Springer Theses

Recognizing Outstanding Ph.D. Research

For further volumes:
http://www.springer.com/series/8790

Aims and Scope

The series "Springer Theses" brings together a selection of the very best Ph.D. theses from around the world and across the physical sciences. Nominated and endorsed by two recognized specialists, each published volume has been selected for its scientific excellence and the high impact of its contents for the pertinent field of research. For greater accessibility to non-specialists, the published versions include an extended introduction, as well as a foreword by the student's supervisor explaining the special relevance of the work for the field. As a whole, the series will provide a valuable resource both for newcomers to the research fields described, and for other scientists seeking detailed background information on special questions. Finally, it provides an accredited documentation of the valuable contributions made by today's younger generation of scientists.

Theses are accepted into the series by invited nomination only and must fulfill all of the following criteria

- They must be written in good English.
- The topic should fall within the confines of Chemistry, Physics, Earth Sciences, Engineering and related interdisciplinary fields such as Materials, Nanoscience, Chemical Engineering, Complex Systems and Biophysics.
- The work reported in the thesis must represent a significant scientific advance.
- If the thesis includes previously published material, permission to reproduce this must be gained from the respective copyright holder.
- They must have been examined and passed during the 12 months prior to nomination.
- Each thesis should include a foreword by the supervisor outlining the significance of its content.
- The theses should have a clearly defined structure including an introduction accessible to scientists not expert in that particular field.

Nodoka Yamanaka

Analysis of the Electric Dipole Moment in the R-parity Violating Supersymmetric Standard Model

Doctoral Thesis accepted by
Osaka University, Osaka, Japan

Springer

Author
Dr. Nodoka Yamanaka
Department of Physics
Graduate School of Science
Kyoto University
Kyoto
Japan

Supervisor
Prof. Toru Sato
Department of Physics
Graduate School of Science
Osaka University
Toyonaka
Japan

ISSN 2190-5053 ISSN 2190-5061 (electronic)
ISBN 978-4-431-56361-7 ISBN 978-4-431-54544-6 (eBook)
DOI 10.1007/978-4-431-54544-6
Springer Tokyo Heidelberg New York Dordrecht London

Printed on acid-free paper

Springer is part of Springer Science+Business Media (www.springer.com)

Parts of this thesis have been published in the following articles:

Nodoka Yamanaka, Toru Sato and Takahiro Kubota, "Reappraisal of two-loop contributions to the fermion electric dipole moments in R-parity violating supersymmetric models," Physical Review D **85**, 117701 (2012).

Nodoka Yamanaka, "R-parity violating supersymmetric contributions to the P, CP-odd electron-nucleon interaction at the one-loop level", Physical Review D **85**, 115012 (2012).

Supervisor's Foreword

The standard gauge theory of the interaction among quarks, leptons, and gauge bosons is extremely successful in describing most of the strong and electroweak phenomena. In spite of the success, there are a few phenomena that suggest the current standard model might be an effective theory of the deeper fundamental theory. One of the leading candidates for such theories is the supersymmetric extension of the standard model (SUSY). The symmetry between fermions and bosons naturally solves one of the theoretical concerns of the standard model, the hierarchy and/or the fine-tuning problem. Since the construction of supersymmetric theory more than some 30 years ago, tremendous efforts to search for its signal have been made. In high-energy collider experiments, the production of supersymmetric particles has been investigated directly. A complementary approach is the low-energy precision test of the symmetry. The electric dipole moment (EDM), which is a CP- and P-violating quantity, is one of the good places to search for the new physics. The standard model, where the Cabibbo–Kobayashi–Maskawa mechanism is the basic source of CP violation, makes an extremely small contribution to flavor diagonal EDM. Currently, the EDM is measured for a variety of systems such as the electron, neutron, atoms, and molecules. Recent precise data of the EDM has started to constrain the models of the new physics.

In this thesis, Nodoka Yamanaka has performed a systematic study of the EDM to disentangle the structure of the Minimal Supersymmetric Standard Model (MSSM) within R-parity-violating (RPV) interactions. The R-parity is defined from the spin, the baryon, and lepton numbers, where a particle has opposite R-parity from its supersymmetric partner. One of the achievements of the thesis is a systematic study of the loop effects on the EDM and CP-violating fermion interactions within RPV-MSSM. The elementary EDM of the fermions are then used to evaluate the EDMs of hadrons, nuclei, and atoms by using the current state-of-the-art procedure. The second feature of the thesis is the development of an analysis tool for the full RPV interaction space of the EDM instead of the conventional single coupling dominance assumption. With this method together with the proposed classification scheme of RPV couplings, complete exploration of the RPV model space has been carried out, and the prospects of the coming EDM experiments have been clarified.

Osaka, March 2013 Prof. Toru Sato

Acknowledgments

I would like to express my sincere gratitude to Prof. Toru Sato and Prof. Takahiro Kubota for their fruitful discussions, invaluable advice, and warm encouragement throughout my research. I also thank Prof. M. Asakawa, Prof. M. Wakamatsu, and Prof. M. Kitazawa for their advice and encouragement. I gratefully acknowledge Prof. M. Oka, Prof. T. Onogi, Prof. E. Hiyama, Prof. H. Fukaya, Prof. H. Iida, Prof. Y. Ikeda, Dr. S. Nasu, and Dr. Y. Yamaguchi for their helpful advice on QCD, chiral perturbation theory, and nuclear physics. In addition, I express my deep appreciation to all the members of the nuclear theory and particle physics groups of Osaka University for their encouragement and assistance. I am grateful, as well, to Osaka University for financial support during my research. Finally, I thank Prof. T.-S. H. Lee for his suggestions for improving this thesis.

Contents

Chapter 1
Introduction

The construction of the quantum chromodynamics [1–3] and the electroweak theory [4] with three generations of fermions [5] lead to the establishment of the standard model (SM) of particle physics. The SM has been able to describe consistently many data from the accelerator experiments, and all particles except the Higgs boson [6–11] within the model have been discovered so far. We can say that the SM is one of the greatest success of modern physics.

However, despite this great success, particle physicists are not always satisfied with the SM. The SM is actually known to have problems with phenomenology:

- The small mass of neutrinos is difficult to explain in the framework of the SM [12].
- The CP violation due to the CKM mechanism is not sufficient to realize the abundance of the matter in our Universe [13].
- The SM does not have candidates for cold dark matter and hence is not consistent with observations [14–16].
- 73 % of the energy the Universe is filled by the unknown dark energy which cannot be explained in the SM. This fact is suggested by the observations of the type Ia supernovae [17, 18].
- The gravity is not included in the SM.
- Recent experimental data, like the anomalous magnetic moment of the muon or the decay asymmetry of the B hadron show discrepancies from the SM predictions [19–21].

In addition, theoretical and convincing arguments against the SM also exist:

- Hierarchical problem due to the radiative corrections of Higgs scalar. The fundamental Higgs scalar poses a serious problem which requires the SM Higgs parameter to be " fine-tuned" (at the level of 10^{-34}!!, if the fundamental scale is the Planck scale).
- The choice of gauge group $(SU(3)_c \times SU(2)_L \times U(1)_Y)$ is ad hoc. Many particle physicists believe that this needs the existence of a "Grand unification" of gauge groups to explain.

N. Yamanaka, *Analysis of the Electric Dipole Moment in the R-parity Violating Supersymmetric Standard Model*, Springer Theses,
DOI: 10.1007/978-4-431-54544-6_1,

- The spontaneous breaking of electroweak symmetry is introduced by adjusting the Higgs potential, which is an ad hoc manipulation. The origin of the Higgs scalar and its potential must be explained.
- The flavor structure and masses (Yukawa couplings) of quarks and leptons are given ad hoc. The flavor seems to be arranged in three generations, but their origin is not known.

All these theoretical arguments strongly suggest the existence of a new physics beyond the SM. Especially, the fine-tuning problem due to the radiative correction of Higgs scalar and the ad hoc choice of the Higgs potential give us a hint that the scale of the new physics is relatively close to that of the electroweak symmetry breaking. The problem of the fine-tuning with fundamental scalar particle merits some explanation. The masses of fermions and gauge bosons (with no scalar in the theory) are protected by symmetries. For example, the radiative corrections to the mass m of fermions is $\delta m \propto m \ln(\Lambda/m)$, where Λ is the cutoff of the effective theory. This can be understood by the fact that the radiative correction cannot flip the chirality without the mass insertion, so that $\delta m \propto m$. The radiative corrections to the mass of the gauge boson is ultimately kept zero by the gauge symmetry. The scale dependence of the theory is then only logarithmic, and we have some stability in fixing parameters, such as the masses of particles. When we insert a scalar particle in the theory, however, the situation changes drastically. An example of the radiative correction of the scalar mass at the one-loop level is shown in Fig. 1.1. After performing loop integrations, we obtain that these one-loop corrections both lead to $\delta m \propto \Lambda^2$. What happens in the case of the SM is that the Higgs scalar with mass around $m_H^2 \sim (100\,\text{GeV})^2$ receives corrections of order $\Lambda_{\text{Planck}}^2 \sim (10^{19}\text{GeV})^2$ if the fundamental scale is taken to the Planck scale. This gives a 10^{34} times larger correction! The expected mass of the Higgs boson is $O(100\,\text{GeV})$, so we must tune the parameter of the theory to 1 part in 10^{34}, which seems to be very unnatural. As said above, this fact suggests the existence of a new physics which incorporates the SM as an effective theory near the scale of electroweak symmetry breaking. Theoretically, the resolution of the fine-tuning problem is the most important requirement in constructing models with new physics.

It is generally beleived that the supersymmetric extension of the SM is an important candidate model with new physics. The supersymmetry was first introduced by Wess and Zumino [22]. Thanks to many works, the supersymmetry acquired

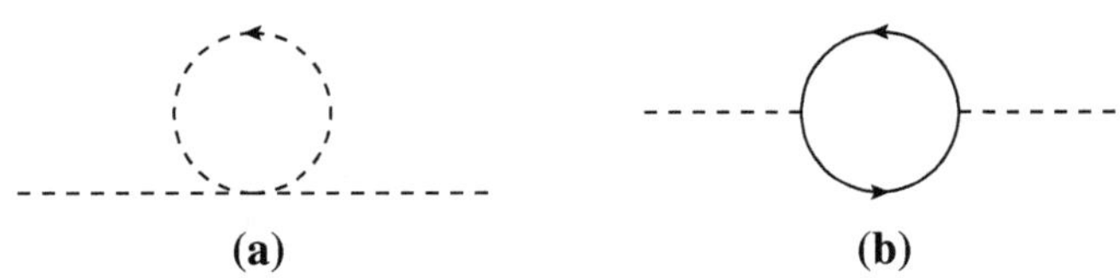

Fig. 1.1 One-loop correction to the scalar mass. *Dashed lines* represent the scalar propagator, and *solid lines* the fermion propagator. **a** Is the correction due to scalar quartic interaction, and (**b**) is the one-loop correction generated by Yukawa interaction

a considerable popularity in particle physics. Compared with the other candidate models with new physics, its phenomenological advantages is particularly interesting [23–26]. Here we can list the following topics:

- The fine-tuning problem due to the radiative corrections of Higgs field can be resolved.
- Soft supersymmetry breaking terms can induce the electroweak symmetry breaking.
- Lightest supersymmetric particles (LSP) can be candidates of dark matters.
- Soft supersymmetry breaking terms can provide new CP violating mechanisms.
- The three running gauge couplings have a better unification at high energy scale.

If the supersymmetry is the true symmetry of the nature, particles discovered so far should have their supersymmetric partners with the same charges and masses. These particles were of course not observed in any past experiments, so we should think that the supersymmetry is a spontaneously broken symmetry. To keep the cancellation of the power divergences in radiative corrections, the supersymmetry must be broken softly. We do not know the definite breaking mechanism of supersymmetry, so the soft breaking terms are introduced by hand. The spontaneous breakdown of supersymmetry is important in phenomenology, since it can provide a new mass scale which is expected to be near above the electroweak scale. The price for introducing the supersymmetry breaking by hand is that we obtain more than 100 soft breaking terms. The soft breaking terms all have mass dimension, and give masses to the particles, which have not been discovered up to 1 TeV [27–31]. The general soft breaking interactions can also have large flavor violation and CP phases, which are also constrained phenomenologically [32]. Other than soft breaking terms, the supersymmetric extension of the SM allows baryon or lepton number violating interactions. These interactions are generated by a set of gauge invariant polynomials of chiral superfields which do not conserve the baryon or lepton numbers, and are called R-parity violating (RPV) interactions. To prevent from such violation, we often assume the conservation of R-parity. This manipulation is however ad hoc. A strong argument to consider RPV interactions comes from the fact that many theoretical physicists believe in the existence of a Grand unified theory of particles and interactions. In Grand unified theories, there are no convincing reasons to distinguish RPV interactions from the R-parity conserving matter fermion-Higgs interactions which give fermion masses, or Higgs self-interactions. Thus it is natural to consider also the violation of R-parity. Many studies of RPV interactions have been done [33–36]. These interactions can generate large baryon number, lepton number, flavor and CP violations, and are therefore strongly constrained by phenomenology.

As discussed above, the models based on the supersymmetric extension of the SM have been studied extensively. To test such models, we can use the available data. But we also need new data which will soon be available from many on-going experiments. The collider experiments at the LHC, through the direct production of new high energy particles, can help to probe the masses of the supersymmetric particles or their lower bounds. There are also many low energy experiments. The Super-K experiments can probe the decay of protons, the mass differences and flavor mixing of neutrinos. The double beta decay experiments will search for the lepton number violations. The muon decay experiments will probe the violation of lepton flavor. Many available experimental results already give significant bounds on parameters of the supersymmetric models, for both R-parity conserving and violating sectors.

In this thesis, we focus on the *electric dipole moments* (EDM), a promising experimental observable which can probe the CP violation originated from the new physics [37–41]. The search for large CP violation is a very important subject in particle physics, since it is known that the CP phase of the CKM matrix cannot provide enough CP violation to realize our matter abundant Universe. In searching for large CP violation, the EDM is an excellent tools for many reasons. First, the EDM is a very "clean" observable. The EDM receives a very small contribution from the SM, due to the higher order effect of the CKM phase. It is also a static observable, so that the final state interaction effects do not disturb the observation. The second important advantage of the EDM is its high accuracy. Due to the strong experimental limits, the EDM has constrained so far many parameters of many candidate models with new physics including the supersymmetry.

The EDM is measurable in a variety of systems, ranging from the elementary particles like the muon to the complex bound states such as molecules. The current experiments already provide very accurate data for each of the systems. The experimental techniques are being improved and many next generation experiments are being prepared, such as the experiments using ultracold neutrons, storage rings, cold molecular beams, ion trap method, etc.

The main objective of this thesis is to investigate the phenomenology of the R-parity violation within the EDM experimental data. There were many previous works in this subject, and many upper limits on the CP violation of the RPV interactions were obtained [42–52]. The RPV interactions contribute to the EDM observables via two leading contributions. The first one is the EDM of quarks and charged leptons, and also the chromo-EDM of quarks. The other contribution is due to the two-body interactions between fermions (lepton-quark, 4-quark and 4-lepton). The trilinear RPV interactions which are the main focus of the study, contribute to the fermion EDM starting from the two-loop level [43, 45]. This is due to the helicity flip of the EDM operator and the structure of the RPV interactions. It was shown by Godbole et al. that the the fermion EDM generated by the RPV contributions at the one-loop level does not exist. The detailed analysis of the two-loop contributions to the fermion EDM due to the R-parity violation was done by Chang et al. They found that the Barr-Zee type two-loop diagrams give the leading contribution, with other suppressed with more than one factor of light quark mass. The P, CP-odd 4-fermion interactions are generated within R-parity violation by sneutrino exchange, and their tree level effects have been studied [46–48].

The two elementary P, CP-odd processes discussed above contribute to experimental EDM observables via intermediate mechanisms, and RPV interactions involved can be constrained by the available experimental data of EDMs of neutron, atoms and molecules. It was shown that the imaginary parts of many bilinears of RPV couplings ($\lambda_{ijj}\lambda^*_{ikk}$, $\lambda_{ijj}\lambda'^*_{ikk}$ and $\lambda'_{ijj}\lambda'^*_{ikk}$, where $i, j, k = 1, 2, 3$) can be constrained via these two processes [43, 45–48]. It is noticed that in those analyses, the dominance of single RPV bilinears is assumed, and the interference between RPV bilinears were neglected.

Recently, we have noticed that the previous calculation of the Barr-Zee type diagram within R-parity violation was not correct [53]. Since the fermion EDM or quark chromo-EDM receive the leading contribution from Barr-Zee type diagram, the entire analysis of EDM within R-parity violation should be revised.

The purpose of this work is to rederive the fermion EDM and the quark chromo-EDM generated from RPV interactions, and analyze the available EDM observables (EDMs of neutron, atoms and molecule) with the corrected formula for Barr-Zee type diagram together with the P, CP-odd 4-fermion interactions. We also predict the EDMs for the planned experiments. As a first step of the extended analysis, we also try to analyze the subleading contribution [54].

In Part I, we briefly review the framework of supersymmetry, the minimal supersymmetric extension of the SM, and the R-parity violation, needed for deriving the P, CP-odd elementary processes contributing to the EDM observables. In order to evaluate the EDM of atoms, nuclei and hadrons, we need to investigate P, CP-odd interactions at the hadronic, nuclear and atomic levels. At each level, we encounter difficult many-body and non-perturbative physics. In this work, we use the best available information on those problems. Part II is a review of the subject on EDM. In Part III, we describe our analysis of the R-parity violation within the EDM-constraints. We first derive the fermion EDM and quark chromo-EDM within the two-loop level Barr-Zee type diagram with detailed explanations of our corrections. Together with the tree level P, CP-odd 4-fermion interaction, we then try to obtain upper bounds on RPV couplings from the atomic, nuclear and hadronic EDM observables using the consequences of the many-body physics presented in Part II. In doing this, we have shown clear classification of RPV bilinears into six types, which clarifies the dependences of the RPV couplings on EDM observables and helps our subsequent analysis. The first step of our phenomenological analysis is to derive upper bounds when single RPV bilinear is considered. This is an update of the previous analyses, including the corrected formula for fermion EDM and quark chromo-EDM. In the next step, we have analyzed RPV contribution to EDMs when all leading RPV bilinears are relevant. In this analysis, the interference between RPV bilinears are also taken into account, within a 10-dimensional parameter space. We also predict the EDM observables for planned experiments. They are also compared between the case of single RPV bilinear dominance and the case where interference can occur. The prospect for each future EDM experiment is discussed from the point of view of the determination of RPV couplings. After that, we present the investigation of one of the subleading contribution to the EDM observables within R-parity violation, the analysis on the P, CP-odd 4-fermion interactions at the one-loop level. This analysis is done by assuming the dominance of one RPV bilinear. This analysis is interesting since the atomic EDMs have a large sensitivity against the P, CP-odd 4-fermion interactions and we can expect that even the subleading RPV contribution can be constrained. It is also the first step of the extended analysis including the subleading RPV contribution to EDM observables. The last chapter is devoted to the summary.

References

1. C.N. Yang, R. Mills, Phys. Rev. **96**, 191 (1954)
2. M. Gell-Mann, Phys. Lett. **8**, 214 (1964)
3. M.Y. Han, Y. Nambu, Phys. Rev. **139**, B1006 (1965)
4. S. Weinberg, Phys. Rev. Lett. **19**, 1264 (1967)
5. M. Kobayashi, T. Maskawa, Prog. Theor. Phys. **49**, 652 (1973)
6. F. Englert, R. Brout, Phys. Rev. Lett. **13**, 321 (1964)
7. P.W. Higgs, Phys. Lett. **12**, 132 (1964)
8. P.W. Higgs, Phys. Rev. Lett. **13**, 508 (1964)
9. P.W. Higgs, Phys. Rev. **145**, 1156 (1966)
10. G.S. Guralnik, C.R. Hagen, T.W.B. Kibble, Phys. Rev. Lett. **13**, 585 (1964)
11. T.W.B. Kibble, Phys. Rev. **155**, 1554 (1967)
12. Y. Fukuda et al., Super-Kamiokande Collaboration. Phys. Rev. Lett. **81**, 1562 (1998)
13. A. D. Sakharov, Pisma Zh. Eksp. Teor. Fiz. **5**, 32 (JETP Lett. 5, 24) (1967)
14. F. Zwicky, Astrophys. J. **86**, 217 (1937)
15. M. Davis, G. Efstathiou, C.S. Frenk, S.D.M. White, Astrophys. J. **292**, 371 (1985)
16. D. Clowe et al., Astrophys. J. **648**, L109 (2006)
17. A.G. Riess et al., Supernova Search Team. Astron. J. **116**, 1009 (1998)
18. S. Perlmutter et al., Supernova Cosmology Project. Astrophys. J. **517**, 565 (1999)
19. G.W. Bennett et al., Muon G-2 Collaboration. Phys. Rev. D **73**, 072003 (2006)
20. V.M. Abazov et al., D0 Collaboration, Phys. Rev. Lett. **105**, 081801 (2010)
21. V.M. Abazov et al., D0 Collaboration. Phys. Rev. D **82**, 032001 (2010)
22. J. Wess, B. Zumino, Phys. Lett. **B49**, 52 (1974)
23. H.E. Haber, G.L. Kane, Phys. Rept. **117**, 75 (1985)
24. J.F. Gunion, H.E. Haber, Nucl. Phys. B **272**, 1 (1986)
25. S.P. Martin, arXiv:hep-ph/9709356
26. H. Baer, X. Tata, *Weak Scale Supersymmetry* (Cambridge Univ, Press , 2006)
27. G. Aad et al., ATLAS Collaboration. Phys. Rev. Lett. **106**, 131802 (2011)
28. G. Aad et al., ATLAS Collaboration. Phys. Lett. B **701**, 186 (2011)
29. G. Aad et al., ATLAS Collaboration, arXiv:1110.6189 [hep-ex]
30. V. Khachatryan et al., CMS Collaboration. Phys. Lett. B **698**, 196 (2011)
31. V. Khachatryan et al., CMS Collaboration, arXiv:1111.2733 [hep-ex]
32. F. Gabbiani, E. Gabrielli, A. Masiero, L. Silvestrini, Nucl. Phys. B **477**, 321 (1996)
33. G. Bhattacharyya, arXiv:hep-ph/9709395
34. H.K. Dreiner, arXiv:hep-ph/9707435
35. R. Barbier et al., Phys. Rept. **420**, 1 (2005)
36. M. Chemtob, Prog. Part. Nucl. Phys. **54**, 71 (2005)
37. W. Bernreuther, M. Suzuki, Rev. Mod. Phys. 63, 313 (1991); Erratum-ibid. 64, 633 (1992)
38. I.B. Khriplovich, S.K. Lamoreaux, *CP Vioaltion Without Strangeness* (Springer, Berlin, 1997)
39. J.S.M. Ginges, V.V. Flambaum, Phys. Rept. **397**, 63 (2004)
40. M. Pospelov, A. Ritz, Ann. Phys. **318**, 119 (2005)
41. T. Fukuyama, Int. J. Mod. Phys. A **27**, 1230015 (2012)
42. R. Barbieri, A. Masiero, Nucl. Phys. B **267**, 679 (1986)
43. R.M. Godbole, S. Pakvasa, S.D. Rindani, X. Tata, Phys. Rev. D **61**, 113003 (2000)
44. S.A. Abel, A. Dedes, H.K. Dreiner, JHEP **0005**, 13 (2000)
45. D. Chang, W.-F. Chang, M. Frank, W.-Y. Keung, Phys. Rev. D **62**, 095002 (2000)
46. P. Herczeg, Phys. Rev. D **61**, 095010 (2000)
47. A. Faessler, T. Gutsche, S. Kovalenko, V. E. Lyubovitskij, Phys. Rev. D 73, 114023
48. A. Faessler, T. Gutsche, S. Kovalenko, V.E. Lyubovitskij, Phys. Rev. D **74**, 074013 (2006)
49. K. Choi, E.J. Chun, K. Hwang, Phys. Rev. D **63**, 013002 (2000)
50. Y.Y. Keum, O.C.W. Kong, Phys. Rev. Lett. **86**, 393 (2001)
51. Y.Y. Keum, O.C.W. Kong, Phys. Rev. D **63**, 113012 (2001)
52. C.-C. Chiou, O.C.W. Kong, R.D. Vaidya, Phys. Rev. D **76**, 013003 (2007)
53. N. Yamanaka, T. Sato, T. Kubota, Phys. Rev. D **85**, 117701 (2012)
54. N. Yamanaka, Phys. Rev. D **85**, 115012 (2012)

Part I
Supersymmetry and R-parity Violation

Chapter 2
The Supersymmetry

2.1 Supersymmetry Algebra

The supersymmetry (SUSY) algebra can be explained by using the Wess-Zumino model [1]. The Wess-Zumino lagrangian can be written as follows:

$$\mathscr{L}_{WZ} = \frac{1}{2}(\partial_\mu A)^2 + \frac{1}{2}(\partial_\mu B)^2 + \frac{i}{2}\bar{\psi}\partial\!\!\!/ \psi - \frac{1}{2}(F^2+G^2) - m\left(\frac{1}{2}\bar{\psi}\psi - GA - FB\right). \tag{2.1}$$

Here, A and B are complex scalar fields, ψ a Majorana spinor field, G and F auxiliary fields (no kinetic term) with mass dimension 2. This Lagrangian is invariant (up to a divergent term) under the following infinitesimal transformation:

$$\begin{cases} \delta A = i\bar{\alpha}\gamma_5\psi \\ \delta B = -\bar{\alpha}\psi \\ \delta\psi = -F\alpha + iG\gamma_5\alpha + \partial\!\!\!/ \gamma_5 A\alpha + i\partial\!\!\!/ B\alpha \\ \delta F = i\bar{\alpha}\partial\!\!\!/ \psi \\ \delta G = \bar{\alpha}\gamma_5\partial\!\!\!/ \psi \end{cases} . \tag{2.2}$$

Here, α is an infinitesimal anti-commuting Majorana spinor parameter, and has mass dimension $-\frac{1}{2}$.

We can generalize the SUSY transformation by using the SUSY generator Q which is a Majorana spinor and obeys the following algebra:

$$\begin{aligned} \left[P_\mu, Q_a\right] &= 0, \\ \left[M_{\mu\nu}, Q_a\right] &= -\frac{1}{2}(\sigma_{\mu\nu})_{ab}Q_b, \\ \left\{Q_a, \bar{Q}_b\right\} &= 2P_\mu(\gamma^\mu)_{ab}, \end{aligned} \tag{2.3}$$

where $M_{\mu\nu}$ are the generators of the Lorentz transformation. These relations are from a direct extension of the Poincaré algebra. The middle line of the above equations

N. Yamanaka, *Analysis of the Electric Dipole Moment in the R-parity Violating Supersymmetric Standard Model*, Springer Theses,
DOI: 10.1007/978-4-431-54544-6_2, © Springer Japan 2014

(commutation relation between Lorentz and supersymmetric generators) means that the supersymmetric partner of a particle has spin which differs by 1/2. This describes well the statement that "supersymmetry is symmetry between bosons and fermions". We must note that the last line is an anti-commutation relation, and gives a result that creates momentum. This is an example of graded Lie-algebra (algebra with anti-commutation), and its elements can mix with the space-time. We can think that supersymmetry is part of the extended space-time geometry (we call it super-space).

2.2 Chiral Superfields and Superpotential

Let us now see how the supersymmetric field theory is formulated. It is known that the general scalar function of the super-space has two terms which always give a total divergence [for example, see the supersymmetric transformation of F and G of Eq. (2.2)]. This property allows us to construct the general supersymmetric lagrangian by defining the scalar super-spatial function with the conditions of interest (gauge invariance, renormalizability, conservation of given parities, etc), and taking these terms invariant up to total derivative. This approach is called the superfield formalism, and the lagrangian of the supersymmetric models is formulated in this manner.

The general scalar field of the super-space can be written as

$$\begin{aligned}\hat{\Phi}(x,\theta) = &\mathcal{S} - i\sqrt{2}\bar{\theta}\gamma_5\psi - \frac{i}{2}(\bar{\theta}\gamma_5\theta)\mathcal{M} + \frac{1}{2}(\bar{\theta}\theta)\mathcal{N} + \frac{i}{2}(\bar{\theta}\gamma_5\gamma_\mu\theta)V^\mu \\ &+ i(\bar{\theta}\gamma_5\theta)[\bar{\theta}(\lambda + \frac{i}{\sqrt{2}}\partial\!\!\!/\,\psi)] - \frac{1}{4}(\bar{\theta}\gamma_5\theta)^2[\mathcal{D} - \frac{1}{2}\partial^2\mathcal{S}],\end{aligned} \tag{2.4}$$

where θ is the Grassmann Majorana spinor variable. Fields in the super-space are called superfields. The supersymmetric transformation can be expressed in terms of Grassmann variable θ as follow:

$$\delta\hat{\Phi} = \left[\bar{\alpha}Q, \hat{\Phi}\right] = \left(-\bar{\alpha}\frac{\partial}{\partial\bar{\theta}} - i\bar{\alpha}\partial\!\!\!/\,\theta\right)\hat{\Phi}. \tag{2.5}$$

The supersymmetric transformation of the above superfield $\hat{\Phi}$ is then

$$\begin{cases} \delta\mathcal{S} &= i\sqrt{2}\bar{\alpha}\gamma_5\psi \\ \delta\psi &= -\frac{1}{\sqrt{2}}\left(\alpha\mathcal{M} + i\gamma_5\alpha\mathcal{N} + i\gamma^\mu\alpha V_\mu + \gamma_5\partial\!\!\!/\,\mathcal{S}\alpha\right) \\ \delta\mathcal{M} &= \bar{\alpha}\left(\lambda + i\sqrt{2}\partial\!\!\!/\,\psi\right) \\ \delta\mathcal{N} &= i\bar{\alpha}\gamma_5\left(\lambda + i\sqrt{2}\partial\!\!\!/\,\psi\right) \\ \delta V^\mu &= -i\bar{\alpha}\gamma^\mu\lambda + \sqrt{2}\bar{\alpha}\partial^\mu\psi \\ \delta\lambda &= -i\gamma_5\alpha\mathcal{D} - \frac{1}{2}[\partial\!\!\!/, \gamma_\mu]V^\mu\alpha \\ \delta\mathcal{D} &= \bar{\alpha}\partial\!\!\!/\,\gamma_5\lambda \end{cases} \tag{2.6}$$

We remark that the general superfield can be separated into three irreducible multiplets which transform independently (up to total derivative). The first two are the *left-* and *right-chiral superfields*, defined as

$$\mathcal{S} = \frac{1}{\sqrt{2}}(A \pm iB), \tag{2.7}$$

$$\psi_{L/R} = \frac{1}{2}(1 \mp \gamma_5)\psi, \tag{2.8}$$

$$\mathcal{F} = \frac{1}{\sqrt{2}}(F \pm iG) = \frac{1}{\sqrt{2}}(\mathcal{M} \mp i\mathcal{N}), \tag{2.9}$$

$$V^\mu = \pm i\partial^\mu \mathcal{S}, \tag{2.10}$$

We see that by setting V^μ, λ and $\mathcal{D}$ to zero, the supersymmetric transformation of the Wess-Zumino model [see Eq. (2.2)] can be obtained. The fields $\mathcal{S}$, $\psi_{L/R}$ and $\mathcal{F}$ thus form an irreducible supermultiplet. The left- (or right-) chiral superfield is used in defining matter fields. The left-chiral superfield can be written as

$$\hat{\mathcal{S}}_L = \mathcal{S} + i\sqrt{2}\bar{\theta}\psi_L + i\bar{\theta}\theta_L\mathcal{F}, \tag{2.11}$$

where $\theta_L \equiv P_L\theta$. The supersymmetric transformation for the left-chiral scalar superfield is

$$\begin{cases} \delta\mathcal{S} &= -i\sqrt{2}\bar{\alpha}\psi_L \\ \delta\psi_L &= -\sqrt{2}\mathcal{F}\alpha_L + \sqrt{2}\partial\!\!\!/\,\mathcal{S}\alpha_R \\ \delta\mathcal{F} &= i\sqrt{2}\bar{\alpha}\partial\!\!\!/\,\psi_L \end{cases}, \tag{2.12}$$

where $\alpha_L \equiv \frac{1}{2}(1-\gamma_5)\alpha$ and $\alpha_R \equiv \frac{1}{2}(1+\gamma_5)\alpha$.

The remaining third type supermultiplet, called the *curl superfield*, consists of $F^{\mu\nu} \equiv \partial^\mu V^\nu - \partial^\nu V^\mu$, λ and $\mathcal{D}$. The supersymmetric transformation of the curl superfield can be written as

$$\begin{cases} \delta F^{\mu\nu} &= -i\bar{\alpha}[\gamma^\nu\partial^\mu - \gamma^\mu\partial^\nu]\lambda \\ \delta\lambda &= -i\gamma_5\alpha\mathcal{D} + \frac{1}{4}[\gamma_\nu, \gamma_\nu]F^{\mu\nu}\alpha \\ \delta\mathcal{D} &= \bar{\alpha}\partial\!\!\!/\,\gamma_5\lambda. \end{cases} \tag{2.13}$$

This supermutiplet is used to express supersymmetric gauge fields. We should note that for the curl superfield, it is not possible to set $\mathcal{S}$, $\psi_{L/R}$ and $\mathcal{F}$ to zero [see the term with V^μ in the second line of Eq. (2.6)]. The supersymmetric transformation of this supermultiplet is in fact not irreducible. In gauge theories however, it is possible to introduce an additional degree of freedom to remove this dependence.

By observing the supersymmetric transformation [Eq. (2.6)], we find that $\mathcal{D}$ is only transformed by a total derivative. This fact suggests that the $\mathcal{D}$ component of a general superfield is supersymmetric. The first idea for the construction of the supersymmetric lagrangian is to take the component $\mathcal{D}$ (we call it the D-term) of a field polynomial with the conditions of interest (gauge invariance, renormalizability, conservation of given parities, etc). By considering the D-term of the bilinear of left- and right chiral superfields ($\hat{\mathscr{S}}_L^\dagger \hat{\mathscr{S}}_L$), we obtain the kinetic term as follows:

$$\mathscr{L}_D = \partial^\mu \mathscr{S}^\dagger \partial_\mu \mathscr{S} + \frac{i}{2}\bar{\psi}\partial\!\!\!/ \psi + \mathscr{F}^\dagger \mathscr{F}. \tag{2.14}$$

The bilinear $\hat{\mathscr{S}}_L^\dagger \hat{\mathscr{S}}_L$ is called the *Kähler potential*. The Kähler potential is generally introduced to generate not only the kinetic terms, but also the gauge interactions which will be explained in the next section.

In a general superfield, there is one more term which leaves only a total derivative after supersymmetric transformation. This is the term with $\mathscr{F}$ in Eq. (2.11) (see Eq. (2.12) for its supersymmetric transformation). As a general rule, polynomials of left-chiral scalar superfields are also left-chiral superfields. It is therefore possible to obtain the supersymmetric lagrangian by taking the F-term of some polynomial of superfields with the conditions of interest. This left-chiral polynomial is called the *superpotential*. In a renormalizable theory, the superpotential is a polynomial with at most three factors of superfields, and the interaction lagrangian generated from it is given by the following formula:

$$\begin{aligned}\mathscr{L}_f = &-\sum_i \left| \frac{\partial \hat{f}}{\partial \hat{\mathscr{S}}_i} \right|^2_{\hat{\mathscr{S}}=\mathscr{S}} \\ &- \frac{1}{2}\sum_{i,j} \bar{\psi}_i \left[\left(\frac{\partial^2 \hat{f}}{\partial \hat{\mathscr{S}}_i \partial \hat{\mathscr{S}}_j} \right)_{\hat{\mathscr{S}}=\mathscr{S}} P_L + \left(\frac{\partial^2 \hat{f}}{\partial \hat{\mathscr{S}}_i \partial \hat{\mathscr{S}}_j} \right)^\dagger_{\hat{\mathscr{S}}=\mathscr{S}} P_R \right] \psi_j,\end{aligned} \tag{2.15}$$

where indices i and j denote the label of each superfield.

We see that the superpotential generates mass terms, mixing bilinears, Yukawa and scalar 4-point interactions. By setting the superpotential $\hat{f} \to \hat{f}_{\mathrm{WZ}} = m\hat{\mathscr{S}}^2$, the Wess-Zumino lagrangian (2.1) is obtained. For the supersymmetric extension of the standard model, the superpotential is composed of quark, lepton, and Higgs superfields.

2.3 Supersymmetric Gauge Theory

The next task is to extend the Kähler potential to include the gauge interaction. The supersymmetric extension of gauge theories is less obvious than that of the matter fields, due to the freedom associated with the gauge transformation. It can be obtained with the superfield formalism using Kähler potential. We will proceed in three steps in constructing the supersymmetrizing gauge theories. The first step is to express the gauge transformation in terms of the gauge potential superfield. The second is to couple gauge superfields to the chiral superfields, and the third is to construct the gauge kinetic term.

We begin by defining the supersymmetrization of the gauge transformation. The gauge transformation of the left-chiral superfield $\hat{\mathscr{S}}$ (with components $\mathscr{S}$, ψ and $\mathscr{F}$) can be written as follow:

$$\hat{\mathscr{S}}_a(\hat{x}, \theta) = \left[e^{-igt_A \hat{\Omega}_A(\hat{x})} \right]_{ab} \hat{\mathscr{S}}_b(\hat{x}, \theta), \tag{2.16}$$

Here, we have defined the gauge transformation parameter $\hat{\Omega}$ as a superfield. Actually, the gauge transformation needs to be a left-chiral superfield (and consequently so does the transformation parameter superfield $\hat{\Omega}$) in order to keep the field after transformation a left-chiral superfield. This is due to the fact that the coordinate $\hat{x}$ ($\equiv x^\mu + \frac{i}{2}\bar{\theta}\gamma_5\gamma^\mu\theta$) also has degrees of freedom in the Grassmann super-space. This is also the result of the supersymmetric extension of the space-time. The gauge coupling g is defined by the gauge covariant derivative $D^\mu \equiv \partial^\mu - igA^\mu_B t_B$ where t_B is the generator of the gauge group. Note that the convention for the sign of the gauge coupling in the Ref. [2] is opposite.

Our next step is to couple gauge superfields to the chiral superfields. For that, we construct the Kähler potential containing the left- and right-chiral scalar superfield in a gauge invariant way, and finally take its D-term to obtain the gauge invariant lagrangian of chiral fields (We can avoid dealing with chiral interaction terms, because these do not contain any derivatives).

We now introduce the gauge potential superfield $\hat{\Phi}$. Its gauge transformation can be written as

$$e^{2gt_A \hat{\Phi}'_A} = e^{igt_P \hat{\Omega}^\dagger_P} e^{2gt_A \hat{\Phi}_A} e^{-igt_Q \hat{\Omega}_Q}. \tag{2.17}$$

By using this property, we can construct the following gauge invariant Kähler potential:

$$K = \hat{\mathscr{S}}^\dagger e^{2gt_A \hat{\Phi}_A} \hat{\mathscr{S}}. \tag{2.18}$$

The gauge potential superfield can be expressed in the form of a general superfield, but its components are not all physical (the scalar, fermion and auxiliary term), so we need to eliminate the unphysical part. Actually, the gauge transformation of the left-chiral scalar superfield can furnish the degree of freedom needed in the cancellation of the unphysical components of the gauge potential. This gauge fixing is called the Wess-Zumino gauge. We can think that the Wess-Zumino gauge fixing is a gauge fixing over the super-space. After this fixing, the gauge potential superfield can be written as

$$\hat{\Phi}_A = \frac{1}{2}(\bar{\theta}\gamma_5\gamma_\mu\theta)(V^\mu_A + \partial^\mu\chi_A) + i\bar{\theta}\gamma_5\theta\bar{\theta}\lambda_A - \frac{1}{4}(\bar{\theta}\gamma_5\theta)^2\mathscr{D}_A. \tag{2.19}$$

Here, V^μ, λ and $\mathscr{D}$ are respectively the gauge field, gaugino (the fermionic supersymmetric partner of gauge boson) and the auxiliary $\mathscr{D}$ field. Since the gauge potential is real, the number of degrees of freedom that can be supplied is exactly one for the abelian gauge theory (N in $SU(N)$ gauge theories). This gives the free parameter of the gauge theory, and the consistency is kept safe. This result is quite general and works as well in the non-abelian case.

If we take the F-term of the Kähler potential in the Wess-Zumino gauge, we find

$$\begin{aligned}
\mathcal{L}_{gauge} = & \frac{i}{2}\bar{\psi}\partial\!\!\!/\,\psi + (\partial_\mu \mathcal{S})^\dagger(\partial^\mu \mathcal{S}) + \mathcal{F}^\dagger \mathcal{F} \\
& - i(\partial_\mu \mathcal{S})^\dagger g(t\cdot V^\mu)\mathcal{S} + i\mathcal{S}^\dagger g(t\cdot V^\mu)\partial_\mu \mathcal{S} + \mathcal{S}^\dagger\left[gt\cdot\mathcal{D} + g^2(t\cdot V)^2\right]\mathcal{S} \\
& + \frac{1}{2}[g\bar{\psi}(t\cdot V\!\!\!/)\psi_L - g\bar{\psi}(t^*\cdot V\!\!\!/)\psi_R] \\
& + \left(\sqrt{2}g\mathcal{S}^\dagger t_A\bar{\lambda}_A\frac{1-\gamma_5}{2}\psi + \text{h.c.}\right),
\end{aligned} \tag{2.20}$$

which is the final form of the gauge interacting supersymmetric chiral lagrangian.

We must now build the gauge kinetic term. For that, we construct a gauge invariant left-chiral superfield from the gauge potential, and then take its F-term. Let us define the following left-chiral (spinor) superfield

$$gt_A\hat{W}_A \equiv \frac{i}{8}\bar{D}D_R\left[e^{-2gt_C\hat{\Phi}_C}D_L e^{2gt_B\hat{\Phi}_B}\right], \tag{2.21}$$

where D, $\bar{D}$, D_L and D_R are supercovariant derivatives defined as follows

$$\begin{cases}
D \equiv \dfrac{\partial}{\partial\bar{\theta}} - i\partial\!\!\!/\,\theta \\
\bar{D} \equiv -\dfrac{\partial}{\partial\theta} + i\bar{\theta}\partial\!\!\!/ \\
D_R \equiv P_R D \\
D_L \equiv P_L D
\end{cases}. \tag{2.22}$$

It is important to note here that the superfield $\hat{W}_A$ is a spinor in the above construction.

This left-chiral superfield is actually gauge covariant in the adjoint representation, like the field strength $F_{\mu\nu A}$ (Indeed, the superfield $e^{2gt_A\hat{\Phi}_A}$ is not gauge covariant and there was a necessity to bring a gauge covariant superfield). Thus we are led to the idea of constructing a gauge invariant superfield by combining $\hat{W}_A$ with another left-chiral gauge covariant superfield which is transformed like $\bar{\psi}$ under Lorentz transformation. The only (renormalizable) possibility of such a combination is

$$\overline{\hat{W}^c_A}\hat{W}_A. \tag{2.23}$$

Note that $\overline{\hat{W}^c_A}$ transforms covariantly under gauge transformation. We can now take its F-term to give the lagrangian. After some manipulation, we obtain the following gauge kinetic lagrangian

$$\mathscr{L}_{GK} = \frac{i}{2}\bar{\lambda}_A \not{D}_{AC}\lambda_C - \frac{1}{4}F_{\mu\nu A}F_A^{\mu\nu} + \frac{1}{2}\mathscr{D}_A\mathscr{D}_A, \tag{2.24}$$

where A and C are gauge indices in the adjoint representation, and

$$F_{\mu\nu A} \equiv \partial_\mu V_{\nu A} - \partial_\nu V_{\mu A} + g f_{ABC} V_{\mu B} V_{\nu C}, \tag{2.25}$$

$$(\not{D}\lambda)_A \equiv \not{\partial}\lambda_A - ig(t_B \not{V}_B)_{AC}\lambda_C. \tag{2.26}$$

The final supersymmetric lagrangian is $\mathscr{L}_{GK} + \mathscr{L}_{gauge}$. The derivation of the supersymmetric gauge theory is carefully presented in the excellent book of Baer and Tata [2].

2.4 Non-Renormalization Theorem

We have seen previously that the quadratic divergence of th e one-loop corrections on the scalar mass is cancelled by the fermion loop. This helpful cancellation is due to the general property of the supersymmetry, in which any loop corrections can be written as a D-term, and F-terms do not receive any loop corrections. This is the *non-renormalization theorem*. It means that the superpotential is not generated from loop corrections. Since the mass of scalar particles is given by the superpotential, loops cannot contribute to their mass correction. Since the wave function renormalization is at most logarithmic, the loop corrections to the scalar propagator (one-loop example shown in Fig. 1.1) can also be at most only logarithmic, and thus leads to the absence of the fine-tuning problem associated with the quadratic divergence of the Higgs boson mass corrections. This aspect was first shown in Ref. [3] using the supergraph methods, and directly in Ref. [4].

References

1. J. Wess, B. Zumino, Phys. Lett. **B49**, 52 (1974)
2. H. Baer, X. Tata, *Weak Scale Supersymmetry* (Cambridge University Press, Cambridge, 2006)
3. M.T. Grisaru, W. Siegel, M. Rocek, Nucl. Phys. B **159**, 429 (1979)
4. N. Seiberg, Phys. Lett. B **318**, 469 (1993)

Chapter 3
Minimal Supersymmetric Standard Model

3.1 Supersymmetric Extension of the Standard Model

The minimal supersymmetric extension of the standard model (SM) can be obtained by taking the F- and D-terms of the Kähler and superpotentials formulated previously. The left-chiral superfields for quarks, leptons and Higgs should be combined with the following conditions:

- $SU(3)_c \times SU(2)_L \times U(1)_Y$ gauge group.
- Three generations of quarks and leptons.
- Higgs potential with anomaly cancellation of the higgsino.
- Renormalizability.

In constructing the superpotential, the chiral-superfields listed in Table. 3.1 are needed. The superpartners of leptons and quarks are named with a prefix s-(for example, the superpartner of the quark is called *squark*). For the superpartners of bosons, we add the suffix-ino (for example, the superpartner of the gauge boson is called *gaugino*).

We must note that there are two Higgs fields in the supersymmetric extension of the SM. This is due to the fact that the superpotential cannot have a complex conjugate Higgs field needed for generating the masses of up and down quarks. Therefore we must introduce a new field which transforms in the conjugate representation of $SU(2)_L$. The second Higgs field is also needed for the cancellation of the chiral anomaly of the higgsinos.

The supermultiplets of the MSSM must be coupled to each other at the level of the superpotential and the Kähler potential, following the gauge invariance and renormalizability. We remark here that if we construct the fully allowed superpotential, some baryon and lepton number violating terms will also be allowed, such as

$$\lambda_{ijk}\varepsilon_{ab}\hat{L}_i^a\hat{L}_j^b\hat{E}_k^c\,, \quad \lambda'_{ijk}\varepsilon_{ab}\hat{L}_i^a\hat{Q}_j^b\hat{D}_k^c\,, \quad \mu'_i\varepsilon_{ab}\hat{L}_i^a\hat{H}_u^b\,, \quad \lambda''_{ijk}\hat{U}_i^c\hat{D}_j^c\hat{D}_k^c\,, \tag{3.1}$$

N. Yamanaka, *Analysis of the Electric Dipole Moment in the R-parity Violating Supersymmetric Standard Model*, Springer Theses,
DOI: 10.1007/978-4-431-54544-6_3, © Springer Japan 2014

Table 3.1 Chiral-superfields in the MSSM

Field	$SU(3)_c$	$SU(2)_L$	$U(1)_Y$
$\hat{L} = \begin{pmatrix} \hat{\nu}_L \\ \hat{e}_L \end{pmatrix}$	1	2	$-\frac{1}{2}$
$\hat{E}^c$	1	1	1
$\hat{Q} = \begin{pmatrix} \hat{u}_L \\ \hat{d}_L \end{pmatrix}$	3	2	$\frac{1}{6}$
$\hat{U}^c$	3*	1	$-\frac{2}{3}$
$\hat{D}^c$	3*	1	$\frac{1}{3}$
$\hat{H}_u = \begin{pmatrix} \hat{h}_u^+ \\ \hat{h}_u^0 \end{pmatrix}$	1	2	$\frac{1}{2}$
$\hat{H}_d = \begin{pmatrix} \hat{h}_d^- \\ \hat{h}_d^0 \end{pmatrix}$	1	2*	$-\frac{1}{2}$

Quark and lepton fields are composed of 3 generations

where i, j, k denote the generation, and a, b are the $SU(2)_L$ indices ($SU(3)_c$ indices were omitted). We can avoid the baryon/lepton number violation if we introduce the following *R-parity*

$$R = (-1)^{3(B-L)+2s} , \tag{3.2}$$

where B, L and s are respectively the baryon, lepton numbers and spin. By imposing this parity conservation, processes generated from the superpotential do not violate baryon and lepton numbers. It is now possible to yield a renormalizable, gauge invariant and R-parity conserving phenomenological supersymmetric extension of the SM. We call it the *minimal supersymmetric standard model* (MSSM).

3.2 SUSY Breaking Terms

The supersymmetry requires the existence of a supersymmetric partner (sparticle) which has the same properties and the same quantum numbers (especially the mass), for each particle in the SM. However, no such sparticles have been observed so far, so we must think that the supersymmetry is a spontaneously broken symmetry. As discussed in the introduction, the mechanism of the spontaneous supersymmetry breaking is not well known, and we need to break it by hand. To avoid the regeneration of the fine-tuning problem, we must break it without generating any quadratic divergences. This is called the "Soft SUSY breaking".

The classification of the soft SUSY breaking terms can be made by inspecting the tadpole diagrams. Girardello and Grisaru [1] listed the following soft SUSY breaking terms

$$\begin{cases} \mu_1^2(A^2+B^2) & \text{(mass shift of the scalar)} \\ \mu_2^2(A^2-B^2) & \text{(mass splitting of the complex scalar)} \\ \mu_3\bar{\lambda}\lambda & \text{(mass shift of the gaugino)} \\ \mu_4(A^3-3AB^2) & \text{(scalar trilinear interaction)} \end{cases} . \quad (3.3)$$

Including these terms in the MSSM, the resulting lagrangian breaks the supersymmetry softly and is of the following form

$$\begin{aligned} \mathscr{L}_{\text{soft}} = &-\Big[\tilde{Q}_i^\dagger m^2_{Q_{ij}}\tilde{Q}_j + \tilde{d}^\dagger_{Ri} m^2_{D_{ij}}\tilde{d}_{Rj} + \tilde{u}^\dagger_{Ri} m^2_{U_{ij}}\tilde{u}_{Rj} \\ &+ \tilde{L}_i^\dagger m^2_{L_{ij}}\tilde{L}_j + \tilde{e}^\dagger_{Ri} m^2_{E_{ij}}\tilde{e}_{Rj} + m^2_{H_u}|H_u|^2 + m^2_{H_d}|H_d|^2\Big] \\ &-\frac{1}{2}\big[M_1\bar{\lambda}_0\lambda_0 + M_2\bar{\lambda}_A\lambda_A + M_3\overline{\tilde{g}_B}\tilde{g}_B\big] \\ &-\frac{i}{2}\big[M_1'\bar{\lambda}_0\gamma_5\lambda_0 + M_2'\bar{\lambda}_A\gamma_5\lambda_A + M_3'\overline{\tilde{g}_B}\gamma_5\tilde{g}_B\big] \\ &+\Big[(a_u)_{ij}\varepsilon_{ab}\tilde{Q}_i^a H_{ub}\tilde{u}^\dagger_{Rj} + (a_d)_{ij}\tilde{Q}_i^a H_{da}\tilde{d}^\dagger_{Rj} + (a_e)_{ij}\tilde{L}_i^a H_{da}\tilde{e}^\dagger_{Rj} + \text{ h.c. }\Big] \\ &+\Big[(c_u)_{ij}\varepsilon_{ab}\tilde{Q}_i^a H^*_{ub}\tilde{u}^\dagger_{Rj} + (c_d)_{ij}\tilde{Q}_i^a H^*_{da}\tilde{d}^\dagger_{Rj} + (c_e)_{ij}\tilde{L}_i^a H^*_{da}\tilde{e}^\dagger_{Rj} + \text{ h.c. }\Big] \\ &+[\, bH_{ua}H_{da} + \text{ h.c. }\,], \end{aligned} \quad (3.4)$$

where λ_0, λ_A and $\tilde{g}_B$ are respectively the gauginos for the $U(1)_Y$, $SU(2)_L$ and $SU(3)_c$ gauge groups. The matrices with flavor indices all have off-diagonal components. All these SUSY breaking terms can have CP violating phases. We must note here that although SUSY is broken softly by the gaugino bilinears (due to the gauge invariance: cancellation of the tadpole diagram), but hardly by the chiral fermion bilinears (i.e. generate quadratic divergence). In fact the chiral scalar field gains (or loses) their mass, but the mass of the chiral fermion field is not shifted in the soft SUSY breaking. The soft SUSY breaking lagrangian enlarges considerably the parameter space of the MSSM. After the introduction of SUSY breaking terms by hand (field redefinition taken into account, c parameters neglected), the model has 124 parameters. To obtain the physical spectrum of the MSSM, we must diagonalize the mass matrices. The detail is presented in Appendix A (see also Ref. [2]).

The study of the spontaneous breakdown of supersymmetry is a very rich subject [3], but it is beyond the scope of this thesis.

3.3 Phenomenological Constraints on Supersymmetric SM

Since we do not know the true mechanism of the SUSY breaking, we are forced to accept the huge number of SUSY breaking parameters. However, as seen previously, the parameter space of the MSSM is restricted by phenomenology. In the following we will review the constraints on the parameter space of soft breaking terms.

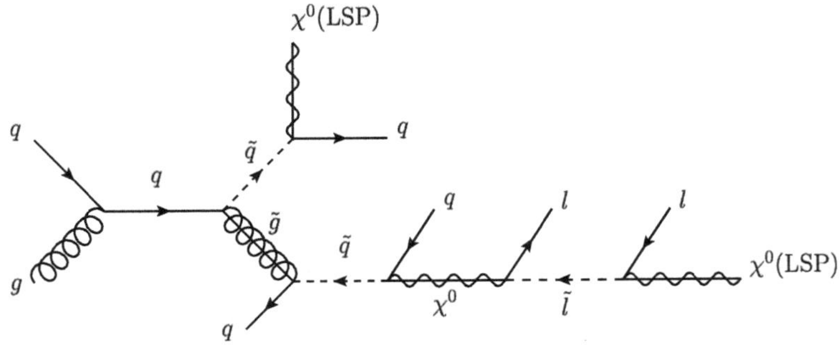

Fig. 3.1 Example of production and cascade decay of sparticles. The final state terminates only with neutral LSPs which bring high missing transverse energy

Constraints on sparticle masses from collider experiments

Sparticles, not discovered so far, are given lower bounds to their masses from high energy collider experiments. The experimental principle for the detection of sparticles is the measurement of high missing energy in collision processes. Particles in the MSSM with R-parity conservation all decay successively (cascade decays, see Fig. 3.1) to final state with the lightest supersymmetric particle (LSP), which yields a large missing energy. The analysis of large missing energy with jet, lepton and photon final states at the LHC has excluded colored sparticle masses (squarks and gluino) less than 1 TeV [4–6].

Constraints on flavor and CP violations

The data from experiments on testing fundemenal symmetries have provided useful constraints on the soft SUSY breaking parameters, especially on the soft breaking mass matrices of fermions [1st, 2nd, 5th and 6th lines of Eq. (3.4)] [7]. The most known constraints are from the non-observation of the flavor changing neutral current processes. The first type of constraint is the suppression of the quark flavor changing neutral current, due to the K $(s\bar{d})$, D $(c\bar{u})$ and B $(b\bar{d})$ oscillations (see Fig. 3.2). Experimental results restrict the flavor off-diagonal components. The ratio between the latter and flavor diagonal terms is approximately restricted to $\frac{\Delta m^2_{\tilde{q}}}{m^2_{\tilde{q}}} \sim \mathcal{O}(10^{-3})$. The CP violation of the K system yields also a constraint of same order to the imaginary part of the squark mass matrices.

Fig. 3.2 Example of supersymmetric $\Delta F = 2$ process: K^0 mixing

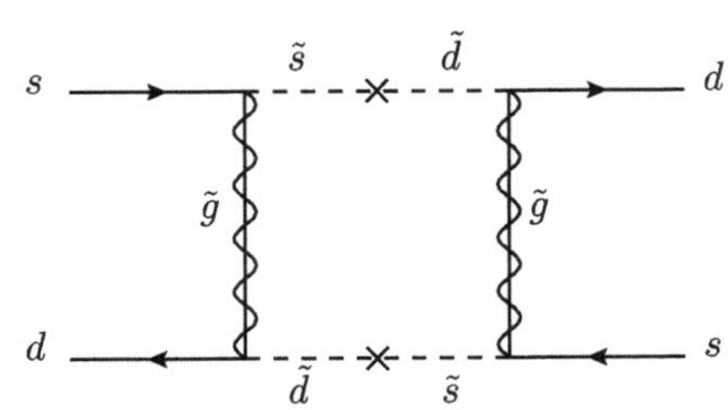

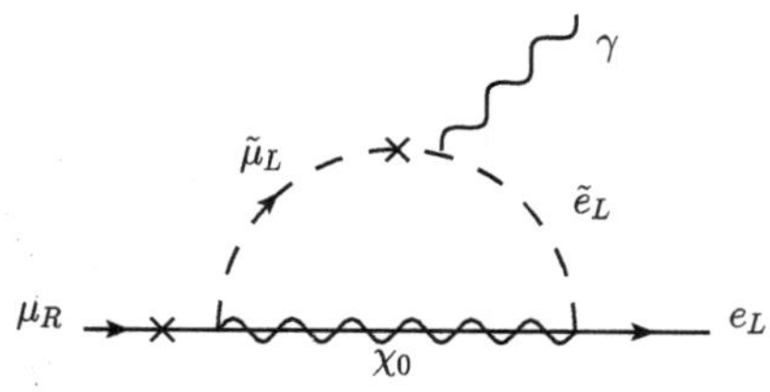

Fig. 3.3 $\Delta F = 1$ lepton flavor violating process

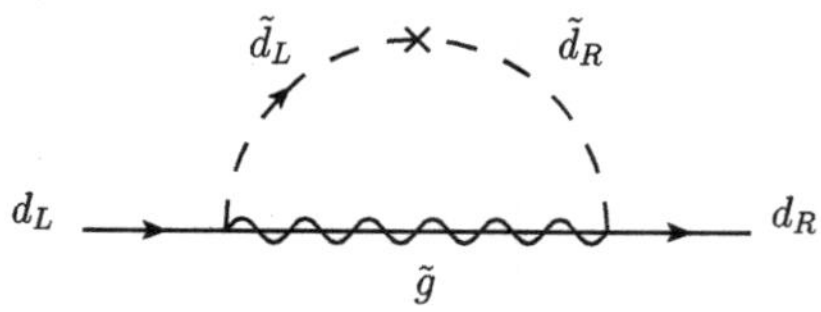

Fig. 3.4 Mass correction to d quark. There are also similar contribution for other light fermion

The second type of constraint is the $\Delta F = 1$ flavor violating process. These constraints apply to the lepton sector, and are obtained from the following flavor violating process:

$$\mu \to e + \gamma . \tag{3.5}$$

Of course, we can replace electron and muon by particles of other flavor. Figure 3.3 shows its typical contribution. The experimental results give tight bounds especially on the flavor off-diagonal component of the trilinear coupling [a_u, a_d, a_e of Eq. (3.4)]. The suppression is approximately $\frac{\Delta m_{\tilde{q}}^2}{m_{\tilde{q}}^2} \sim \mathcal{O}(10^{-3})$ (processes involving the third generation do not give strong constraint).

Flavor conserving processes can also yield restrictions. The third type of restriction comes from the mixing of fermion with their chiral partner (intra-generation mixing, $L \leftrightarrow R$). This constraint applies only to the flavor diagonal components of the trilinear interaction [terms with a_u, a_d, a_e in Eq. (3.4)], for both quarks and leptons. The typical process is depicted in Fig. 3.4. This is a mass correction to fermions, and it must not exceed their masses, so only the first and second generations suffer this suppression.

Finally, measurement of the electric dipole moment constrains the imaginary part of the flavor diagonal components. Figure 3.5 shows its mechanism. The experimental results strongly constrain the imaginary part of the trilinear Higgs-sfermion-sfermion

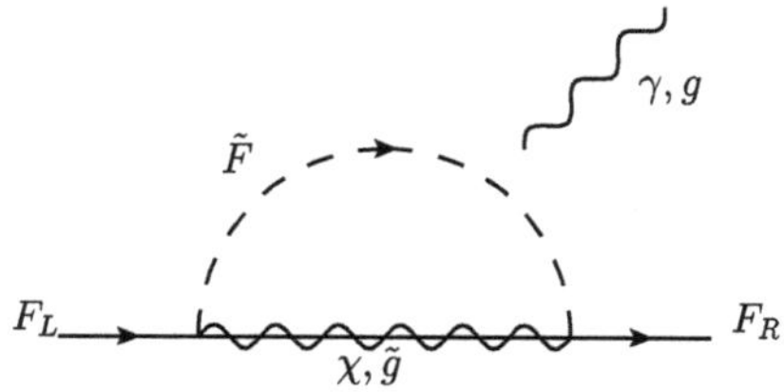

Fig. 3.5 Contribution to the EDM. F is an arbitrary fermion

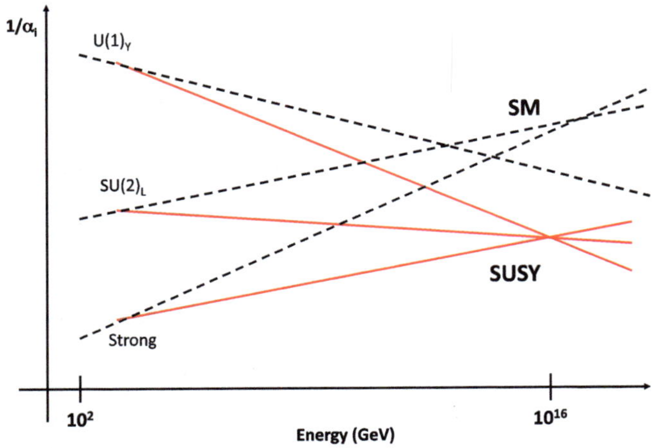

Fig. 3.6 Unification of gauge coupling is better in MSSM than in the SM

interactions [terms with a_u, a_d, a_e in Eq. (3.4)]. The first generation (electron, u and d quarks) is tightly constrained because the available experimental data depend mainly on the first generation fermions. The EDM is the main subject of this thesis, and will be studied in detail.

Renormalization group analysis

The supersymmetric extension of the SM has also an important impact on the renormalization group analysis. Since the quadratic divergences are removed by supersymmetry, it is possible to extrapolate the parameters of the MSSM to very high energy scale using the renormalization group. The inclusion of additional supersymmetric degrees of freedom can deviate significantly the renormalization flow in the high energy region. Surprisingly, the three gauge couplings have a very good unification at 10^{16} GeV(= Λ_{GUT}) in the MSSM [8] (see Fig. 3.6). This fact strongly suggests that the MSSM is an effective theory of some underlying grand unified theory.

The soft breaking terms were also analyzed with renormalization group based on the grand unification [9–13]. We suppose that the soft SUSY breaking terms are also unified at the same scale as gauge coupling unification (Λ_{GUT}) by imposing the following universality relations

$$g_{\mathrm{GUT}} \equiv g_1 = g_2 = g_3 \,, \tag{3.6}$$

$$m_{1/2} \equiv M_1 = M_2 = M_3 \,, \tag{3.7}$$

$$m_0^2 \equiv m_{Q_i}^2 = m_{U_i}^2 = m_{D_i}^2 = m_{L_i}^2 = m_{E_i}^2 = m_{H_u}^2 = m_{H_d}^2 \,, \tag{3.8}$$

$$A_0 \equiv A_t = A_b = A_\tau \,, \tag{3.9}$$

and evolve them along the renormalization group equations to obtain the sparticle masses at the electroweak scale. The supersymmetric model obtained with these boundary conditions are called the constrained-MSSM (cMSSM). The gauge charges of the particles have a strong influence on the evolution of the running of the parameters. As a consequence of renormalization group, the masses of colored particles grows very fast, and the mass m_{E_i} varies the least between $\Lambda_{\rm GUT}$ and the electroweak scale. One important feature of this renormalization group approach is that the Yukawa coupling decreases the growth of the soft breaking masses. It was discovered that for sufficiently large Yukawa coupling of the top quark, the Higgs mass parameter becomes negative, and thus induces the spontaneous breaking of the electroweak symmetry [14–19]. This is actually one of the strong argument which supports the supersymmetric extension of the SM. In this thesis, the analysis will be done without constraining the relative size of sparticle masses. In future work, the result of renormalization group analysis will be taken into account.

References

1. L. Girardello, M.T. Grisaru, Nucl. Phys. **B194**, 65 (1982)
2. J. Rosiek, Phys. Rev. D **41**, 3464 (1990)
3. M. Dine, J.D. Mason, Rept. Prog. Phys. **74**, 056201 (2011)
4. ATLAS Collaboration (Georges Aad et al.), Phys. Rev. Lett. **106**, 131802 (2011)
5. ATLAS Collaboration (Georges Aad et al.), Phys. Lett. B **701**, 186 (2011). arXiv:1110.6189 [hep-ex]
6. CMS Collaboration (Vardan Khachatryan et al.), Phys. Lett. B **698**,196 (2011). arXiv:1111.2733 [hep-ex]
7. F. Gabbiani, E. Gabrielli, A. Masiero, L. Silvestrini, Nucl. Phys. B **477**, 321 (1996)
8. U. Amaldi, W. de Boer, H. Fürstenau, Phys. Lett. B **260**, 447 (1991)
9. N.K. Falck, Z. Phys. C **30**, 247 (1986)
10. S. Martin, M. Vaughn, Phys. Rev. D **50**, 2282 (1994)
11. Y. Yamada, Phys. Rev. D **50**, 3537 (1994)
12. I. Jack, D.R.T. Jones, Phys. Lett. B **333**, 372 (1994)
13. J. Ellis, A. Mustafayev, K. Olive, Eur. Phys. J. C **69**, 201 (2010)
14. L. Ibáñez, G.G. Ross, Phys. Lett. B **110**, 215 (1982)
15. K. Inoue et al., Prog. Theor. Phys. **68**, 927 (1982)
16. K. Inoue et al., Prog. Theor. Phys. **71**, 413 (1984)
17. L. Ibáñez, Phys. Lett. B **118**, 73 (1982)
18. J. Ellis, J. Hagelin, D. Nanopoulos, M. Tamvakis, Phys. Lett. B **125**, 275 (1983)
19. L. Alvarez-Gaumé, J. Polchinski, M. Wise, Nucl. Phys. B **221**, 495 (1983)

Chapter 4
R-parity Violation and Phenomenological Constraints

4.1 R-parity Violating Interactions

The R-parity violating (RPV) interactions are generated by the following superpotential:

$$W_{\not R} = \mu'_i \varepsilon_{ab} \hat{L}_i^a \hat{H}_u^b + \frac{1}{2}\lambda_{ijk}\varepsilon_{ab}\hat{L}_i^a \hat{L}_j^b (\hat{E}^c)_k + \lambda'_{ijk}\varepsilon_{ab}\hat{L}_i^a \hat{Q}_j^b (\hat{D}^c)_k + \frac{1}{2}\lambda''_{ijk}(\hat{U}^c)_i (\hat{D}^c)_j (\hat{D}^c)_k \,, \tag{4.1}$$

with $i, j, k = 1, 2, 3$ indicating the generation, and $a, b = 1, 2$ the $SU(2)_L$ indices. The sum is taken for each index. For the baryon number violating interactions (terms with λ''), the $SU(3)_c$ indices have been omitted. The lepton left-chiral superfields $\hat{L}$ and $\hat{E}^c$ are respectively the $SU(2)_L$ doublet and singlet. The quark superfields $\hat{Q}$, $\hat{U}^c$ and $\hat{D}^c$ denote respectively the quark $SU(2)_L$ doublet, up quark singlet and down quark singlet left-chiral superfields, and $\hat{H}_u$ the up type Higgs left-chiral superfield. The RPV superpotential gives rise to the following baryon or lepton number violating interactions (see Fig. 4.1).

$$\begin{aligned}
\mathscr{L}_{\not R} =& \mu'_i \left[\bar{\nu}_i P_L \left(\tilde{h}_u^0\right)^c - \bar{e}_i P_L \left(\tilde{h}_u^+\right)^c \right] + \text{h.c.} \\
&- \frac{1}{2}\lambda_{ijk}\left[\tilde{\nu}_i \bar{e}_k P_L e_j + \tilde{e}_{Lj}\bar{e}_k P_L \nu_i + \tilde{e}^\dagger_{Rk}\bar{\nu}_i^c P_L e_j - (i \leftrightarrow j) \right] + \text{h.c.} \\
&- \lambda'_{ijk}\Big[\tilde{\nu}_i \bar{d}_k P_L d_j + \tilde{d}_{Lj}\bar{d}_k P_L \nu_i + \tilde{d}^\dagger_{Rk}\bar{\nu}_i^c P_L d_j \\
&\qquad - \tilde{e}_{Li}\bar{d}_k P_L u_j - \tilde{u}_{Lj}\bar{d}_k P_L e_i - \tilde{d}^\dagger_{Rk}\bar{e}_i^c P_L u_j \Big] + \text{h.c.} \\
&- \frac{1}{2}\lambda''_{ijk}\left[\tilde{u}^\dagger_{Ri}\bar{d}_j P_L d_k^c + \tilde{d}^\dagger_{Rj}\bar{u}_i P_L d_k^c + \tilde{d}^\dagger_{Rk}\bar{u}_i P_L d_j^c \right] + \text{h.c.} \,,
\end{aligned} \tag{4.2}$$

N. Yamanaka, *Analysis of the Electric Dipole Moment in the R-parity Violating Supersymmetric Standard Model*, Springer Theses,
DOI: 10.1007/978-4-431-54544-6_4, © Springer Japan 2014

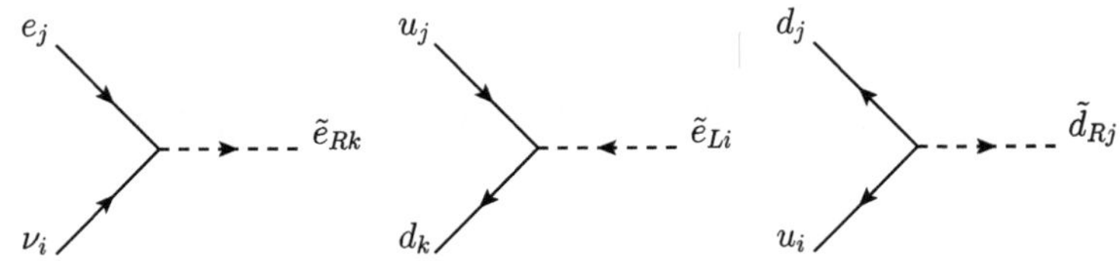

Fig. 4.1 Example of Yukawa interactions generated from R-parity violation. Arrows indicate the lepton or baryon number flow

where $P_L \equiv \frac{1}{2}(1-\gamma_5)$, and $\tilde{h}_u$ denotes the up type higgsino. The first three terms in Eq. (4.2) are lepton number violating and the last term is baryon number violating. There are also RPV scalar quartic interactions, but we do not consider them since these interactions have less effects on observable.

We can also add the general soft SUSY breaking lagrangian in the RPV sector:

$$\begin{aligned}\mathscr{L}_{\not{R}\mathrm{soft}} = &\mu'^2_{ui}\tilde{L}_i H_u + \tilde{m}^2_{di} H^\dagger_d \tilde{L}_i \\ &+ m_{\tilde{G}}\left[\frac{1}{2}A^\lambda_{ijk}\lambda_{ijk}\tilde{L}_i\tilde{L}_j\tilde{E}^c_k + A^{\lambda'}_{ijk}\lambda'_{ijk}\tilde{L}_i\tilde{Q}_i\tilde{D}^c_k + \frac{1}{2}A^{\lambda''}_{ijk}\lambda''_{ijk}\tilde{U}^c_i\tilde{D}^c_j\tilde{D}^c_k\right] \\ &+ \text{h.c.}\,,\end{aligned} \tag{4.3}$$

where the field operators with tildes are the scalar component of the chiral-superfields. The $SU(2)_L$ and $SU(3)_c$ indices were omitted, but fields must be combined in a gauge invariant way.

As mentioned before, the motivation of eliminating the RPV interactions is mainly to prevent the proton decay in the theory. However, there is no definite reasons to forbid all RPV interactions. On the phenomenological ground, there is actually no reasons to prefer R-parity conserved models than RPV models. Furthermore, the RPV interactions can play roles in the grand unification. If we believe the grand unification, the quarks and leptons should be embedded in the same multiplets, and the conservation of R-parity seems to be incompatible. Many grand unified models which effectively give RPV interactions at low energy have been studied [1–6]. From the point of view of the grand unification, there is no preference between the R-parity conserving and RPV models, and all grand unification models do not have any generic prediction for the size of the RPV interactions. (In string theories, it is also possible to construct models with or without R-parity [7, 8]). We can say that the study of RPV interactions has potential to provide us with knowledge about the grand unification.

4.2 Bilinear R-parity Violation

The study of bilinear RPV interactions (mixing between lepton and Higgs) is interesting by itself. The bilinear RPV interactions can be rotated away by redefining the Higgs field as

$$\hat{H}'_{da} = \frac{\mu \hat{H}_{da} + \sum_i \varepsilon_{ba} \mu'_i \hat{L}^b_i}{\sqrt{\mu'^2 + \mu'^2_1 + \mu'^2_2 + \mu'^2_3}}, \tag{4.4}$$

where the massive parameter μ of the first term in the numerator is the coefficient of the mixing between up type and down type Higgs (the so-called μ-term of the superpotential). This redefinition also converts the Higgs-fermion-fermion (standard Yukawa) terms of the superpotential to the RPV superpotential (for example, $\varepsilon_{ab} \hat{Q}^a_i \hat{H}^b_u \hat{U}^c_j \to \varepsilon_{ab} \hat{Q}^a_i \hat{L}^b_k \hat{U}^c_j$). This is the reason why we often treat bilinear and trilinear RPV interactions separately. We must note that this rotation cannot get rid of the bilinear RPV soft breaking terms. In the case where these soft breaking terms are present, the sneutrinos also develop vacuum expectation value.

By using the above properties, it is possible to construct a scenario with R-parity which breaks spontaneously using the following superpotential [9]:

$$\hat{f} = \sum_{i,j=1,2,3} \Big[(\mathbf{f}_u)_{ij} \varepsilon_{ab} \hat{Q}^a_i \hat{H}^b_u \hat{U}^c_j + (\mathbf{f}_d)_{ij} \hat{Q}^a_i \hat{H}_{da} \hat{D}^c_j + (\mathbf{f}_e)_{ij} \hat{L}^a_i \hat{H}_{da} \hat{E}^c_j$$
$$+ (\mathbf{f}_\nu)_{ij} \varepsilon_{ab} \hat{L}^a_i \hat{H}^b_u \hat{\nu}^c_j + (\mathbf{f})_{ij} \hat{\Phi} \hat{S}_i \hat{\nu}^c_j \Big] + (f_0 \hat{H}_u \hat{H}_d - \varepsilon^2) \hat{\Phi}, \tag{4.5}$$

where $\hat{\Phi}$, $\hat{S}_i$ and $\hat{\nu}^c_i$ are the new chiral superfields with lepton numbers 0, −1 and +1, respectively, and they are all with baryon number 0. The mechanism goes as follows. First, the scalar potential gets vacuum expectation values in the directions of $\tilde{\nu}_{Ri}$, $\tilde{S}_i$, $\tilde{\Phi}$, h^0_u and h^0_d. These vacuum expectation values break the lepton number, thus generating the effective bilinear RPV interaction (both the superpotential and the soft breaking lagrangian). This can also be redefined in a basis with the RPV trilinear superpotential and the vacuum expectation value of the sneutrinos. Note that this spontaneous breakdown of R-parity does not change the proton life time since the superpotential of Eq. (4.5) does not minimize to baryon number violating vacuum.

4.3 Phenomenological Constraints on Trilinear RPV Interactions

Many of the trilinear RPV interactions are constrained phenomenologically. We will review in detail the most important ones.

Constraints from the Non-Observation of the Proton Decay

The simultaneous presence of lepton and baryon number violating RPV interactions leads to proton decays (see Fig. 4.2). As the proton decay is not observed in experiments, the combination of λ'', λ or λ'', λ' are strongly constrained [10–12].

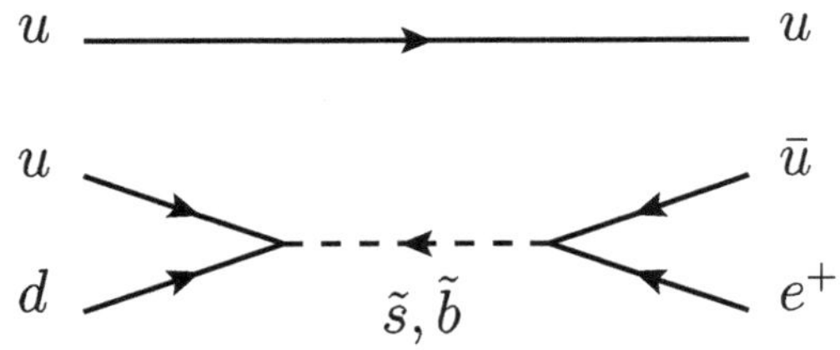

Fig. 4.2 Existence of both baryon and lepton number violating RPV interactions induces proton decay

Table 4.1 Upper limits on combinations of RPV couplings from double beta decay experiments

RPV couplings	λ'^2_{111}	$\lambda'_{112}\lambda'_{121}$	$\lambda'_{113}\lambda'_{131}$
Upper limits ($[m_{\rm SUSY}]^3$)	7.7×10^{-6}	4.0×10^{-7}	1.7×10^{-8}

$[m_{\rm SUSY}]^3$ is the mass of the SUSY particle (d squark involved in the decay) in unit of 100 GeV

Constraints from proton life time can be written as

$$|\lambda'_{ijk}\lambda''_{lmn}| < 10^{-26} \sim 10^{-11},$$
$$|\lambda_{ijk}\lambda''_{lmn}| < 10^{-11} \sim 10^{-3}, \tag{4.6}$$

which set very strong upper limits [10–16]. Due to this result, we often assume in theoretical analysis that baryon and lepton number violating RPV interactions do not co-exist.

Constraints from Lepton Number Violating Processes

The non-observation of the neutrinoless double beta decay sets also strong constraints on RPV couplings [13–21] (see Fig. 4.3).

The combinations of RPV couplings λ'^2_{111}, $\lambda'_{112}\lambda'_{121}$ and $\lambda'_{113}\lambda'_{131}$ are constrained as shown in Table 4.1.

The effective Majorana mass of the neutrino can also be generated by lepton number violating combination of RPV interactions [13–16, 22–24] (see Fig. 4.4). As the neutrino (Majorana) mass is constrained by observation, it is possible to limit

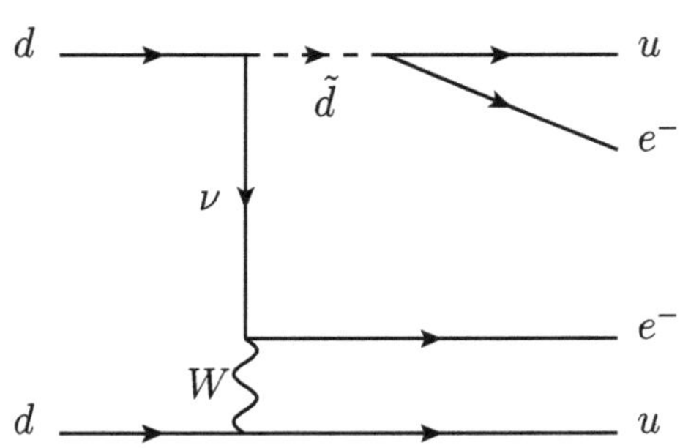

Fig. 4.3 Example of neutrinoless double beta decay amplitude induced from RPV interactions

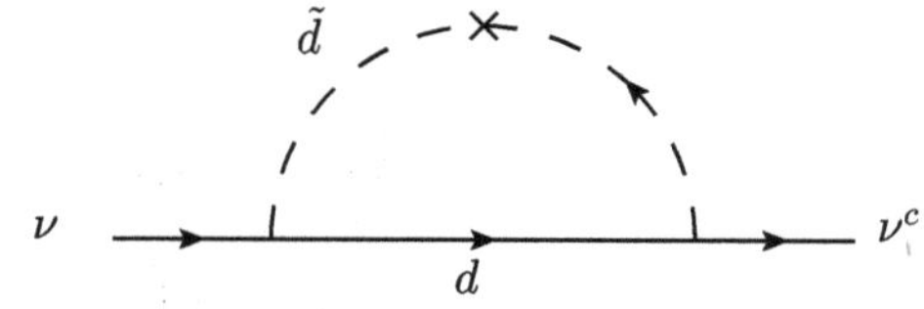

Fig. 4.4 RPV contribution to the Majorana mass of the neutrino

RPV couplings contributing to the process. This gives a relatively tight constraint on the RPV coupling λ'_{133}: $|\lambda'_{133}| < 3.5 \times 10^{-3}$.

Constraints from Precision Tests: Lepton Flavor Violating Process

Some combinations of RPV interactions are tightly constrained by lepton flavor violating processes. Here we present the example of the flavor changing radiative decay of charged lepton $l \to l'\gamma$ [13–16, 25] (see Fig. 4.5).

This process gives the following constraints on RPV interactions:

$$
\begin{aligned}
|\lambda^*_{121}\lambda_{122}| &< 5.7 \times 10^{-5}, \\
|\lambda^*_{131}\lambda_{132}| &< 5.7 \times 10^{-5}, \\
|\lambda^*_{23k}\lambda_{131}| &< 1.1 \times 10^{-4}, \\
|\lambda'^*_{2mk}\lambda'_{1mk}| &< 4.5 \times 10^{-4}, \\
|\lambda'^*_{23n}\lambda'_{13n}| &< 7.7 \times 10^{-3}, \\
|\lambda'^*_{233}\lambda'_{133}| &< 1.0 \times 10^{-2}, \\
|\lambda'^*_{1jk}\lambda'_{3jk}| &< 1.2 \times 10^{-2},
\end{aligned}
\tag{4.7}
$$

where k $(= 1, 2, 3)$ and n $(= 1, 2)$ denote the generation of charged leptons.

Constraints from Precision Tests: Rare Hadron Decays

Hadron decays are very sensitive probe of RPV interactions since they receive contribution from the four-fermion interaction generated from R-parity violation [13–16,

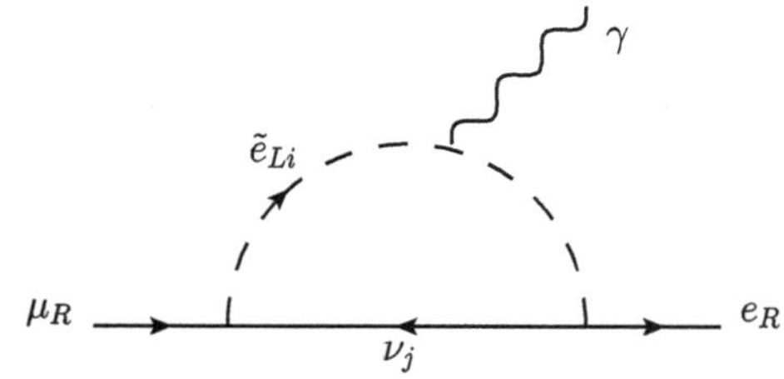

Fig. 4.5 $\mu \to e\gamma$ process within RPV interactions

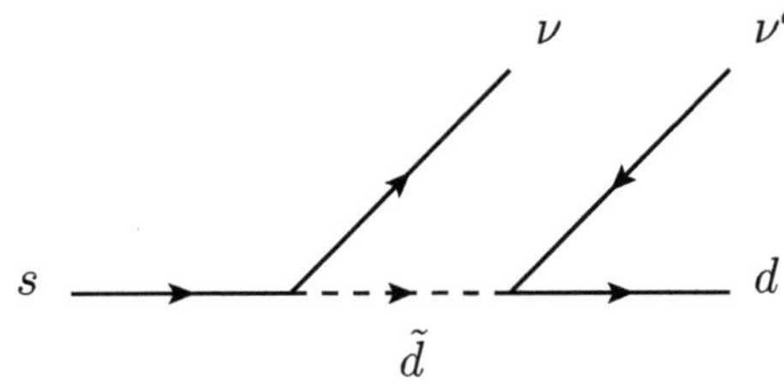

Fig. 4.6 RPV contribution to the $K \to \pi\nu\bar{\nu}$

26–31]. Here we will present the example of the semi-leptonic $K^+ \to \pi^+\nu\bar{\nu}$ decay. This K meson decay is induced by the following effective interaction (see Fig. 4.6)

$$\mathscr{L}_K = \frac{\lambda'^*_{ijk}\lambda'_{i'j'k}}{2m^2_{\tilde{d}_{Rk}}}\bar{d}_j\gamma^\mu P_L d_{j'}\,\bar{\nu}_i\gamma_\mu\nu_{i'} - \frac{\lambda'^*_{ijk}\lambda'_{i'jk'}}{2m^2_{\tilde{d}_{Lj}}}\bar{d}_k\gamma^\mu P_R d_{k'}\,\bar{\nu}_{i'}\gamma_\mu\nu_i + \text{h.c.}\,. \quad (4.8)$$

The branching ratio of the purely RPV $K^+ \to \pi^+\nu\bar{\nu}$ decay is given by [28–31]

$$B_{\text{RPV}}(K^+ \to \pi^+\nu\bar{\nu}) = \frac{r_+ B(K^+ \to \pi^0 e^+\nu_e)}{16|V_{us}|^2 G_F^2}\left|\frac{\lambda'^*_{i2n}\lambda'_{j1n}}{m^2_{\tilde{d}_{Rn}}} - \frac{\lambda'^*_{in1}\lambda'_{jn2}}{m^2_{\tilde{d}_{Ln}}}\right|^2, \quad (4.9)$$

where $r_+ = 0.901$ is the isospin correction factor. The branching ratio of the K decay into isospin partner is given by $B(K^+ \to \pi^0 e^+\nu_e) = (5.07 \pm 0.04) \times 10^{-2}$. The above RPV branching ratio should not excess the discrepancy between experimental data [32–35] and the standard model prediction [36]. Recently, experiment has observed the rare $K^+ \to \pi^+\nu\bar{\nu}$ decay, and the result is consistent with the standard model prediction. The RPV contribution is therefore constrained and should not excess the error of $B_{\text{exp}} - B_{\text{SM}}$, the difference between experimental value and the standard model contribution. We do not consider the interference between RPV and standard model contributions. The experimental value of the branching ratio of the decay $K^+ \to \pi^+\nu\bar{\nu}$ is [32–35]

$$B_{\text{exp}}(K^+ \to \pi^+\nu\bar{\nu}) = (1.73^{+1.15}_{-1.05}) \times 10^{-10}. \quad (4.10)$$

The theoretical estimation of the standard model contribution is [36]

$$B_{\text{SM}}(K^+ \to \pi^+\nu\bar{\nu}) = (7.81^{+0.80}_{-0.71} \pm 0.29) \times 10^{-11}, \quad (4.11)$$

where the first error is related to the uncertainty of the input parameters and the second one to the theoretical uncertainty. We obtain then the following inequality for the bilinear of RPV couplings:

$$4.1 \times \left| \frac{\lambda'^*_{i2n}\lambda'_{j1n}}{[m_{\tilde{d}_{Rn}}]^2} - \frac{\lambda'^*_{in1}\lambda'_{jn2}}{[m_{\tilde{d}_{Ln}}]^2} \right|^2 < 2.2 \times 10^{-10}, \tag{4.12}$$

where the expression with $[\cdots]$ denotes the mass of the sparticle in unit of 100 GeV. The right-hand side of the above equation is the error of $B_{\rm exp} - B_{\rm SM}$. This gives then the following bounds to the RPV couplings:

$$|\lambda'^*_{i2n}\lambda'_{j1n}| < 7.3 \times 10^{-6}[m_{\tilde{d}_{Rn}}]^2 , \qquad |\lambda'^*_{in1}\lambda'_{jn2}| < 7.3 \times 10^{-6}[m_{\tilde{d}_{Ln}}]^2, \tag{4.13}$$

where the dominance of the single bilinear of RPV couplings was assumed. Note that we have not considered the interference of the RPV amplitudes with standard model and R-parity conserving supersymmetric contributions.

This limit can be used to constrain RPV interactions with other flavors via flavor mixing. The change from the current basis to the mass basis yields the change $\bar{d}_k d_{k'} \to \bar{d}'_k d'_{k'} \simeq V_{k'1} V^*_{k2} \bar{s} d + \cdots$ for the quark bilinear in the effective lagrangian. We then obtain the following upper limits

$$\begin{aligned} |\lambda'_{imk}| &< 5.7 \times 10^{-3}[m_{\tilde{d}_{Rk}}], \\ |\lambda'_{i3k}| &< 0.14\,[m_{\tilde{d}_{Rk}}], \end{aligned} \tag{4.14}$$

where $i, k = 1, 2, 3$ and $m = 1, 2$.

Similar analysis holds for the B meson decays [37–40]. The effective lagrangian of Eq. (4.8) involving b quark generates the following purely RPV decay $B^+ \to X_s \nu_j \bar{\nu}_i$, and can be expressed as follows [37]

$$\frac{B_{\rm RPV}(B^+ \to X_s \nu_j \bar{\nu}_i)}{B(B^+ \to X_c e^+ \nu_e)} = \frac{1}{8G_F^2 |V_{cb}|^2 f_{PS}(m_c^2/m_b^2)} \sum_k \left\{ \left| \frac{\lambda'_{i2k}\lambda'^*_{j3k}}{2m^2_{\tilde{d}_{Rk}}} \right|^2 + \left| \frac{\lambda'_{ik2}\lambda'^*_{jk3}}{2m^2_{\tilde{d}_{Lk}}} \right|^2 \right\}, \tag{4.15}$$

where $f_{PS}(x) = 1 - 8x + 8x^3 - x^4 - 12x^2 \ln x^2$ is the phase space factor. By using the quark masses $m_c = 1.29^{+0.05}_{-0.11}$ GeV and $m_b = 4.19^{+0.18}_{-0.06}$ GeV [39, 40], we obtain $f_{PS}(m_c^2/m_b^2) \approx 0.5$. The notations X_s and X_c denote the strange and charmed hadronic final states respectively. From the review of Particle data group, we have [39, 40]

$$B(B^+ \to K^+ \nu\bar{\nu}) < 1.3 \times 10^{-5}, \tag{4.16}$$

$$B(B^+ \to K^*(892)^+ \nu\bar{\nu}) < 8 \times 10^{-5} , \tag{4.17}$$

$$B(B^+ \to X_c e^+ \nu_e) = (10.8 \pm 0.4) \times 10^{-2}, \tag{4.18}$$

From the upper two inequalities, we have $B(B^+ \to X_s \nu\bar{\nu}) < 1 \times 10^{-4}$. The standard model prediction is $B_{\rm SM}(B^+ \to X_s \nu\bar{\nu}) < 5 \times 10^{-5}$ [37]. By neglecting the standard model contribution, we obtain the following bound to the RPV interactions:

$$|\lambda'_{i2k}\lambda'^{*}_{j3k}| < 5.8 \times 10^{-4}[m_{\tilde{d}_{Rk}}]^2, \qquad |\lambda'_{ik2}\lambda'^{*}_{jk3}| < 5.8 \times 10^{-4}[m_{\tilde{d}_{Lk}}]^2. \quad (4.19)$$

As for the $K \to \pi\nu\bar{\nu}$ decay, the change from the current basis to the mass basis can set limits to other combination of RPV couplings. The mixing between s and b quarks gives the following bound

$$|\lambda'_{i3k}| < 0.12[m_{\tilde{d}_{Rk}}]. \quad (4.20)$$

where $i, k = 1, 2, 3$.

Constraints from Precision Tests: Electric Dipole Moments

The electric dipole moments of neutron, YbF molecule, ^{205}Tl and ^{199}Hg atoms can set severe constraints on the CP phases between RPV couplings. This topic is the main subject of this thesis and will be discussed in detail in Part III.

Constraints from Precision Tests: Universalities

The universality of the gauge coupling is an important tool to rule out the interactions of new physics. If the universality of the weak coupling holds, the contribution from the RPV must be embedded in the uncertainty of the standard model. This can be applied to the RPV interactions by noticing that the weak decay of leptons (or hadrons) can be mimicked by RPV amplitude with the same Lorentz structure [41, 42].

Let us examine the decays of leptons. The lepton number violating RPV interaction λ_{ijk} also contributes to the process. In Fig. 4.7, an example of the RPV muon decay process is shown. This contribution can interfere with the muon beta decay, and leads to the following redefinition of the Fermi weak coupling constant for the muon decay

$$\frac{G_F}{\sqrt{2}} = \frac{g_2^2}{8m_W^2}\left[1 + \frac{m_W^2}{g_2^2 m_{\tilde{e}_{R_k}}^2}|\lambda_{21k}|^2\right]. \quad (4.21)$$

Similarly, the Fermi constants for the decays of τ lepton to electron and to muon will be shifted by $(m_W^2/g_2^2 m_{\tilde{e}_{R_k}}^2)|\lambda_{31k}|^2$ and $(m_W^2/g_2^2 m_{\tilde{e}_{R_k}}^2)|\lambda_{32k}|^2$, respectively. The decay ratios $R_{\tau\mu} \equiv \Gamma(\tau^- \to \mu^-\bar{\nu}_\mu\nu_\tau)/\Gamma(\mu^- \to e^-\bar{\nu}_e\nu_\mu)$ and $R_\tau \equiv \Gamma(\tau^- \to e^-\bar{\nu}_e\nu_\tau)/\Gamma(\tau^- \to \mu^-\bar{\nu}_\mu\nu_\tau)$ will then be shifted by

$$R_{\tau\mu} = [R_{\tau\mu}]_{SM}\left\{1 + 2[(m_W^2/g_2^2 m_{\tilde{e}_{R_k}}^2)|\lambda_{32k}|^2 - (m_W^2/g_2^2 m_{\tilde{e}_{R_k}}^2)|\lambda_{21k}|^2]\right\}, \quad (4.22)$$

$$R_{\tau} = [R_{\tau}]_{SM}\left\{1 + 2[(m_W^2/g_2^2 m_{\tilde{e}_{R_k}}^2)|\lambda_{31k}|^2 - (m_W^2/g_2^2 m_{\tilde{e}_{R_k}}^2)|\lambda_{32k}|^2]\right\}. \quad (4.23)$$

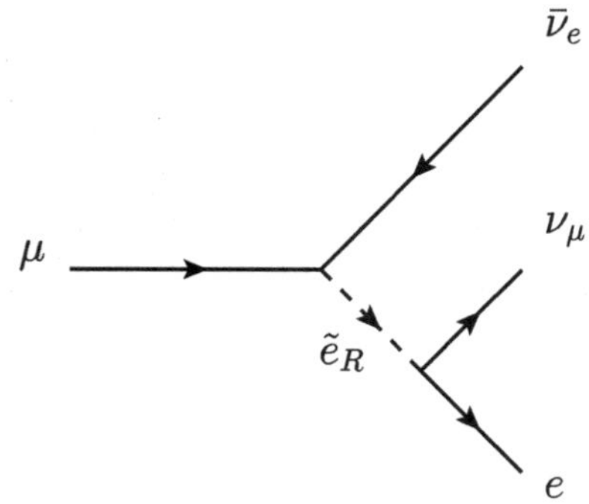

Fig. 4.7 Example of process with RPV interactions interfering to the SM weak process (example of muon decay)

The standard model prediction of these ratios are

$$[R_{\tau\mu}]_{\rm SM} = 1.309 \times 10^6,$$
$$[R_\tau]_{\rm SM} = 1.028. \tag{4.24}$$

These values were calculated by taking into account the radiative corrections and the running coupling [42]. The experimental values listed by the review of Particle data group are [39, 40]

$$[R_{\tau\mu}] = (1.315 \pm 0.006) \times 10^6,$$
$$[R_\tau] = 1.025 \pm 0.003. \tag{4.25}$$

The consistency between the experimental values and the standard model predictions implies that the RPV couplings must be within the experimental errors. The following constraints can then be given:

$$|\lambda_{21k}| < 0.05[m_{\tilde{e}_{Rk}}],$$
$$|\lambda_{31k}| < 0.03[m_{\tilde{e}_{Rk}}],$$
$$|\lambda_{32k}| < 0.05[m_{\tilde{e}_{Rk}}], \tag{4.26}$$

where $k = 1, 2, 3$. $[\cdots]$ denotes the mass of sparticles in unit of 100 GeV. Here we have also assumed the dominance of single RPV couplings.

Similar analysis holds for the decay ratios $\Gamma(\pi^- \to e^- \bar{\nu}_e)/\Gamma(\pi^- \to \mu^- \bar{\nu}_\mu)$ and $\Gamma(\tau^- \to \pi^- \nu_\tau)/\Gamma(\pi^- \to \mu^- \bar{\nu}_\mu)$, and also for the decay ratios $\Gamma(D^0 \to \mu^+ \nu_\mu K^-)/\Gamma(D^0 \to e^+ \nu_e K^-)$, $\Gamma(D^+ \to \mu^+ \nu_\mu \bar{K}^0)/\Gamma(D^+ \to e^+ \nu_e \bar{K}^0)$ and $\Gamma(D^+ \to \mu^+ \nu_\mu \bar{K}^*(892)^0)/\Gamma(D^+ \to e^+ \nu_e \bar{K}^*(892)^0)$. These processes receive contribution from lepton number violating RPV interactions λ'_{ijk}, so it is possible to constrain them. This method has the advantage to cancel the theoretical uncertainty due to the meson form factors, and the ratio can be fully calculated in the standard model. In these cases, the uncertainty of the CKM matrix elements V_{ud} and V_{cs} has to be taken into account. For detailed discussion, see Ref. [42].

RPV at Colliders

The RPV processes can also be observed in collider experiments [22, 41, 43, 44]. The first type of manifestation of the R-parity violation is the resonance of the sneutrino in pp collision, as shown in Fig. 4.8. The analysis of the data accumulated at the LHC provides a stringent constraint on the coupling λ'_{311}, if the τ sneutrino is the LSP. The non-observation of such resonance gives $\lambda'_{311} < 10^{-2}$ for $m_{\tilde{\nu}_\tau} = 1$ TeV and $\lambda'_{311} < 10^{-3}$ for $m_{\tilde{\nu}_\tau} = 100$ GeV [43, 44].

The other way to constrain RPV interactions is to analyze the displaced vertices of the heavy particle decay. The typical process is the decay of the lightest neutralino as shown in Fig. 4.9. No significant result was found at the LHC, and it was concluded that the product between the production cross-section and the decay branching fraction of the neutralino is less than 5 pb for $m_{\tilde{q}} = 1.5$ TeV [43, 44].

Constraints from Cosmology

The non-conservation of R-parity leads to the decays of lightest sparticles (LSPs) (example of the decay is shown in Fig. 4.9) [13–16, 22]. This fact can disturb the current picture of the Universe in two ways, and we need to constrain RPV interactions in each case.

The first topic is the stability of the supersymmetric dark matters. The candidates of dark matter in supersymmetric models are the lightest neutralino or sneutrino. To

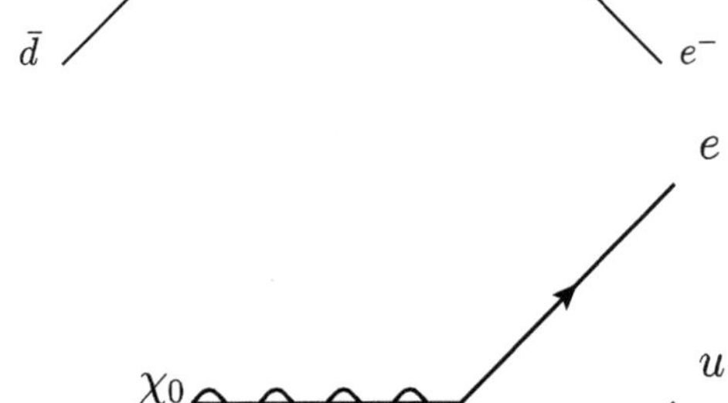

Fig. 4.8 Example of resonance process with RPV interactions in collider search

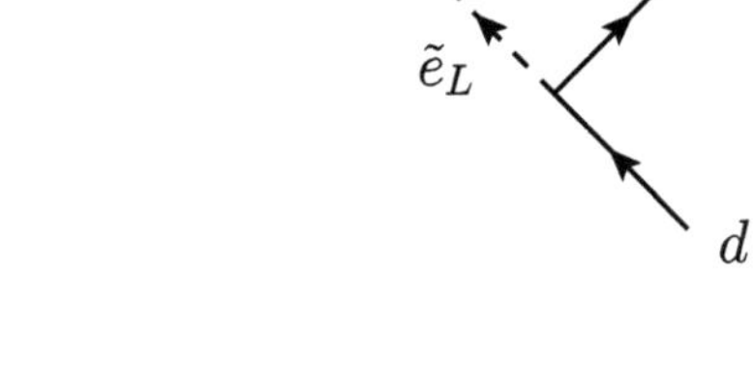

Fig. 4.9 Neutralino decay within RPV interactions

reconcile the scenario of supersymmetric dark matters with the existence of RPV interactions, the following constraint must hold:

$$|\lambda_{ijk}|, |\lambda'_{ijk}|, |\lambda''_{ijk}| < 10^{-21}, \tag{4.27}$$

for $m_{\rm SUSY} = O(100\,{\rm GeV})$, where $i, j, k = 1, 2, 3$. Of course these limits do not apply when the dark matter is not composed of supersymmetric particles.

If unstable LSPs with RPV interactions is larger than 10^{-20}, there is a second type of constraints. The LSP must actually decay before the nucleosynthesis to not disturb the nucleosynthesis [45]. The decay of LSPs before nucleosynthesis is translated to the following *lower* bounds

$$|\lambda_{ijk}|, |\lambda'_{ijk}|, |\lambda''_{ijk}| > 10^{-12}. \tag{4.28}$$

The other cosmological source of strong constraints on RPV interactions is the dilution of the baryon asymmetry [46–49]. The RPV interactions can dilute the baryon number asymmetry of the Universe when the expansion rate of the Universe is smaller than the decay rate of matter. The non-dilution of the matter imposes also constraints on RPV interactions as follows:

$$\begin{aligned} |\lambda_{ijk}|, |\lambda'_{ijk}|, |\lambda''_{ijk}| &< 1.6 \times 10^{-7} [m_{\tilde{f}}]^{1/2} \quad (T > m_{\tilde{f}}), \\ |\lambda_{ijk}|, |\lambda'_{ijk}|, |\lambda''_{ijk}| &< 1.6 \times 10^{-6} [m_{\tilde{f}}]^{1/2} \quad (T < m_{\tilde{f}}), \end{aligned} \tag{4.29}$$

where $[m_{\tilde{f}}]$ is the mass of sfermions in unit of 100 GeV. We must note that these bounds do not apply simultaneously to all RPV interactions. It is possible to protect the baryon number asymmetry by imposing the above constraints to a fixed generation.

References

1. L.J. Hall, M. Suzuki, Nucl. Phys. B **231**, 419 (1984)
2. D. Brahm, L. Hall, Phys. Rev. D **40**, 2449 (1989)
3. K. Tamvakis, Phys. Lett. B **382**, 251 (1996)
4. R. Hempfling, Nucl. Phys. B **478**, 3 (1996)
5. G.F. Giudice, R. Rattazzi, Phys. Lett. B **406**, 321 (1997)
6. B.C. Allanach, A. Dedes, H.K. Dreiner, Phys. Rev. D **60**, 075014 (1999)
7. M.C. Bento, L. Hall, G.G. Ross, Nucl. Phys. B **292**, 400 (1987)
8. N. Ganoulis, G. Lazarides, Q. Shafi, Nucl. Phys. B **323**, 374 (1989)
9. A. Masiero, J.W.F. Valle, Phys. Lett. B **251**, 273 (1990)
10. A. Yu, Smirnov and F. Vissani. Phys. Lett. B **380**, 317 (1996)
11. G. Bhattacharyya, P.B. Pal, Phys. Lett. B **439**, 81 (1998)
12. G. Bhattacharyya, P.B. Pal, Phys. Rev. D 59 (1999)
13. G. Bhattacharyya, arXiv:hep-ph/9709395
14. H.K. Dreiner, arXiv:hep-ph/9707435
15. R. Barbier et al., Phys. Rept. **420**, 1 (2005)

16. M. Chemtob, Prog. Part. Nucl. Phys. **54**, 71 (2005)
17. K.S. Babu, R.N. Mohapatra, Phys. Rev. Lett. **75**, 2276 (1995)
18. M. Hirsch, H.V. Klapdor-Kleingrothaus, S.G. Kovalenko, Phys. Rev. Lett. **75**, 17 (1995)
19. A. Faessler, S. Kovalenko, F. Simkovic, J. Schwieger, Phys. Rev. Lett. **78**, 183 (1997)
20. M. Hirsch, H.V. Klapdor-Kleingrothaus, S.G. Kovalenko, Phys. Rev. D **57**, 1947 (1998)
21. A. Faessler, T. Gutsche, S. Kovalenko, F. Simkovic, Phys. Rev. D **77**, 113012 (2008)
22. S. Dawson, Nucl. Phys. B **261**, 297 (1985)
23. S. Dimopoulos, L.J. Hall, Phys. Lett. B **207**, 210 (1988)
24. R.M. Godbole, P. Roy, X. Tata, Nucl. Phys. B **401**, 67 (1993)
25. M. Chaichian, H. Huitu, Phys. Lett. B **384**, 157 (1996)
26. H.K. Dreiner, M. Krämer, B. O'Leary, Phys. Rev. D **75**, 114016 (2007)
27. Y. Kao, T. Takeuchi, arXiv:0909.0042 [hep-ph]
28. K. Agashe, M. Graesser, Phys. Rev. D **54**, 4445 (1996)
29. D. Choudhury, P. Roy, Phys. Lett. B **378**, 153 (1996)
30. N.G. Deshpande, D.K. Ghosh, X.-G. He, Phys. Rev. D **70**, 093003 (2004)
31. A. Deandrea, J. Welzel, M. Oertel, JHEP **0410**, 38 (2004)
32. A.V. Artamonov et al., BNL-E949 collaboration. Phys. Rev. D **79**, 092004 (2009)
33. S. Adler et al., Phys. Rev. D **77**, 052003 (2008)
34. S. Adler et al., Phys. Rev. D **70**, 037102 (2004)
35. S. Adler et al., Phys. Lett. B **537**, 211 (2002)
36. J. Brod, M. Gorbahn, E. Stamou, Phys. Rev. D **83**, 034030 (2011)
37. Y. Grossman, Z. Ligeti, E. Nardi, Nucl. Phys. B **465**, 369 (1996); [Erratum] Nucl. Phys. B **480**, 753 (1996)
38. G. Buchalla, A.J. Buras, Nucl. Phys. B **400**, 225 (1993)
39. K. Nakamura et al., Particle data group. J. of Phys. G **37**, 075021 (2010) and 2011 partial update for the 2012 edition (URL: http://pdg.lbl.gov)
40. J. Beringer et al., Particle data group. Phys. Rev. D **86**, 010001 (2012)
41. V. Barger, G.F. Giudice, T. Han, Phys. Rev. D **40**, 2987 (1989)
42. Y. Kao, T. Takeuchi, arXiv:0910.4980 [hep-ph]
43. J.R. Ellis, G. Gelmini, C. Jarlskog, G.G. Ross, J.W.F. Valle, Phys. Lett. B **150**, 142 (1985)
44. H.K. Dreiner, G.G. Ross, Nucl. Phys. B **365**, 597 (1991); ATLAS Collaboration (presented by P. D. Jackson), arXiv:1112.0369 [hep-ex]
45. J.E. Kim, B. Kyae, J.D. Park, arXiv:hep-ph/9810503
46. A. Bouquet, P. Salati, Nucl. Phys. B **284**, 557 (1987)
47. W. Fischler, G. Giudice, R.S. Leigh, S. Paban, Phys. Lett. B **258**, 45 (1991)
48. B.S. Campbell, S. Davidson, J. Ellis, K. Olive, Phys. Lett. B **256**, 457 (1991)
49. H.K. Dreiner, G.G. Ross, Nucl. Phys. B **410**, 188 (1993)

Part II
Electric Dipole Moments

Chapter 5
Backgrounds and Motivations for EDM Search

The study of the electric dipole moment (EDM) began in the measurement of neutron EDM done by Purcell and Ramsey in 1949 [1]. Attentions were paid to EDM, known as parity-odd observable, when parity violation was discovered in the decay of polarized ^{60}Co [2, 3]. Landau pointed out however that EDM can not probe the parity violation since it is also time reversal-odd [4–6]. The time reversal symmetry (or equivalently CP, within the assumption of CPT theorem) of the theory must be broken to observe finite EDM. The parity (P) and time reversal (T) violations of the EDM can be explicitly shown in the following way. Consider the EDM interaction hamiltonian $H = -d\langle\sigma\rangle \cdot \mathbf{E}$ (the EDM is proportional to the spin unit vector $\langle\sigma\rangle$ since it is the only vector quantity in the system). We then have

$$-d\langle\sigma\rangle \cdot \mathbf{E} \xrightarrow{P} -d\langle\sigma\rangle \cdot (-\mathbf{E}) = -H, \tag{5.1}$$

$$-d\langle\sigma\rangle \cdot \mathbf{E} \xrightarrow{T} -d(-\langle\sigma\rangle) \cdot \mathbf{E} = -H. \tag{5.2}$$

After the CP violation in the K meson system was discovered [7], the importance of EDM search was then fully recognized.

The EDM is an observable measurable in a variety of systems. It was tested also for many atoms from 60's. In 1963, Schiff gave the famous *Schiff's screening theorem* [8] which states that the *EDM of non-relativistic, point-like components is screened in a neutral electrostatically bound system.* The EDM of components in atoms suffers then from the screening due to the rearrangement of the system. Schiff also pointed out the loopholes of the theorem. Atomic EDM can arise due to the following mechanisms:

- Relativistic effect to electrons with intrinsic EDM.
- P, T-odd interactions between electron and nucleus.
- Finite size effect of the nucleus: the residual effect, the *nuclear Schiff moment* is not screened by electron rearrangement.

The 70's was an era of a great development for CP violation in theoretical physics. Kobayashi and Maskawa predicted the existence of 6 quarks with the famous

N. Yamanaka, *Analysis of the Electric Dipole Moment in the R-parity Violating Supersymmetric Standard Model*, Springer Theses,
DOI: 10.1007/978-4-431-54544-6_5, © Springer Japan 2014

Cabbibo-Kobayashi-Maskawa (CKM) mixing and successfully included the CP violation in renormalizable theory [9]. The EDMs of neutron and electron were then evaluated, but exhibited a very small contribution from the CKM matrix. This will be shown in the subsequent chapter.

The potential role of the Strong CP lagrangian was also pointed out. The QCD lagrangian can also have the *Strong CP* θ-term, which contributes to the P, CP violation of the hadron sector. The θ-term has a large contribution to the neutron EDM [10]. It is then possible to use the experimental data to constrain the size of the θ-term. An extremely fine-tuned coupling θ (to better than one part in 10^9) has been obtained. This unusual fine-tuning is known as the *Strong CP problem*. This fact goes to the development of the dynamical relaxation of the Strong CP problem introduced by Peccei and Quinn [11], which solves the problem by making the θ-term unphysical while giving rise to the pseudo-Nambu-Goldstone boson, the *axion*. The axions are still not discovered in experiment, and this subject is one of the most important experimental challenges in particle physics.

While the CKM theory was very successful in describing CP violation in experiments, it has also a serious problem in interpreting our matter abundant Universe. Our Universe is thought to be initially empty of matter, and need satisfy some conditions to dynamically generate baryon and lepton number asymmetries in a sufficient manner. Sakharov pointed out the famous three criteria for realizing matter abundant Universe [12, 13]: *(a)* The existence of baryon number violating interactions; *(b)* Sufficient C and CP violations; *(c)* Departure from thermal equilibrium. In the CKM theory, the CP violation (condition *(b)*) was estimated to be too weak to explain the matter abundant Universe. The baryon number asymmetry of our Universe is now known to be one of the serious problem of the standard model (SM). We need therefore some new physics beyond the SM. The supersymmetric extension of the SM is known to be one of the candidates of new physics which has the potential to solve such problems and has been developed and analyzed by many theorists. The supersymmetric extension of the SM can provide many CP violation via the soft breaking terms, thus being a very interesting target of the EDM analysis. The CP violation given by SUSY also has a large contribution to the EDM, and current experimental data give constraints to the CP violating phases of soft breaking terms. It is then possible to constrain the CP phases of the supersymmetric SM from experimental data of neutron and atomic EDMs by more than 2 orders of magnitude (when supersymmetric particles have masses around TeV, which is preferred by the hierarchy between the breakdowns of supersymmetry and electroweak symmetry). This unusual tuning is called the *SUSY CP problem*. Studies of other models, such as the multi Higgs doublet models [14–16], left-right symmetric models [17, 18], give also constraints on the CP violation of the model considered.

Up to now the EDM searches give only null results in every system studied. However, as seen above, the search and analysis of EDMs gave us many hints and visions of what new physics look like, and stimulated the development of many theoretical investigations. The search for EDMs are still one of the most important challenges in particle physics, with the main goal: *beyond the standard model*.

EDM has many advantages in searching new physics. Its experimental study can be categorized in the low energy precision test of the SM, with lower cost than high energy experiments using colliders. In General, the low energy precision tests including EDM experiments and high energy experiments are complementary in new physics search. Moreover, the EDM can be measured in a variety of systems: the EDMs of each system (neutron, atoms, molecules, muon, nuclei, etc) are sensitive to their own parameter space of new physics, and each EDM is complementary. This is one of the important motivations for the EDM search. Many EDM experiments are now being planned or prepared to search for EDMs in a large variety of systems [19]. The small SM contribution must also be emphasized. The CKM contribution to the EDM is known to be be very small, due to the suppression in low orders of perturbation. These more or less accidental suppressions give a very small SM background on EDM, which makes the EDM to be a very good probe of new physics. This topic will be discussed in detail later.

The sensitivity of the EDM on new physics in comparison with other experimental approaches is also remarkable. This can be illustrated by making a very naïve comparison of the two experimental approaches. Let us assume that the contribution of the new phyics to the EDM is from the one-loop mechanism with a $O(1)$ coupling with CP phase. For the neutron EDM d_n, we then obtain

$$d_n \sim \frac{Y_q e}{4\pi^2 M_{\rm NP}} \sim \frac{10^{-21}}{M_{\rm NP}/\mathrm{GeV}} e\,\mathrm{cm}, \tag{5.3}$$

where $M_{\rm NP}$ stands for the mass of the particles in the loop, and Y_q is the quark Yukawa coupling. If we interpret the mass of the intermediate particles to be the scale of new physics, for $d_n < 10^{-26}$ e cm, we are probing new physics of the scale of 100 TeV. This value is well beyond the current sensitivity of the present high energy experiments (for LHC, $E_{\rm CM} = 14$ TeV). Of course, small CP phases or accidental suppression of interactions contributing to EDM can upset this argument, and probing these scenarios is more or less limited, but this naïve comparison nevertheless tells us the power and the potential of EDM experiments.

With all these arguments seen above, we can affirm that EDM is a very interesting topic of particle physics, and this subject must be discussed.

There are now many EDM experimental data available for many systems. The present experimental upper bounds of EDMs are given in Table 5.1. This table shows the importance of the EDM search. The future plans of many experiments on EDM have been made. The present planned neutron EDM experiments targets $O(10^{-28})$ e cm level sensitivity by increasing the ultracold neutron density stored in cells (PSI, Los Alamos, RCNP-TRIUMF, J-PARC, etc). For paramagnetic atoms, the search for the EDM of Francium is prepared in Tohoku university, aiming at $O(10^{-28})$ e cm. For the EDM of diamagnetic atoms, ^{129}Xe (Tokyo Inst. Tech., $O(10^{-30})$ e cm), ^{225}Ra (Argonne, $O(10^{-28})$ e cm), radon (TRIUMF, $O(10^{-29})$ e cm) are planned. Searching for molecular EDMs is also planned in many facilities (with ThO, PbO molecules, etc). The atomic and molecular EDM measurements have known significant improvements with laser trap techniques. On the side of charged

Table 5.1 Current experimental limits from direct EDM search and methods used in measurement

System	Experimental limit (*e* cm)	Techniques
Neutron	$d_n < 2.9 \times 10^{-26}$	Storage of Ultracold neutron [24]
^{205}Tl atom	$d_{\mathrm{Tl}} < 9.4 \times 10^{-25}$	Beam of Tl atoms [25]
^{199}Hg atom	$d_{\mathrm{Hg}} < 3.1 \times 10^{-29}$	Storage of Hg atoms in vapor cell [26]
^{129}Xe atom	$d_{\mathrm{Xe}} < 4.0 \times 10^{-27}$	Spin exchange pumped maser [27]
YbF molecule	$(d_e < 1.05 \times 10^{-27})$	Beam of YbF molecules [28]
Muon	$d_\mu < 1.8 \times 10^{-19}$	Storage ring [29]

For the case of YbF molecule experiment, the corresponding limit to the electron EDM is shown

particles, muon (prospected sensitivity $O(10^{-25})$ *e* cm), proton ($O(10^{-29})$ *e* cm), deuteron ($O(10^{-29})$ *e* cm) EDM measurements are prepared at J-PARC and BNL. The measurement of charged particles uses the new technique with storage ring [20–23]. This new technological innovation opens up the possibility for the measurement of light nuclei, such as the ^{3}He nucleus.

To give constraints on parameters of new physics from EDM experimental data, we need to analyze the hadronic, nuclear and atomic many-body physics [30–34]. The EDM of a composite system depends in fact on many underlying P, CP-odd mechanisms and the relative sensitivity varies between systems. This is the reason why the EDM data for different composite systems are complementary. The dependence of EDMs on the underlying mechanisms are illustrated in Fig. 5.1.

The accuracy of the hadronic, nuclear and atomic level calculations is therefore crucial for the search of new physics through EDMs. Details of these calculations are explained in the subsequent chapters.

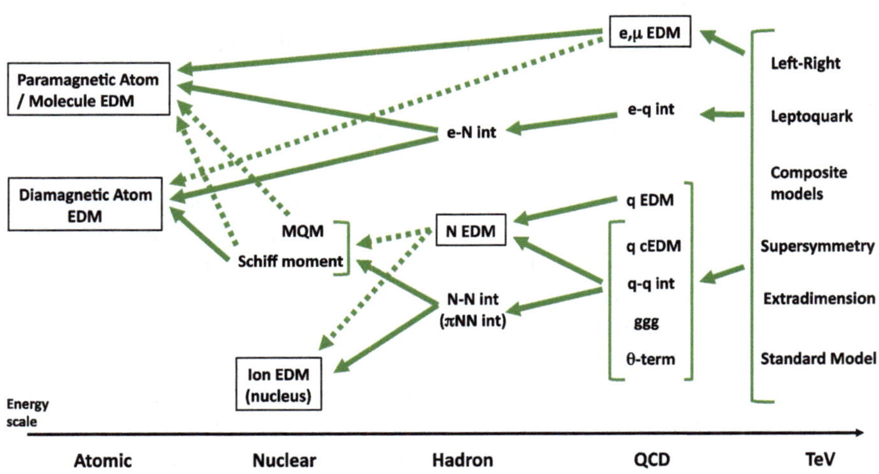

Fig. 5.1 Dependence of the EDM observables on the underlying P, CP-odd mechanisms [32]

We will review in detail the many-body physics contributing to EDMs, the prediction of the EDMs within the standard model and the constraints on supersymmetry. In the next three chapters, we will see the detail of the hadronic, nuclear and atomic many-body calculations. In Chap. 9, we will review the prediction of the small EDM in the standard model. In Chap. 10 we will present the constraints on supersymmetry.

References

1. E.M. Purcell, N.F. Ramsey, Phys. Rev. **78**, 807 (1950)
2. T.D. Lee, C.N. Yang, Phys. Rev. **104**, 254 (1956)
3. C. S. Wu el al. Phys. Rev. **105**, 1413 (1957)
4. L.D. Landau, Zh Eksp, Teor. Fiz. **32**, 405 (1957)
5. L.D. Landau, Zh Eksp, Sov. Phys. JETP 5, 405 (1957)
6. L.D. Landau, Zh Eksp,. Nucl. Phys. **3**, 127 (1957)
7. J.H. Christensen, J.W. Cronin, V.L. Fitch, R. Turlay, Phys. Rev. Lett. **13**, 138 (1964)
8. L.I. Schiff, Phys. Rev. **132**, 2194 (1963)
9. M. Kobayashi, T. Maskawa, Prog. Theor. Phys. **49**, 652 (1973)
10. R.J. Crewther, P. Di Vecchia, G. Veneziano, E. Witten, Phys. Lett. B **88**, 123 (1979)
11. R.D. Peccei, H.R. Quinn, Phys. Rev. Lett. **38**, 1440 (1977)
12. A.D. Sakharov, Pisma Zh. Eksp. Teor. Fiz.**5**, 32 (1967)
13. A.D. Sakharov, Zh. JETP Lett. **5**, 24 (1967)
14. S. Weinberg, Phys. Rev. Lett. **63**, 2333 (1989)
15. S.M. Barr, A. Zee, Phys. Rev. Lett. **65**, 21 (1990)
16. V. Barger, A. Das, C. Kao, Phys. Rev. D **55**, 7099 (1997)
17. G. Beall, A. Soni, Phys. Rev. Lett. **47**, 552 (1981)
18. F. Xu, H. An, X. Ji, JHEP **1003**, 088 (2010)
19. N. Sasao, in *5th International conference on "Fundamental Physics Using Atoms"*, Okayama University, 2011. (URL:http://xqw.hep.okayama-u.ac.jp/kakenhi/index.php/fpua2011_home/fpua2011_top_e/). Accessed Oct 2011
20. I.B. Khriplovich, Phys. Lett. B **444**, 98 (1998)
21. F. J. M. Farley et al., Phys. Rev. Lett. **93**, 052001 (2004)
22. Y. K. Semertzidis et al., AIP Conf. Proc. **698**, 200 (2004)
23. Y.F. Orlov, W.M. Morse, Y.K. Semertzidis, Phys. Rev. Lett. **96**, 214802 (2006)
24. C.A. Baker et al., Phys. Rev. Lett. **97**, 131801 (2006)
25. B.C. Regan et al., Phys. Rev. Lett. **88**, 071805 (2002)
26. W.C. Griffith et al., Phys. Rev. Lett. **102**, 101601 (2009)
27. M.A. Rosenberry et al., Phys. Rev. Lett. **86**, 22 (2001)
28. J.J. Hudson et al., Nature **473**, 493 (2011)
29. G.W. Bennett et al., Muon $(g-2)$ Collaboration. Phys. Rev. D **80**, 052008 (2009)
30. W. Bernreuther, M. Suzuki, Rev. Mod. Phys. **63**, 313 (1991); Erratum-ibid. **64**, 633 (1992)
31. I.B. Khriplovich, S.K. Lamoreaux, *CP Vioaltion Without Strangeness* (Springer, Berlin, 1997)
32. J.S.M. Ginges, V.V. Flambaum, Phys. Rept. **397**, 63 (2004)
33. M. Pospelov, A. Ritz, Ann. Phys. **318**, 119 (2005)
34. T. Fukuyama, Int. J. Mod. Phys. A **27**, 1230015 (2012)

Chapter 6
Hadron Level Calculation

The first many-body physics relevant to the evaluation of the electric dipole moment (EDM) is the effects on the hadron level. In this chapter we present the calculations of the contributions from the leading P, CP-odd hadronic mechanisms. The processes considered are the nucleon EDM, P, CP-odd nucleon-nucleon interaction, P, CP-odd electron-nucleon interaction due to the P, CP-odd quark level operators, the quark EDM, quark chromo-EDM, gluon chromo-EDM (i.e. Weinberg operator), θ-term, P, CP-odd 4-quark interaction, and the P, CP-odd electron-quark interaction. The schematic dependences of the hadronic scale operators on the quark level operators are shown in Fig. 6.1.

The P, CP-odd lagrangian on the hadron level

$$\mathscr{L}_{\text{hadron}} = \mathscr{L}_{\text{edm}} + \mathscr{L}_{\pi NN} + \mathscr{L}_{eN}, \tag{6.1}$$

with

- the nucleon EDM

$$\mathscr{L}_{\text{edm}} = -\frac{i}{2} \sum_{N=p,n} d_N \bar{N} \sigma_{\mu\nu} \gamma_5 N F^{\mu\nu}, \tag{6.2}$$

- the P, CP-odd pion-nucleon interaction

$$\mathscr{L}_{\pi NN} = \sum_{N=p,n} \sum_{a=1}^{3} \Big[\bar{g}^{(0)}_{\pi NN} \bar{N} \tau^a N \pi^a + \bar{g}^{(1)}_{\pi NN} \bar{N} N \pi^0 + \bar{g}^{(2)}_{\pi NN} (\bar{N} \tau^a N \pi^a - 3 \bar{N} \tau^3 N \pi^0) \Big], \tag{6.3}$$

where a denotes the isospin index, and
- the P, CP-odd electron-nucleon interaction

$$\mathscr{L}_{eN} = -\frac{G_F}{\sqrt{2}} \sum_{N=p,n} \Big[C_N^{\text{SP}} \bar{N} N \, \bar{e} i \gamma_5 e + C_N^{\text{PS}} \bar{N} i \gamma_5 N \, \bar{e} e + \frac{1}{2} C_N^{\text{T}} \epsilon^{\mu\nu\rho\sigma} \bar{N} \sigma_{\mu\nu} N \, \bar{e} \sigma_{\rho\sigma} e \Big]. \tag{6.4}$$

N. Yamanaka, *Analysis of the Electric Dipole Moment in the R-parity Violating Supersymmetric Standard Model*, Springer Theses,
DOI: 10.1007/978-4-431-54544-6_6, © Springer Japan 2014

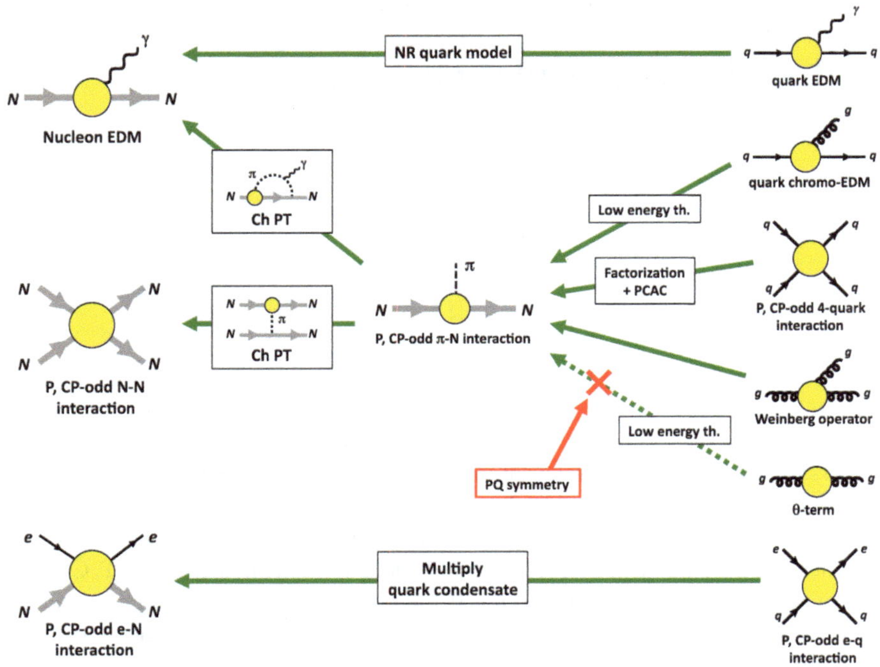

Fig. 6.1 Detailed flow diagram of the dependence of the hadron level P, CP-odd processes on leading quark level P, CP-odd operators. The hadron level processes are shown in the *left side* and the quark level operators in the *right side*. The Weinberg operator contribution is not discussed in our analysis

The P, CP-odd nucleon-nucleon interaction receives the leading contribution from one-pion exchange as shown in the flow chart of Fig. 6.1. It can be obtained by combining the above P, CP-odd pion-nucleon interaction $\mathscr{L}_{\pi NN}$ with the standard P, CP-even pion-nucleon interaction ($\mathscr{L} = g_{\pi NN} \bar{N} i \gamma_5 \tau N \cdot \pi$).

P, CP-odd quark level operators include the θ-term, quark EDM, quark chromo-EDM, four-fermion interaction and the Weinberg operator (gluon chromo-EDM). These operators with the lowest mass dimension, give the leading contribution to the hadron level P and CP violations. The leading quark level P, CP-odd interactions are the followings:

- θ-term:

$$\mathscr{L}_\theta = \frac{g_s^2}{64\pi^2} \bar{\theta}\, \epsilon^{\mu\nu\rho\sigma} G^a_{\mu\nu} G^a_{\rho\sigma}, \tag{6.5}$$

- quark EDM:

$$\mathscr{L}_{\rm qEDM} = -\frac{i}{2} \sum_{q=u,d,s} d_q\, \bar{q} \sigma_{\mu\nu} \gamma_5 q\, F^{\mu\nu}, \tag{6.6}$$

- quark chromo-EDM:

$$\mathscr{L}_{\mathrm{cEDM}} = -\frac{i}{2} \sum_{q=u,d,s} d_q^c \, g_s \bar{q} \sigma_{\mu\nu} t_a \gamma_5 q G_a^{\mu\nu}, \tag{6.7}$$

- P, CP-odd 4-quark interactions:

$$\mathscr{L}_{4\mathrm{q}} = \frac{G_F}{\sqrt{2}} \sum_{q,q'=u,d,s,c,b} \left[C_{qq'} (\bar{q}q)(\bar{q}' i \gamma_5 q') + \frac{1}{2} C'_{qq'} \epsilon^{\mu\nu\rho\sigma} (\bar{q}\sigma_{\mu\nu} q)(\bar{q}' \sigma_{\rho\sigma} q') \right], \tag{6.8}$$

- P, CP-odd electron-quark interactions:

$$\mathscr{L}_{eq} = -\frac{G_F}{\sqrt{2}} \sum_{q=u,d,s,c,b} \left[C_{eq}^{\mathrm{SP}} \bar{q}q \bar{e} i \gamma_5 e + C_{eq}^{\mathrm{PS}} \bar{q} i \gamma_5 q \bar{e} e + \frac{1}{2} C_{eq}^{\mathrm{T}} \varepsilon^{\mu\nu\rho\sigma} \bar{q} \sigma_{\mu\nu} q \bar{e} \sigma_{\rho\sigma} e \right], \tag{6.9}$$

- Weinberg operator [1]:

$$\mathscr{L}_w = \frac{1}{6} w \frac{G_F}{\sqrt{2}} f^{abc} \varepsilon^{\alpha\beta\gamma\delta} G^a_{\mu\alpha} G^b_{\beta\gamma} G_\delta^{\mu,c}, \tag{6.10}$$

  where f^{abc} is the $SU(3)$ structure constant of the Lie algebra. (The Weinberg operator will not be discussed in this thesis, since it is not relevant in our analysis.)

To obtain the P, CP-odd hadron level interactions, we need the results from model calculations which need many inputs, such as hadron matrix elements, quark masses, etc. Strong interaction processes should *in principle* be calculated within the framework of the Quantum chromodymanics (QCD). However, we still do not have fully analytic non-perturbative methods to calculate hadron level processes starting from quark level interactions. One way of calculating them in QCD is the Lattice QCD simulation, which consists of numerical analysis using Montecarlo techniques in discretized space-time, but not many data for hadron matrix elements needed in EDM analysis are currently available, and their calculations remain one of the important homework. On the side of model calculations, many results are available. Hadron matrix elements needed in EDM analysis were essentially evaluated using non-relativistic quark models, low energy theorems, chiral techniques and QCD sum rules [2]. In this thesis, we mainly focus on the chiral approach using low energy theorems for the evaluation of hadron matrix elements. The first reason of this choice is that many calculations of hadron matrix elements are available. Since different models have different source of errors, it is very important to use the same model for every hadron matrix elements needed in the analysis. The second reason is the systematic estimation of the precision of hadron matrix elements. We will also use the results from *ab initio* lattice QCD calculations if available. In the following, we first examine the quark contents of nucleon which are needed in many subsequent analyses. We will then briefly review the method using low energy theorems which provides us with the relation between quark level P, CP-odd operators and the P, CP-odd meson-baryon couplings. After obtaining meson-baryon couplings, we will discuss the calculation of nucleon EDMs using the chiral method. The Peccei-Quinn mechanism is then reviewed. We finally summarize the results.

6.1 Quark Contents of the Nucleon

In the evaluation of the hadron level effective P, CP-odd interactions, we often need the quark contents of the nucleon with suitable Lorentz structure. (for example, the P, CP-odd electron-nucleon (e-N) interactions $C_N^{\rm SP}\bar N N\,\bar e i\gamma_5 e = \sum_q C_{eq}^{\rm SP}\langle N|\bar q q|N\rangle\,\bar q q\,\bar e i\gamma_5 e$.) Explicitly, we must calculate $\langle N|\bar q q|N\rangle$, $\langle N|\bar q i\gamma_5 q|N\rangle$ and $\langle N|\bar q\sigma^{\mu\nu}q|N\rangle$ $(N = p, n\,; q = u, d, s, c, b)$. We begin first with the evaluation of the scalar content of nucleon $\langle N|\bar q q|N\rangle$. The physical meaning of these matrix elements is the fraction of the quark mass over the nucleon mass. In the classic phenomenological approach using SU(3) symmetry and breaking with baryon mass splitting [3, 4], and the u, d quark content of the nucleon (the so-called σ term, $\sigma \equiv \frac{m_u+m_d}{2}\langle p|\bar u u + \bar d d|p\rangle = 55 \sim 75\,\text{MeV}$ [5]), we obtain the following values:

$$\langle p|\bar u u|p\rangle = 7.7, \tag{6.11}$$

$$\langle p|\bar d d|p\rangle = 6.9, \tag{6.12}$$

$$\langle p|\bar s s|p\rangle = 4.2, \tag{6.13}$$

where we have equated the proton-neutron mass splitting $\langle p|\bar u u - \bar d d|p\rangle = \frac{m_p^0-m_n^0}{m_u-m_d}$, the Ξ-Λ mass splitting $\langle p|\bar u u+\bar d d - 2\bar s s|p\rangle = 3\frac{m_\Xi-m_\Lambda}{m_s}$ and the σ-term above. The recent values of the quark and baryon masses have been used, with $m_u \approx 2.5\,\text{MeV}$, $m_d \approx 4.9\,\text{MeV}$, $m_s \approx 100\,\text{MeV}$, $m_p^0 - m_n^0 = -2.05\,\text{MeV}$ (nucleon masses without electromagnetic contribution), $m_\Xi = 1321$ MeV and $m_\Lambda = 1116$ MeV [6–8]. The quark contents of the nucleon were also evaluated in lattice QCD [9–13]. The u and d quark contents are in agreement with lattice QCD simulations, and the result shows that the chiral expansion in u and d quark masses works well. The strange quark content of the nucleon merits a short discussion. The earlier quenched lattice QCD simulations for $\langle N|\bar s s|N\rangle$ showed agreement with the classic value (Eq. 6.13). However, recent lattice analyses with dynamical quarks indicate that the strange content of the nucleon is suppressed about one order compared with the classic values. This can be understood by the suppression of disconnected quark loop (sea quark) contribution. In this analysis, we adopt the results of the lattice QCD analyses with dynamical quarks [10–13]. Calculations present a small result with

$$\langle p|\bar s s|p\rangle = \langle N|\bar s s|N\rangle \approx 0.1\ , \tag{6.14}$$

derived with $m_s = 100\,\text{MeV}$. This disagreement shows that the chiral expansion in terms of strange quark mass is rather difficult. The result of Eq. (6.13) was from using the first order expansion of the strange quark mass. However, higher order corrections can be important since m_s is far from the chiral limit. For these reasons, we will use the result of lattice QCD calculations ($\langle p|\bar{s}s|p\rangle \approx 0.1$) [10–13] in this thesis.

The charm and bottom quark contributions can be calculated by using the heavy quark expansion [3, 4]. The leading order contribution of the heavy quark condensate is given by

$$\langle N|\bar{Q}Q|N\rangle \approx -\left\langle N\left|\frac{\alpha_s}{12\pi m_Q}G_{\mu\nu,a}G_a^{\mu\nu}\right|N\right\rangle + O\left(\frac{1}{m_Q^2}\right), \tag{6.15}$$

where Q stands for heavy quark. By neglecting higher order terms of $O(1/m_Q^2)$, we obtain

$$\langle N|\bar{c}c|N\rangle \approx 3\times 10^{-2}, \tag{6.16}$$

$$\langle N|\bar{b}b|N\rangle \approx 1\times 10^{-2}, \tag{6.17}$$

where the quark masses $m_c = 1.3\,\text{GeV}$ and $m_b = 4.2\,\text{GeV}$ were used [6–8].

We now calculate the matrix element $\langle N|\bar{q}i\gamma_5 q|N\rangle$ by using PCAC and axial anomaly [14] to get

$$\langle p|\bar{q}i\gamma_5 q|p\rangle = \frac{m_N}{m_q}\left(\Delta q' + \frac{\alpha_s}{2\pi}\Delta g\right), \tag{6.18}$$

where $\Delta q'$ is the fraction of the axial vector current of the quark q in the proton, Δg is defined by $\langle p|\text{Tr}\, G_{\mu\nu}\tilde{G}^{\mu\nu}|p\rangle = -2m_N \Delta g \bar{u}_p i\gamma_5 u_p$ [14], where $G_{\mu\nu}$ is the gluon field strength and $\tilde{G}_{\mu\nu}$ its dual. We use $\Delta u' = 0.82$, $\Delta d' = -0.44$, $\Delta s' = -0.11$ [15, 16], $(\alpha_s/2\pi)\Delta g = -0.16$ [14] and the recent values of quark masses cited previously. This gives

$$\langle p|\bar{u}i\gamma_5 u|p\rangle = 248, \tag{6.19}$$

$$\langle p|\bar{d}i\gamma_5 d|p\rangle = -115, \tag{6.20}$$

$$\langle p|\bar{s}i\gamma_5 s|p\rangle = -2.5. \tag{6.21}$$

The pseudoscalar condensate of the bottom quark can also be calculated in heavy quark expansion [3, 4]. The leading order contribution is given by

$$\langle N|\bar{Q}i\gamma_5 Q|N\rangle \approx -\left\langle N\left|\frac{\alpha_s}{8\pi m_Q}G_{\mu\nu,a}\tilde{G}_a^{\mu\nu}\right|N\right\rangle + O\left(\frac{1}{m_Q^2}\right). \tag{6.22}$$

This gives

$$\langle N|\, \bar{c} i\gamma_5 c\,|N\rangle \approx -9\times 10^{-2},$$
$$\langle N|\, \bar{b} i\gamma_5 b\,|N\rangle \approx -3\times 10^{-2}. \qquad (6.23)$$

Again, higher order corrections were neglected.

The calculation of the tensor matrix element was done in quenched lattice QCD with the Wilson quark [17]. The result is

$$\langle p|\bar{u}\sigma^{\mu\nu}u|p\rangle = (0.839\pm 0.060)\bar{p}\sigma^{\mu\nu}p,$$
$$\langle p|\bar{d}\sigma^{\mu\nu}d|p\rangle = -(0.231\pm 0.055)\bar{p}\sigma^{\mu\nu}p,$$
$$\langle p|\bar{s}\sigma^{\mu\nu}s|p\rangle = -(0.046\pm 0.034)\bar{p}\sigma^{\mu\nu}p. \qquad (6.24)$$

Here the strange quark tensor charge receives purely disconnected contributions. All these results suffer from a large finite volume effect, and it is difficult to determine the strange quark tensor charge.

6.2 The PCAC Techniques and the P, CP-Odd Meson-Baryon Interactions

We now explain several calculational techniques. The first one is the *PCAC technique*. The quark sector of QCD at a few hundred of MeV is known to have approximately chiral flavor $SU(3)_L\times SU(3)_R\times U(1)_V$ symmetry. Quarks have small masses, which explicitly break the axial $SU(3)$ symmetry, but this symmetry breaking is quite small compared to the scale of confinement, so an approximate relation for axial currents is possible. This is the *partially conserved axial vector current* (PCAC) relation. One of the important consequence of PCAC is the soft-pion theorem:

$$\langle \pi^c B_a|\, O_i\mid B_b\rangle \approx \frac{i}{f_\pi}\langle B_a|\,[O_i, (J_5^c)_0]\mid B_b\rangle, \qquad (6.25)$$

where B is an $SU(3)$ octet baryon state and $(J_5^c)_0$ the axial charge. The baryon field and the Nambu-Goldstone boson (meson) field π^c are defined as

$$\pi^c = M = \begin{pmatrix} \frac{\pi^0}{\sqrt{2}}+\frac{\eta^0}{\sqrt{6}} & \pi^+ & K^+ \\ \pi^- & -\frac{\pi^0}{\sqrt{2}}+\frac{\eta^0}{\sqrt{6}} & K^0 \\ K^- & \bar{K}^0 & -2\frac{\eta^0}{\sqrt{6}} \end{pmatrix}, \qquad (6.26)$$

and

$$B = \begin{pmatrix} \frac{\Sigma^0}{\sqrt{2}} + \frac{\Lambda^0}{\sqrt{6}} & \Sigma^+ & p \\ \Sigma^- & -\frac{\Sigma^0}{\sqrt{2}} + \frac{\Lambda^0}{\sqrt{6}} & n \\ \Xi^- & \Xi^0 & -2\frac{\Lambda^0}{\sqrt{6}} \end{pmatrix} . \tag{6.27}$$

The right-hand side of the PCAC relation receives corrections from meson rescattering. Since we are interested in very low energy meson exchange, this contribution can be ignored. This relation is part of the low energy theorem, and is often said to be model independent. It can be used to reduce the P and CP violating hadron matrix elements with external meson to the P and CP-even hadron matrix elements without external meson, and so it will be used here to calculate the P, CP-odd meson-baryon couplings (for example, $\bar{g}^{(0)}_{\pi NN}, \bar{g}^{(1)}_{\pi NN}, \bar{g}^{(2)}_{\pi NN}$).

An important example is the calculation of the θ-term contribution to the neutron EDM [18]. The θ-term (Eq. 6.5) is related via the chiral anomaly to the CP-odd quark mass $\mathscr{L}_{\text{CPVM}} = -\bar{\theta} m_* (\bar{u} i\gamma_5 u + \bar{d} i\gamma_5 d + \bar{s} i\gamma_5 s)$ (with $m_* \equiv \frac{m_u m_d}{m_u + m_d}$), and the parameter $\bar{\theta}$ can be transferred to each other. By applying the PCAC relation to $\langle N\pi^a | \mathscr{L}_{\text{CPVM}} | N\rangle$, we obtain

$$\bar{g}^{(0)}_{\pi NN}(\bar{\theta}) = \frac{\bar{\theta} m_*}{f_\pi} \langle N | \bar{q}\tau^3 q | N\rangle \approx 0.015\bar{\theta} , \tag{6.28}$$

where $q = u, d$. The hadron matrix elements $\langle p | \bar{u}u | p\rangle$ and $\langle p | \bar{d}d | p\rangle$ were derived from the proton and neutron mass splitting (see previous chapter). The strange quark contribution was neglected since it is of the order of m_*/m_s.

Another important example of calculation using the PCAC relation is the calculation of the chromo-EDM contribution to the meson-baryon couplings. The discussion below follows the paper of Hisano and Shimizu [19, 20]. After using the PCAC relation, we obtain the following contribution to the P, CP-odd pion-nucleon couplings:

$$\langle B_a \pi^c | \mathscr{L}_{\text{cEDM}} | B_b\rangle = \sum_{q,q'} \frac{1}{f_\pi} d^c_q \langle B_a | \bar{q} g_s G^{\mu\nu}_a \sigma_{\mu\nu} t_a T_c q | B_b\rangle , \tag{6.29}$$

where T_c is the generator of the flavor $SU(3)$ symmetry, and t_a the one of the color $SU(3)$. The values of these condensates are not calculable with chiral techniques. For the evaluation of the mixed condensate in nucleon $\langle B_a | \bar{q}' g_s G^{\mu\nu}_a \sigma_{\mu\nu} t_a T_c q | B_b\rangle$, we use the relation based on QCD sum rules [21–23] with the saturation of the lightest 0^{++} state and the low energy theorem [24] adopted in Refs. [3, 4, 25–27] to write

$$\langle B_a | \bar{q} g_s G^{\mu\nu}_a \sigma_{\mu\nu} t_a T_c q | B_b\rangle \approx \frac{5}{3} m_0^2 \langle B_a | \bar{q} T_c q | B_b\rangle, \tag{6.30}$$

where $m_0^2 \equiv \langle 0 | \bar{q} g_s G^{\mu\nu}_a \sigma_{\mu\nu} t_a q | 0\rangle / \langle 0 | \bar{q} q | 0\rangle \approx 0.8\,(\text{GeV})^2$ [21–23, 28, 29]. For the derivation, see Appendix C.

The result of the θ-term and chromo-EDM contributions can then be regrouped to get the following expression:

$$\langle B_a \pi^c | \mathscr{L}_{\theta+\text{cEDM}} | B_b \rangle \approx \sum_q \frac{1}{f_\pi} \left(2\alpha_q m_q + \frac{5}{3} d_q^c m_0^2 \right) \langle B_a | \bar{q} T_c q \mid B_b \rangle$$
$$= \sum_q \frac{1}{f_\pi} \langle B_a | \bar{q} \{T_c, A\} q \mid B_b \rangle, \tag{6.31}$$

where $\sum_q \alpha_q = \bar{\theta}$. $A = \text{diag}(A_u, A_d, A_s)$ is the flavor $SU(3)$ breaking effect, with components

$$A_q = \alpha_q m_q + \frac{5}{6} m_0^2 d_q^c \, . \tag{6.32}$$

If we assume that the Peccei-Quinn symmetry and the axion mechanism hold, the θ-term contribution will become unphysical. We can then expect the term with α_q to vanish. However, it is known that the chromo-EDM can induce the θ-term even in the presence of the axion mechanism [30, 31]. The induced chromo-EDM contribution is

$$\alpha_q = -\frac{m_0^2}{2} \frac{d_q^c}{m_q}, \tag{6.33}$$

so that

$$A_q = -\frac{m_0^2}{2} d_q^c + \frac{5}{6} m_0^2 d_q^c = \frac{m_0^2}{3} d_q^c \approx 0.27 d_q^c \, \text{GeV}. \tag{6.34}$$

We will see the derivation of the contribution of the chromo-EDM to the induced θ-term under the axion mechanism later in Sect. 6.5.

By considering the flavor $SU(3)$ breaking matrix element of Eq. (6.31), we obtain the following P, CP-odd meson-baryon interaction:

$$\mathscr{L}_{CPV} = \frac{1}{\sqrt{2} f_\pi} \big[\langle p | \bar{s}s - \bar{d}d | p \rangle \text{Tr}(\bar{B} B \{M, A\})$$
$$+ \langle p | \bar{u}u - \bar{d}d | p \rangle \text{Tr}(\bar{B} \{M, A\} B)$$
$$+ 2 \langle p | \bar{d}d | p \rangle \text{Tr}(AM) \text{Tr}(\bar{B} B) \big]. \tag{6.35}$$

The relevant P, CP-odd meson-baryon lagrangian is

$$\mathscr{L}_{CPV} = \frac{\langle p | \bar{s}s - \bar{d}d | p \rangle}{\sqrt{2} f_\pi} \left[(A_d + A_s) \bar{\Sigma}^+ p \bar{K}^0 + (A_u + A_s) \left(\frac{\bar{\Sigma}^0}{\sqrt{2}} + \frac{\bar{\Lambda}}{\sqrt{6}} \right) p K^- \right.$$
$$\left. + (A_u + A_s) \bar{\Sigma}^- n K^- + (\text{h.c.}) - \frac{4}{\sqrt{6}} A_s \bar{p} p \eta^0 \right]$$

$$+\frac{\langle p|\bar{u}u-\bar{d}d|p\rangle}{\sqrt{2}f_\pi}\left[(A_u+A_d)\bar{p}n\pi^+ - \frac{2}{\sqrt{6}}(A_u+A_s)\bar{p}\Lambda K^+ + (\text{h.c.}) + 2A_u\bar{p}p\left(\frac{\pi^0}{\sqrt{2}}+\frac{\eta^0}{\sqrt{6}}\right) - \frac{2}{\sqrt{2}}A_d\bar{n}n\pi^0\right]$$
$$+\frac{2\langle p|\bar{d}d|p\rangle}{\sqrt{2}f_\pi}\left(\frac{A_u-A_d}{\sqrt{2}}\pi^0 + \frac{A_u+A_d-2A_s}{\sqrt{6}}\eta^0\right)(\bar{p}p+\bar{n}n), \tag{6.36}$$

where +(h.c.) means that we add the hermitian conjugates of the terms in its left side.

From this we can derive the P, CP-odd pion-nucleon interactions:

$$\mathscr{L}_{\pi NN} = \frac{1}{2f_\pi}(A_u+A_d)\langle p|\bar{u}u-\bar{d}d|p\rangle\bar{N}\tau^a N\pi^a + \frac{1}{2f_\pi}(A_u-A_d)\langle p|\bar{u}u+\bar{d}d|p\rangle)\bar{N}N\pi^0. \tag{6.37}$$

The P, CP-odd pion-nucleon couplings can be explicitly written as

$$\bar{g}^{(0)}_{\pi NN}(\bar{\theta}, d^c_q) = \frac{1}{2f_\pi}(A_u+A_d)\langle p|\bar{u}u-\bar{d}d|p\rangle, \tag{6.38}$$

$$\bar{g}^{(1)}_{\pi NN}(\bar{\theta}, d^c_q) = \frac{1}{2f_\pi}(A_u-A_d)\langle p|\bar{u}u+\bar{d}d|p\rangle. \tag{6.39}$$

For the case of P, CP-odd 4-quark interaction, the vacuum factorization approximation can be used. Combined with the PCAC method, we get the following relation:

$$\begin{aligned}\bar{g}^{(1)}_{\pi NN}(C_{qd}) &= \sum_{q=u,d,s,c,b}\langle\pi^0 N|C_{qd}\frac{G_F}{\sqrt{2}}\bar{q}q\,\bar{d}i\gamma_5 d|N\rangle \\ &\approx \sum_{q=u,d,s,c,b} C_{qd}\frac{G_F}{\sqrt{2}}\langle\pi^0|\bar{d}i\gamma_5 d|0\rangle\langle N|\bar{q}q|N\rangle \\ &= -\frac{F_\pi m_\pi^2 G_F}{2\sqrt{2}m_d}\sum_{q=u,d,s,c,b} C_{qd}\langle N|\bar{q}q|N\rangle.\end{aligned} \tag{6.40}$$

We should note that in this discussion, we have not considered the vacuum condensation of the neutral mesons, which, combined with the CP-even pion-nucleon interaction, can generate an additional contribution to the CP-odd pion-nucleon interaction. This effect may be important as it contributes to the same chiral order as the CP-odd pion-nucleon coupling calculated in this section (see Fig. 3 of Ref. [2] and the corresponding discussion).

6.3 Nucleon EDMs with Chiral Method

Let us move to the evaluation of the nucleon EDM. The one-loop term in the expansion of the nucleon EDM has "chiral logarithm" $\ln(m_\pi/\Lambda)$ (Λ is some hadron scale cutoff). This term appears to be independent of the chiral model chosen [32, 33]. The idea is to consider this term as the leading term in the chiral expansion. The expansion of the nucleon EDM in the chiral perturbation theory has also divergent terms at the one-loop level, which give the renormalization of the low energy constant of the tree level and have to be determined for accurate calculation. In the present case we consider the leading chiral logarithmic terms to be the dominant contribution, although the theoretical uncertainty is as large as order one. The improvement of the accuracy of the nucleon EDM calculation is an important subject which must be treated in the near future. We will return to this subject briefly in the final summary. With the above assumption, the chiral contribution to the nucleon EDM can be evaluated using the following chiral lagrangian.

$$\begin{aligned}\mathscr{L}_{mBB} \approx & - g_{\pi NN}\bar{p}i\gamma_5 p\pi^0 - \sqrt{2}g_{\pi NN}(\bar{p}i\gamma_5 n\pi^+ + \text{h.c.}) - g_{\eta NN}\bar{p}i\gamma_5 p\eta^0 \\ & + g_{K\Lambda N}(\bar{p}i\gamma_5 \Lambda K^+ + \text{h.c.}) \\ & - g_{K\Sigma N}(\bar{\Sigma}^+ i\gamma_5 p\bar{K}^0 + \bar{\Sigma}^- i\gamma_5 n K^- + \frac{1}{\sqrt{2}}\bar{\Sigma}^0 i\gamma_5 p K^- + \text{h.c.}),\end{aligned} \tag{6.41}$$

where $g_{\pi NN} = \frac{m_N}{f_\pi}(D+F) \approx 12.6$, $g_{\eta NN} = \frac{m_N}{\sqrt{3}f_\pi}(3F-D) \approx 3.0$, $g_{K\Lambda N} = \frac{m_N+m_\Lambda}{2\sqrt{3}f_\pi}(D+3F) \approx 6.4$ and $g_{K\Sigma N} = \frac{m_N+m_\Sigma}{\sqrt{2}f_\pi}(D-F) \approx 6.0$, with $D = 0.81$ and $F = 0.44$. In deriving these values, we neglected the isospin splitting. The detail of the derivation is explained in Appendix D.

The P, CP-odd meson-baryon interactions contribute to the single nucleon EDM through one-loop diagrams as shown in Figs. 6.2 and 6.3.

The calculation of the one-loop contribution of the nucleon EDM to the leading chiral logarithm gives the following results:

$$d_n = \frac{e}{8\pi^2 f_\pi^2}(A_u x_u^{(n)} + A_d x_d^{(n)} + A_s x_s^{(n)}), \tag{6.42}$$

$$d_p = \frac{e}{8\pi^2 f_\pi^2}(A_u x_u^{(p)} + A_d x_d^{(p)} + A_s x_s^{(p)}), \tag{6.43}$$

with

$$\begin{aligned}x_u^{(n)} &= (D+F)\langle p|\bar{u}u - \bar{d}d|p\rangle \log\frac{m_N}{m_\pi} + (D-F)\langle p|\bar{d}d - \bar{s}s|p\rangle \log\frac{m_\Sigma}{m_K} \approx 4.1, \\ x_d^{(n)} &= (D+F)\langle p|\bar{u}u - \bar{d}d|p\rangle \log\frac{m_N}{m_\pi} \approx 1.9, \\ x_s^{(n)} &= (D-F)\langle p|\bar{d}d - \bar{s}s|p\rangle \log\frac{m_\Sigma}{m_K} \approx 2.2,\end{aligned}$$

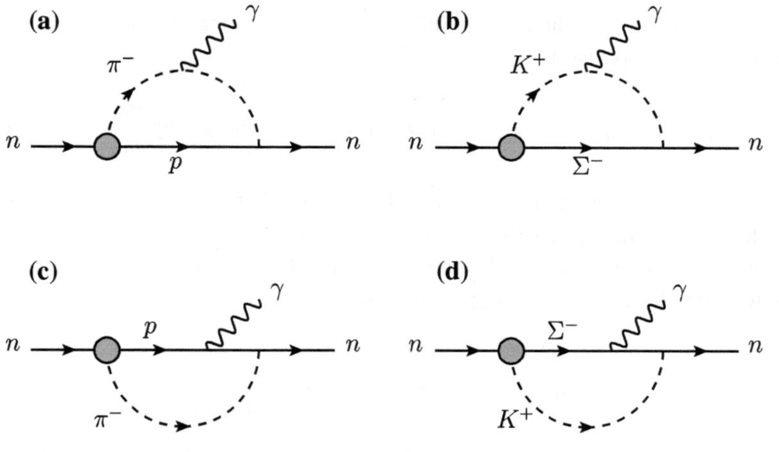

Fig. 6.2 Meson-loop contribution to the neutron EDM. The *grey blob* denotes the P, CP-odd meson-baryon coupling. Graphs (**b**) and (**d**) are strangeness contribution

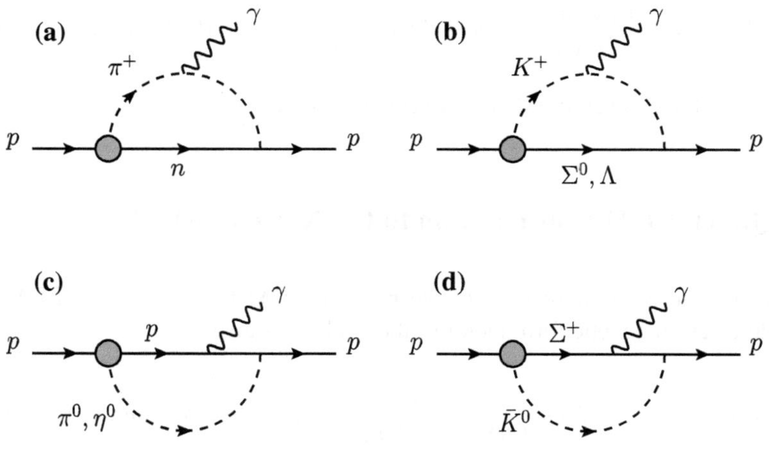

Fig. 6.3 Meson-loop contribution to the proton EDM. The *grey blob* denotes the P, CP-odd meson-baryon coupling. Graphs (**b**) and (**d**) are strangeness contribution

$$
\begin{aligned}
x_u^{(p)} &= (D+F)\langle p|\bar{d}d - \bar{u}u|p\rangle \log\frac{m_N}{m_\pi} + \frac{D-F}{2}\langle p|\bar{d}d - \bar{s}s|p\rangle \log\frac{m_\Sigma}{m_K} \\
&\qquad + \frac{D+3F}{6}\langle p|\bar{d}d + \bar{s}s - 2\bar{u}u|p\rangle \log\frac{m_\Lambda}{m_K} \approx -3.2, \\
x_d^{(p)} &= (D+F)\langle p|\bar{d}d - \bar{u}u|p\rangle \log\frac{m_N}{m_\pi} \approx -1.9, \\
x_s^{(p)} &= \frac{D-F}{2}\langle p|\bar{d}d - \bar{s}s|p\rangle \log\frac{m_\Sigma}{m_K} + \frac{D+3F}{6}\langle p|\bar{d}d + \bar{s}s - 2\bar{u}u|p\rangle \log\frac{m_\Lambda}{m_K} \approx -1.3.
\end{aligned}
\tag{6.44}
$$

We also write down the case without strangeness contribution (K, Λ and Σ neglected) for comparison.

$$x_u^{\prime(n)} = x_d^{\prime(n)} = -x_u^{\prime(p)} = -x_d^{\prime(p)} = (D+F)\langle p|\bar{u}u - \bar{d}d|p\rangle \log \frac{m_N}{m_\pi} \approx 1.9. \quad (6.45)$$

In this case, we see that the proton and neutron EDMs have the same size and opposite sign. From Eqs. (6.38) and (6.42), we can derive the well-known formula of the neutron EDM generated by the isoscalar P, CP-odd pion-nucleon interaction in the leading chiral logarithm approximation:

$$d_n(\bar{g}_{\pi NN}^{(0)}) \approx \frac{e g_{\pi NN} \bar{g}_{\pi NN}^{(0)}}{4\pi^2 m_N} \ln \frac{m_N}{m_\pi}. \quad (6.46)$$

The detailed derivation of the above formula is given in Appendix D. By using Eq. (6.28), we obtain the dependence of the neutron EDM on the θ-term [18]:

$$d_n(\bar{\theta}) \approx \bar{\theta} \frac{e g_{\pi NN} m_*}{4\pi^2 m_N f_\pi} \langle p|\bar{u}u - \bar{d}d|p\rangle \ln \frac{m_N}{m_\pi} \approx 2 \times 10^{-16} \bar{\theta}\, e\,\text{cm}. \quad (6.47)$$

We see that the neutron EDM is very sensitive to the θ-term.

6.4 Quark EDM Contribution to the Nucleon EDM

The next topic is the derivation of the nucleon EDM from the quark EDM. The CP-odd electromagnetic form factor of the nucleon is given by

$$\left\langle N(p-q) \middle| j_\mu^{(\text{em})} \middle| N(p) \right\rangle_{\text{CPV}} = -\frac{g_2(q^2)}{2m_N} \bar{u}_N(p-q) \sigma_{\mu\nu} q^\nu \gamma_5 u_N(p), \quad (6.48)$$

where u_N is the nucleon spinor and q^ν is the momentum transfer. The EDM of the nucleon d_N is defined by the limit of zero momentum transfer of the CP-odd nucleon form factor:

$$\frac{d_N}{e} \equiv \lim_{q^2 \to 0} \frac{g_2(q^2)}{2m_N}. \quad (6.49)$$

In the presence of the quark EDM operator, the leading contribution to the CP-odd nucleon matrix element is

$$\left\langle N(p-q) \middle| j_\mu^{(\text{em})} \middle| N(p) \right\rangle_{\text{CPV}} = \sum_{q=u,d,s} -\frac{1}{e} \langle N(p-q) | d_q \bar{q} \sigma_{\mu\nu} q^\nu \gamma_5 q | N(p) \rangle, \quad (6.50)$$

where d_q is the quark EDM. By taking the limit $q^\nu \to 0$ and the nucleon on-shell after combining Eqs. (6.48) and (6.50), we see that the momentum transfer q^ν can be factorized. The dependence of the nucleon EDM on the quark EDM is therefore related to the tensor charge by

$$d_N \bar{u}_N \sigma_{\mu\nu} u_N = d_q \langle N | \bar{q} \sigma_{\mu\nu} q | N \rangle. \tag{6.51}$$

We can thus use the result of the lattice QCD calculation of the quark tensor charge in nucleon (see Eq. (6.24)) to give the dependence of the nucleon EDM on the quark EDMs. The QCD sum rules calculation gives also a consistent result within the theoretical uncertainty [34].

It should be noted that the non-relativistic reduction of the tensor current gives the spin density. Such a situation is realized in the non-relativistic constituent quark model, which assumes that the nucleon is formed of three massive constituent quarks with confining inter-quark potential and small spin/isospin dependent interactions. The constituent quark model is known to work well in the hadron spectroscopy. The tensor charge of the constituent quark in nucleon (or the quark EDM contribution to the nucleon EDM) can be calculated with the $SU(2)$ algebra, and is given as

$$d_n(d_u, d_d) = \frac{4}{3} d_d - \frac{1}{3} d_u, \tag{6.52}$$

$$d_p(d_u, d_d) = \frac{4}{3} d_u - \frac{1}{3} d_d, \tag{6.53}$$

with d_u and d_d the EDMs of the u and d quarks, respectively. For the detailed derivation, see Appendix E. We see that the non-relativistic constituent quark model predicts a larger EDM than the lattice QCD result (6.24).

In watching this discrepancy, two sources of deviation can naïvely be inferred. The first possibility is the dressing of the bare quark tensor charge (or the bare quark EDM) by gluons, and the second one is the spin-dependent bound state effect. If we assume that the non-relativistic quark model works well, this discrepancy should originate in the gluon dressing of the tensor vertex of the quark.

The difference between the constituent quark model prediction and the lattice QCD result is of $O(1)$. As the hadronic level evaluation of other CP-odd quark level operators involves much more theoretical uncertainties, we do not need to be very sensitive on this discrepancy. In this thesis, we will use the results of the non-relativistic constituent quark model (6.52) and (6.53) for simplicity.

6.5 Theta Term and Strong CP Problem

In QCD, there are many distinct classes of gluon field configurations which are not connected by continuous gauge transformations. It is actually possible to assign to each of them a distinct topological charge called *winding number*. There are also gauge transformations which can shift the winding number. This so-called "large gauge transformation" can be generated with the following function [35]

$$\Lambda_1(\mathbf{x}) = \frac{\mathbf{x}^2 - d^2}{\mathbf{x}^2 + d^2} + \frac{2id\mathbf{x}\cdot\tau}{\mathbf{x}^2 + d^2}, \tag{6.54}$$

where $\mathbf{x}$ and τ are respectively the spatial and internal $SU(2)$ subgroup gauge coordinates, and d an arbitrary parameter. This gauge transformation has an impressive characteristic, it actually mixes the external and internal coordinates. To obtain gauge transformations which change the winding number arbitrarily, we simply have to gauge transform the gluon field using the gauge function $\Lambda_n(\mathbf{x}) = [\Lambda_1(\mathbf{x})]^{\mathbf{n}}$. To obtain a fully gauge invariant vacuum, we have to sum up all classes of vacuum states with different winding number. The following coherent superposition of states $|\, m \,\rangle$ with winding number m satisfies the large gauge invariance:

$$|\, \theta \,\rangle = \sum_m e^{-im\theta} |\, m \,\rangle, \tag{6.55}$$

with θ an arbitrary real parameter which specifies the QCD vacuum. This is the *θ-vacuum*. Under gauge transformation, the θ-vacuum behaves as

$$U_n |\, \theta \,\rangle = e^{in\theta} |\, \theta \,\rangle, \tag{6.56}$$

where U_n is the large gauge transformation associated with the gauge function Λ_n. With non-zero θ, we can reexpress the action by adding a new term. The generic amplitude with QCD θ-vacuum can be written as

$$\langle\, \theta \,|\, X \,|\, \theta \,\rangle = \sum_{m_+, m_-} e^{i(m_+ - m_-)\theta} \langle\, m_+ \,|\, X \,|\, m_- \rangle, \tag{6.57}$$

where X is some operator. The shift of the winding number $(m_+ - m_-)$ between asymptotic initial and final states can be measured by the topological charge operator $\int \tilde{G}G = \frac{g_s^2}{64\pi^2} \int d^4x \, \epsilon^{\mu\nu\rho\sigma} G^a_{\mu\nu} G^a_{\rho\sigma}$, as $\langle m_- |\int \tilde{G}G \,|\, m_+ \rangle = m_+ - m_-$. Applying it to the generic amplitudes, we obtain

$$\langle\, \theta \,|\, X \,|\, \theta \,\rangle = \sum_{m_+, m_-} \langle\, m_+ \,|\, X e^{i\theta \int \tilde{G}G} \,|\, m_- \,\rangle = \langle\, \theta = 0 \,|\, X e^{i\theta \int \tilde{G}G} \,|\, \theta = 0 \,\rangle. \tag{6.58}$$

The above expression means that we can translate the general dynamics in the θ-vacuum by the dynamics in $\theta = 0$ vacuum with a new term $S_\theta = \theta \int \tilde{G}G$ in the action. This term is called the θ-term (see Eq. (6.5)), and has an important status in the physics of CP violation. The θ-term is a total derivative, so it should not be relevant in perturbation theory, but it is actually not invariant under the "large gauge transformation", and can have an important role at the non-perturbative level.

Moreover, the existence of finite θ-term induces a sizable P and CP violation, as we have seen previously.

We will now see that the θ-term is related to the chiral transformation. The θ-term has the same form as the anomaly term relevant in the Ward identity of the axial (or chiral) $U(1)_A$ current. The 4-divergence of the axial $U(1)_A$ current can be written as

$$\partial_\mu j_5^\mu = 2\sum_q^{N_f} m_q \bar{q} i\gamma_5 q + \frac{N_f g_s^2}{16\pi^2} G_{\mu\nu}^a \tilde{G}^{\mu\nu,a}, \tag{6.59}$$

where $\tilde{G}_a^{\mu\nu} \equiv \frac{1}{2}\epsilon^{\mu\nu\rho\sigma} G_{\rho\sigma}^a$, $j_5^\mu \equiv \sum_q^{N_f} \bar{q}\gamma^\mu\gamma_5 q$ and N_f is the number of flavors for quarks. This is the concrete expression which describes the anomaly effect. The physical meaning of the anomaly is the breakdown of the classical level symmetry at the quantum level. Let us define the following "modified" chiral charge:

$$\tilde{Q}_5 = \int d^3x \left\{ j_5^0 - K^0 \right\}, \tag{6.60}$$

where K^μ is defined such that $\partial_\mu K^\mu \equiv 2N_f \tilde{G}G$. The chiral transformation with this modified chiral charge is conserved for massless quarks. Note that this chiral charge is not gauge invariant. When we apply the gauge transformation U_n which changes the winding number by n, we have

$$U_n \tilde{Q}_5 U_n^{-1} = n(\tilde{Q}_5 - 2N_f). \tag{6.61}$$

This means that the θ-vacuum is modified by chiral rotations. The large gauge transformation of the chirally rotated θ-vacuum (with angle α) is

$$U_n e^{i\alpha\tilde{Q}_5}|\,\theta\,\rangle = U_n e^{i\alpha\tilde{Q}_5} U_n^{-1} U_n |\,\theta\,\rangle = e^{in(\theta-2N_f\alpha)} e^{i\alpha\tilde{Q}_5}|\,\theta\,\rangle = e^{in(\theta-2N_f\alpha)}|\,\theta - 2N_f\alpha\rangle. \tag{6.62}$$

The chiral phase α and the parameter θ of the θ-vacuum can actually be transferred to each other.

Let us now see the chiral transformation of quarks. The bare quark mass matrix is in general not diagonal and may have complex phase. We assume that the quark mass term has the following form

$$\mathscr{L}_m = -\bar{q}'_L \hat{m}' q'_R - \bar{q}'_R \hat{m}'^\dagger q'_L, \tag{6.63}$$

where $q'_{L/R}$ and $\hat{m}'$ are the bare quark and its mass matrix, respectively, with all considered flavors. As the physical mass matrix $\hat{m}$ should be real to have no tachyonic quarks, the chiral rotation of the basis $q'_{L/R} = e^{\mp i\alpha} U_{L/R} q_{L/R}$, where $U_{L/R}$ is the $SU(N_f)$ unitary matrix diagonalizing $\hat{m}'$, is needed to eliminate the axial $U(1)_A$ phase α as

$$\mathcal{L}_m = -e^{2i\alpha}\bar{q}_L U_L^\dagger \hat{m}' U_R q_R - e^{-2i\alpha}\bar{q}_R U_R^\dagger \hat{m}'^\dagger U_L q_L = -\bar{q}_L \hat{m} q_R - \bar{q}_R \hat{m}^\dagger q_L, \tag{6.64}$$

such that $\hat{m}$ is diagonal and $\text{Im}[\det \hat{m}] = \text{Im}[\det(e^{2i\alpha}\hat{m}')] = 0 \Leftrightarrow 2N_f\alpha = -\arg(\det \hat{m}')$. The total relevant θ-term is then

$$\bar{\theta} = \theta + \arg(\det \hat{m}'). \tag{6.65}$$

The physical effect of the $\bar{\theta}$ can be calculated with its contribution entirely transferred to the "P, CP-odd mass" $\eta \bar{q} i\gamma_5 q$ via the chiral rotation. The quark mass lagrangian can thus be expressed as

$$\mathcal{L}_m = -\bar{q}_L \hat{M} q_R - \bar{q}_R \hat{M}^\dagger q_L = -\sum_q^{N_f} m_q \bar{q} q - \eta \sum_q^{N_f} \bar{q} i\gamma_5 q, \tag{6.66}$$

where $\hat{M}$ is the quark mass matrix with the entire $\bar{\theta}$ contribution transferred, and η the corresponding P, CP-odd quark mass. The parameter η should not depend on the quark flavor since the effect of $\bar{\theta}$ was transferred via axial $U(1)_A$ transformation. From this, it is evident that the P, CP-odd effects are suppressed for heavier quarks. For QCD with 3 quark flavors, η satisfies the following relation

$$\bar{\theta} = \arg(\det \hat{M}) = \arg[(m_u + i\eta)(m_d + i\eta)(m_s + i\eta)]. \tag{6.67}$$

We thus have for small $\bar{\theta}$

$$\eta \approx \bar{\theta} \frac{m_u m_d m_s}{m_u m_d + m_u m_s + m_d m_s} \approx \bar{\theta} \frac{m_u m_d}{m_u + m_d} \equiv \bar{\theta} m_*, \tag{6.68}$$

where the second equality is the approximation for heavy strange quark mass. We obtain finally the following replacement between the θ-term and the P, CP-odd quark mass:

$$\bar{\theta} \frac{g_s^2}{32\pi^2} G_{\mu\nu}^a \tilde{G}^{\mu\nu,a} \leftrightarrow -\bar{\theta} m_* \sum_q^{N_f} \bar{q} i\gamma_5 q. \tag{6.69}$$

This replacement now gives us the possibility to calculate the hadron matrix element of the P, CP-odd pion-nucleon vertex with the PCAC techniques, as described in previous sections. It is therefore possible to estimate the physical contribution to the observables such as the neutron EDM. We have seen that the θ-term of the QCD was strongly constrained by EDM experimental data. Crewther et al. have shown that the neutron EDM experimental data constrain the parameter $\bar{\theta}$ to be less than 1 part of 10^{10} [18]. This remarkable fine-tuning can also be seen in many experimental data:

- Neutron EDM:
  From the previous analysis [18] (see Eq. (6.47)) and the experimental upper limit on neutron EDM ($d_n < 2.9 \times 10^{-26} e\,\text{cm}$) [36], we obtain

$$|\bar{\theta}| < 1 \times 10^{-10}. \tag{6.70}$$

- Decay of η meson [18, 37]:

$$\mathrm{Br}(\eta \to \pi^+\pi^-) = 1.8 \times 10^2 \bar{\theta}. \tag{6.71}$$

With the experimental data $\mathrm{Br}(\eta \to \pi^+\pi^-) < 1.3 \times 10^{-5}$eV [38], we obtain

$$|\bar{\theta}| < 2.7 \times 10^{-4}. \tag{6.72}$$

- EDMs of diamagnetic atoms [39, 40]:

Like the neutron EDM, these puts also via P, CP-odd pion-nucleon interactions severe constraints on the θ-parameter.
From the experimental data of the EDM of ^{199}Hg atom we have [39]

$$|\bar{\theta}| < 3 \times 10^{-10}. \tag{6.73}$$

From the experimental data of the EDM of ^{129}Xe atom we have [40]

$$|\bar{\theta}| < 7 \times 10^{-7}. \tag{6.74}$$

All these results present strong arguments for the absence of physical contribution of the strong CP lagrangian. This is in contrast to the physical contribution of the anomalous $\int \tilde{G}G$ contribution to the heavy η' meson, which is believed to be the solution to the $U(1)_A$ problem [41–44]. This problem is called the *Strong CP problem.* These data, although leaving a little possibility of miraculous accidental cancellation, make us think of some mechanism that renders the parameter $\bar{\theta}$ to be unphysical.

One possible resolution to this problem was proposed by Peccei and Quinn, by introducing a new field coupled to the Strong CP lagrangian, the *axion* [45]. Its lagrangian (together with the Strong CP lagrangian) is given as follows:

$$\mathscr{L}_a = \frac{1}{2}\partial_\mu a \partial^\mu a + \left(\bar{\theta} + \frac{a(x)}{f_a}\right) \frac{\alpha_s}{8\pi} G^a_{\mu\nu} \tilde{G}^{\mu\nu,a}, \tag{6.75}$$

where $a(x)$ is the axion field. This field is assumed to be the pseudo-Nambu-Goldstone boson of some chiral $U(1)_{\mathrm{PQ}}$ symmetry spontaneously broken at some high energy scale f_a. From this lagrangian, we see that if a vacuum expectation value $\langle a \rangle = -f_a \bar{\theta}$ is developed by the axion, the Strong CP lagrangian becomes irrelevant in the dynamics. We will see that this is exactly what happens. Let us give the following effective lagrangian

$$\mathscr{L}_a^{\mathrm{eff}} = \frac{1}{2}\partial_\mu a \partial^\mu a - K_1 \left(\bar{\theta} + \frac{a}{f_a}\right) - \frac{1}{2} K \left(\bar{\theta} + \frac{a}{f_a}\right)^2 + \cdots . \tag{6.76}$$

Coefficients K and K_1 can be determined by the calculation of the correlators involving the topological charge $G\tilde{G}$. First, we consider the case with only $\mathscr{L}_a + \mathscr{L}_{\mathrm{QCD}}$. In this case, K_1 vanishes since there are no way to annihilate the odd number of axions. The coefficient K is called the topological susceptibility, and can be obtained by calculating the following correlator [41–44, 46, 47]

$$\begin{aligned} K &= -i \lim_{k \to 0} \int d^4x\, e^{ik(x-y)} \left\langle 0 \left| T \left\{ \frac{\alpha_s}{8\pi} G^a_{\mu\nu} \tilde{G}^{\mu\nu,a}(x) \frac{\alpha_s}{8\pi} G^b_{\rho\sigma} \tilde{G}^{\rho\sigma,b}(y) \right\} \right| 0 \right\rangle \\ &\approx -m_* \langle 0 | \bar{q} q | 0 \rangle + O(m_*^2), \end{aligned} \tag{6.77}$$

where $\langle 0|\bar{q}q|0\rangle \equiv -f_\pi^2 m_\pi^2/(m_u + m_d) \simeq -(0.280\,\mathrm{MeV})^3$ is a negative number. For the derivation of Eq. (6.77), see Appendix C. We will then obtain a system which dynamically chooses the vacuum such that $\bar{\theta} + \langle a \rangle / f_a = 0$, which eliminates the effect of the Strong CP term. This is one scenario which solves the Strong CP problem, called the *axion mechanism*. This spontaneous choice of the vacuum will of course give massive excitations around $\langle a \rangle = -f_a \bar{\theta}$ with the mass of $\frac{1}{f_a}\sqrt{-m_* \langle 0|\bar{q}q|0\rangle}$. The search for this axion particle gives null result at the present time, and the axion is thought to be very light, constraining the scale of the $U(1)_{\mathrm{PQ}}$ symmetry breaking to be $f_a > 10^{10}$ GeV.

We now consider the linear term of the axion potential with K_1 (the second term of the right-hand side of Eq. (6.76)). This term is generated by the correlation between the topological charge $G\tilde{G}$ and the P, CP-odd operator O_{CP}. The coefficient K_1 is given as follows [30, 31]:

$$\begin{aligned} K_1(O_{CP}) &= -i \lim_{k \to 0} \int d^4x e^{ik(x-y)} \left\langle 0 \left| T \left\{ \frac{\alpha_s}{8\pi} G^a_{\mu\nu} \tilde{G}^{\mu\nu,a}(x) O_{CP}(y) \right\} \right| 0 \right\rangle \\ &= -i \int d^4x \left\{ -\frac{1}{2N_f} \delta(x_0 - y_0) \langle 0 | [\, j_5^0(x), O_{CP}(y)\,] | 0 \rangle \right. \\ &\qquad \left. - \frac{1}{N_f} \sum_q \langle 0 | T\{ m_q \bar{q} i \gamma_5 q(x) O_{CP}(y) \} | 0 \rangle \right\}. \end{aligned} \tag{6.78}$$

For instance, the chromo-EDM contribution ($O_{CP} = -\frac{i}{2} d_q^c \bar{q} g_s \sigma^{\mu\nu} G^a_{\mu\nu} t_a \gamma_5 q$) can be written as

$$K_1 = -\frac{m_*}{2} \sum_{q=u,d,s} \frac{d_q^c}{m_q} \langle 0 | \bar{q} g_s \sigma^{\mu\nu} G^a_{\mu\nu} t_a q | 0 \rangle. \tag{6.79}$$

The derivation of K_1 for the chromo-EDM is presented in Appendix C. The presence of the linear term with K_1 is important since the minimum of the axion potential receives a shift, and thus generates an induced θ-term $\theta_{\mathrm{ind}} = -K_1(O_{CP})/K$ [30, 31]. For the chromo-EDM, this gives the following induced θ-term:

$$\theta_{\text{ind}} = -\frac{m_0^2}{2} \sum_{q=u,d,s} \frac{d_q^c}{m_q}. \tag{6.80}$$

This result has been used in Sect. 6.2.

For the Weinberg operator, the first term of the second equality of Eq. (6.78) gives no contribution, since it involves no quark field operators. The second term is also small, since isospin violation must occur to generate a pion from the isoscalar Weinberg operator in the intermediate state. The θ-term induced by the Weinberg operator is therefore suppressed by at least a factor of light quark mass.

6.6 Summary of Hadron Level Calculation

Here we summarize the dependences of the hadron level P, CP-odd interactions on P, CP-odd quark level operators. These results will be used in our discussion.

- P, CP-odd πNN interactions:

$$\bar{g}_{\pi NN}^{(0)} = 5.9 \times 10^{13} \frac{d_u^c + d_d^c}{\text{cm}} \tag{6.81}$$

$$\bar{g}_{\pi NN}^{(1)} = 1.0 \times 10^{15} \frac{d_u^c - d_d^c}{\text{cm}} - \sum_{q'=u,d,s,c,b} C_{q'd} \frac{F_\pi m_\pi^2 G_F}{2\sqrt{2} m_d} \langle N | \bar{q}' q' | N \rangle \tag{6.82}$$

- Neutron EDM:

$$d_n = \frac{4}{3} d_d - \frac{1}{3} d_u + 4.9 e d_u^c + 2.3 e d_d^c + 2.6 e d_s^c, \tag{6.83}$$

- Proton EDM:

$$d_p = -\frac{1}{3} d_d + \frac{4}{3} d_u - 3.8 e d_u^c - 2.3 e d_d^c - 1.6 e d_s^c, \tag{6.84}$$

- P, CP-odd electron-nucleon interactions:

$$C_p^{\text{SP}} = \sum_{q=u,d,s,c,b} C_{eq}^{\text{SP}} \langle p | \bar{q} q | p \rangle \tag{6.85}$$

$$C_n^{\text{SP}} = \sum_{q=u,d,s,c,b} C_{eq}^{\text{SP}} \langle n | \bar{q} q | n \rangle \tag{6.86}$$

$$C_p^{\text{PS}} = \sum_{q=u,d,s,c,b} C_{eq}^{\text{PS}} \langle p | \bar{q} i \gamma_5 q | p \rangle \tag{6.87}$$

$$C_n^{\text{PS}} = \sum_{q=u,d,s,c,b} C_{eq}^{\text{PS}} \langle n | \bar{q} i \gamma_5 q | n \rangle \tag{6.88}$$

References

1. S. Weinberg, Phys. Rev. Lett. **63**, 2333 (1989)
2. M. Pospelov, A. Ritz, Ann. Phys. **318**, 119 (2005)
3. J.F. Donoghue, C.R. Nappi, Phys. Lett. B **168**, 105 (1986)
4. A.R. Zhitnitsky, Phys. Rev. D **55**, 3006 (1997)
5. J. Gasser, H. Leutwyler, M.E. Sainio, Phys. Lett. B **253**, 252 (1991)
6. K. Nakamura et al. (Particle Data Group), J. Phys. G **37**, 075021 (2010)
7. K. Nakamura et al. (Particle Data Group), Partial update for the 2012 edition (URL: http://pdg.lbl.gov) (2011)
8. J. Beringer et al. (Particle Data Group), Phys. Rev. D **86**, 010001 (2012)
9. R.D. Young, A.W. Thomas, Nucl. Phys. A **844**, 266 (2010)
10. K. Takeda, S. Aoki, S. Hashimoto, T. Kaneko, T. Onogi, N. Yamada (JLQCD Collaboration), PoS **LATTICE2010**, 160 (2010) [arXiv:1012.1907 [hep-lat]]
11. H. Ohki et al. (JLQCD Collaboration), Phys. Rev. D **78**, 054502 (2008)
12. K. Takeda et al., Phys. Rev. D **83**, 114506 (2011)
13. G.S. Bali et al. (QCDSF Collaboration), Phys. Rev. D **85**, 054502 (2012)
14. T.P. Cheng, L.F. Li, Phys. Rev. Lett. **62**, 1441 (1989)
15. A. Airapetian et al. (HERMES Collaboration), Phys. Rev. D **75**, 012007 (2007)
16. J.R. Ellis, M. Karliner, arXiv:hep-ph/9601280
17. S. Aoki, M. Doui, T. Hatsuda, Y. Kuramashi, Phys. Rev. D **56**, 433 (1997)
18. R.J. Crewther, P. Di Vecchia, G. Veneziano, E. Witten, Phys. Lett. B **88**, 123 (1979)
19. J. Hisano, Y. Shimizu, Phys. Lett. B **581**, 224 (2004)
20. J. Hisano, Y. Shimizu, Phys. Rev. D **70**, 093001 (2004)
21. M.A. Shifman, A.I. Vainshtein, V.I. Zakharov, Nucl. Phys. B **147**, 385 (1979)
22. M.A. Shifman, A.I. Vainshtein, V.I. Zakharov, Nucl. Phys. B **147**, 448 (1979)
23. L.J. Reinders, H. Rubinstein, S. Yazaki, Phys. Rept. **127**, 1 (1985)
24. V.A. Novikov, M.A. Shifman, A.I. Vainshtein, V.I. Zakharov, Nucl. Phys. B **191**, 301 (1981)
25. T. Falk, K.A. Olive, M. Pospelov, R. Roiban, Nucl. Phys. B **560**, 3 (1999)
26. V.M. Khatsymovsky, I.B. Khriplovich, A.R. Zhitnitsky, Z. Phys. C **36**, 455 (1987)
27. V.M. Khatsymovsky, I.B. Khriplovich, A.S. Yelkhovsky, Ann. Phys. **186**, 1 (1987)
28. V.M. Belyaev, B.L. Ioffe, Zh Eksp Teor. Fiz. **83**, 876 (1982)
29. V.M. Belyaev, B.L. Ioffe, Sov. Phys. JETP **56**, 493 (1982)
30. I. Bigi, N.G. Uraltsev, Sov. Phys. JETP **100**, 198 (1991)
31. I. Bigi, N.G. Uraltsev, Nucl. Phys. B **353**, 321 (1991)
32. A. Pich, E. de Rafael, Nucl. Phys. B **367**, 313 (1991)
33. B. Borasoy, Phys. Rev. D **61**, 114017 (2000)
34. M. Pospelov, A. Ritz, Phys. Rev. D **63**, 073015 (2001)
35. R. Jackiw, C. Rebbi, Phys. Rev. Lett. **37**, 172 (1976)
36. C.A. Baker et al., Phys. Rev. Lett. **97**, 131801 (2006)
37. K. Kawarabayashi, N. Ohta, Nucl. Phys. B **175**, 477 (1980)
38. F. Ambrosino et al. (KLOE Collaboration), Phys. Lett. B **606**, 276 (2005)
39. W.C. Griffith et al., Phys. Rev. Lett. **102**, 101601 (2009)
40. M.A. Rosenberry et al., Phys. Rev. Lett. **86**, 22 (2001)
41. R.J. Crewther, Phys. Lett. B **70**, 349 (1977)
42. E. Witten, Nucl. Phys. B **156**, 269 (1979)
43. G. Veneziano, Nucl. Phys. B **159**, 213 (1979)
44. S. Coleman, *Aspect of Symmetry* (Cambridge University Press, Cambridge, 1988)
45. R.D. Peccei, H.R. Quinn, Phys. Rev. Lett. **38**, 1440 (1977)
46. M.A. Shifman, A.I. Vainshtein, V.I. Zakharov, Nucl. Phys. B **166**, 493 (1980)
47. P. Di Vecchia, G. Veneziano, Nucl. Phys. B **171**, 253 (1980)

Chapter 7
Nuclear Level Calculation

The nuclear level P, CP-odd effects arise in the form of P, CP-odd nuclear moments, induced by P, CP-odd hadron level interactions. The leading P, CP-odd nuclear moment is of course the nuclear electric dipole moment (EDM), which enhances the contribution of P, CP-odd pion-nucleon interaction via many-body effects. The nuclear EDM is measurable in light nuclear systems such as the deuteron and the ^{3}He nucleus, with recent experimental developments using storage rings [1–4]. As the experimental prospects show a very high sensitivity of $O(10^{-29})e$ cm, it is of primary interest to evaluate the nuclear EDM for the deuteron and ^{3}He nucleus due to the P, CP-odd hadron level processes.

For atomic systems, the situation is different. There the nuclear EDM is actually screened by atomic electron rearrangement, and only a minor effect due to the finite size of the nucleus can contribute to the atomic EDM. This phenomenon first pointed by Schiff [5], leads to the suppression of the nuclear P, CP-odd effects in atoms. The relevant P, CP-odd nuclear moment for the atomic EDM is then the *nuclear Schiff moment*. In this discussion, we are also interested in the atomic EDMs of heavy nuclei where, in spite of the suppression due to Schiff's screening, the P, CP-odd hadronic effects are expected to be sufficiently enhanced via nuclear many-body physics.

To evaluate these P, CP-odd nuclear moments, the calculation of the nuclear wave functions are needed. As nuclei are made of protons and neutrons, the nuclear level calculation involves difficulties due to the many-body problem [7, 8]. For light nuclear systems, *ab initio* methods can be used. For heavy nuclei, however, the *ab initio* methods cannot be applied, due to the calculational cost increasing exponentially in nucleon number. For heavy nuclear systems of interest (^{129}Xe, ^{199}Hg, ^{211}Rn and ^{225}Ra), we will use the results from the sophisticated mean-field approach.

In this chapter, we will first see the *ab initio* calculations of the EDMs of light nuclei (deuteron and ^{3}He). We will then review in detail the screening of the P, CP-odd nuclear EDM in atomic system pointed by Schiff and the formula of the nuclear Schiff moment as important P, CP-odd mechanism contributing to the atomic EDM. We will next present the derivation of the leading P, CP-odd nuclear moments (the nuclear

N. Yamanaka, *Analysis of the Electric Dipole Moment in the R-parity Violating Supersymmetric Standard Model*, Springer Theses,
DOI: 10.1007/978-4-431-54544-6_7,

EDM, Schiff moment and magnetic quadrupole moment) for heavy nuclei within a simple model. After this simple calculation, a sophisticated mean-field approach is reviewed, and its results for the evaluation of the nuclear Schiff moments of heavy nuclei (^{129}Xe, ^{199}Hg, ^{211}Rn and ^{225}Ra) are presented. We finally summarize the nuclear level P, CP-odd moments for all nuclei relevant to the discussions in this chapter. Note that the electric charge e is defined as $e = |e| > 0$ in this chapter, in contrast to the previous chapters.

7.1 Ab Initio Calculations: Deuteron and ^{3}He EDM

Let us first present the *ab initio* calculations of the EDMs in the deuteron and the ^{3}He nucleus. The motivation for the study of these light nuclei is as follows. Recently, new experimental techniques of EDM measurements using magnetic storage ring are in preparation [1–4], and the EDM of light nuclei is the main focus. The EDM of light nuclei has the following advantages. The first advantage is that the system can be measured with high precision. The projected experiment of BNL is aimed at reaching the sensitivity of $O(10^{-29})e$ cm for the deuteron EDM. The second advantage is the absence of electrons which suppress the P, CP-odd nuclear effect through Schiff's screening phenomenon. These arguments indicate that the detection of EDMs in light nuclei has more significant sensitivity than the experiments on measuring hadron level P and CP violations. Although the small number of nucleons in the system, the EDM of light nuclei is actually a very competitive probe of new physics. Here we present the calculation of the deuteron EDM [6] and ^{3}He EDM [9].

The calculation of the deuteron EDM was first done by Khriplovich and Korkin with old strong potential [10]. A more complete analysis was made by Liu and Timmermans using three phenomenological nuclear potentials [6]: Argonne v_{18} [11], Nijmegen models Reid93 and Nijm II [12]. The dependence of the P, CP-odd pion-nucleon, ρ-meson-nucleon, η-meson-nucleon and ω-meson-nucleon interactions were studied. They found the dominance of the isovector pion-nucleon coupling for P, CP-odd interactions while all P, CP-odd hadronic interactions have similar magnitudes. The contribution to the deuteron EDM is composed of (a) single nucleon contribution ($d_A^{(Nedm)}$), (b) polarization contribution ($d_A^{(pol)}$), (c) contribution from exchanged current (in this thesis, we neglect its effect). In the case of the deuteron, the single nucleon contribution is simply given by the sum of proton and neutron EDMs ($d_d^{(Nedm)} = d_n + d_p$), since this is the allowed isoscalar combination. The polarization contribution is determined by the spin/isospin selection rules as

$$
\begin{aligned}
d_d^{(pol)} &= \sum_{i=1}^{A=2} \Big\langle \tilde{d} : j = 1, j_z = 1 \,\Big|\, e_i z_i \,\Big|\tilde{d} : j = 1, j_z = 1 \Big\rangle \\
&= \frac{e}{\sqrt{6}} \langle d || \mathbf{r} || d' \rangle \cdot \frac{1}{2} \langle I = 0 | \tau_1^z - \tau_2^z | I = 1, I_z = 0 \rangle = \frac{e}{\sqrt{6}} \langle d || \mathbf{r} || d' \rangle \qquad (7.1)
\end{aligned}
$$

where $z \equiv z_1 - z_2$. Here $|\tilde{d} : j = 1, j_z = 1\rangle$ is the deuteron ground state wave function polarized in the z-axis, with $|d\rangle$ and $|d'\rangle$ its respective P, CP-even and P, CP-odd components. In the second line, we have factored out the isospin space, with the expression $\langle d||\mathbf{r}||d'\rangle$ denoting the reduced matrix element, and the second element is the matrix element of the isospin part with τ_1^z and τ_2^z the generator of the isospin $SU(2)$ group acting on the first and the second nucleons of the deuteron, respectively. The center of mass frame $\sum_i^A \mathbf{r}_i = 0$ was assumed. The wave functions $|d\rangle$ and $|d'\rangle$ were calculated by using the following P, CP-odd nucleon-nucleon (N-N) potential:

$$W(\mathbf{r}_a - \mathbf{r}_b) = -\frac{g_{\pi NN}}{8\pi m_p}\Big[\Big(\bar{g}^{(0)}_{\pi NN}\tau_a \cdot \tau_b + \bar{g}^{(2)}_{\pi NN}(\tau_a \cdot \tau_b - 3\tau_a^z\tau_b^z)\Big)(\boldsymbol{\sigma}_a - \boldsymbol{\sigma}_b) + \bar{g}^{(1)}_{\pi NN}(\tau_a^z \boldsymbol{\sigma}_a - \tau_b^z \boldsymbol{\sigma}_b)\Big] \cdot \nabla_a \frac{e^{-m_\pi r_{ab}}}{r_{ab}} \tag{7.2}$$

where a and b denote the indices of the two interacting nucleons, and $r_{ab} \equiv |\mathbf{r}_a - \mathbf{r}_b|$. The contribution from the exchanged current is suppressed compared to the polarization contribution with the P, CP-odd pion-nucleon couplings of the same order of magnitude [6]. If we neglect the contribution from ω, ρ and η-mesons and consider the pion exchange as the leading contribution, we obtain the following expression for the deuteron EDM:

$$d_d = d_n + d_p - 0.015 g_{\pi NN} \bar{g}^{(1)}_{\pi NN} \times 10^{-13} e\,\text{cm}. \tag{7.3}$$

The sign for isovector coupling was reversed from the result of Ref. [6] due to the difference of convention.

The calculation of the ^{3}He EDM was done by Stetcu et al. The ^{3}He EDM is also given by the single nucleon, polarization and exchanged current contributions. In this work, the exchanged current was neglected, following the small result of the deuteron EDM [6]. The single nucleon contribution is

$$d^{(Nedm)}_{\text{He}} = \langle \text{He}| \sum_{i=1}^{A=3} \frac{1}{2}\left[(d_p + d_n) + (d_p - d_n)\tau_i^z\right]\sigma_i^z |\text{He}\rangle \tag{7.4}$$

where $|\text{He}\rangle$ is the ground state of ^{3}He nucleus within the P and CP conserving hamiltonian. The contribution from the polarization can be written in the second order of perturbation as

$$\mathbf{d}^{(pol)}_{\text{He}} \approx \langle \text{He}| \sum_{i=1}^{A=3} \frac{e}{2}(1 + \tau_i^z)\mathbf{r}_i \sum_{n \neq 0} \frac{1}{E_0 - E_n} |n\rangle\langle n|W|\text{He}\rangle \tag{7.5}$$

where n are the opposite parity states, E_n their corresponding energy, and W the P, CP-odd interactions of Eq. (7.2).

The evaluation of the ^{3}He wave functions was done in the *ab initio* no-core shell model approach [13, 14] with truncated harmonic oscillator basis. The nuclear forces used in the calculation are the Argonne $v18$ [11], the nonlocal Bonn [15] potentials, with Coulomb interaction and isospin violation taken into account, and two-, three-body interactions derived from EFT [16–19].

The calculation of the dependence of the EDM of the ^{3}He nucleus on P, CP-odd pion-nucleon couplings gives the following result [9]:

$$\begin{aligned} d_{\rm He} =& g_{\pi NN}\left(-0.015\bar{g}^{(0)}_{\pi NN} - 0.023\bar{g}^{(1)}_{\pi NN} + 0.036\bar{g}^{(2)}_{\pi NN}\right) \times 10^{-13} e\,{\rm cm} \\ & - 0.04 d_p + 0.90 d_n \end{aligned} \tag{7.6}$$

where we have not explicitly added the contribution of the P, CP-odd pion-nucleon interactions coming from the single nucleon EDM (pion-loop contribution). Here again, the sign for isoscalar and isovector couplings was reversed from the result of Ref. [9] due to the difference of convention. This result shows that the ^{3}He EDM has higher sensitivity than the deuteron and nucleon EDMs to the P, CP-odd pion-nucleon interactions. All measurements of EDMs of nucleon, deuteron and ^{3}He are complementary.

7.2 Schiff's Screening Phenomenon and the Nuclear Schiff Moment

We now move to the evaluation of nuclear P and CP violations for heavy nuclei in atoms, where nuclear EDM is shielded by atomic electrons. The screening phenomenon of the intrinsic EDM of components in atomic system was first shown by Schiff [5]. Actually, the theorem of Schiff states that the EDM of non-relativistic point-like particle in a neutral electrostatically bound system is completely shielded. The effect of the nuclear EDM in atomic systems is therefore suppressed. Schematically, the screening of the nuclear EDM can be described as shown in Fig. 7.1.

Let us see this phenomenon in detail. The atomic EDM receives contribution from the sum of the intrinsic EDMs of its components, and the polarization of the system induced by the mixing of opposite parity states due to P, CP-odd interactions ($\langle s_{1/2}|H_{\not P \not T}|p_{1/2}\rangle$). The atomic EDM (and more generally the EDM of neutral electrostatically bound systems) can then be expressed as

$$\mathbf{d}_{\rm atom} = \sum_i \langle \Psi | d_i \gamma_0 \boldsymbol{\sigma}_i | \Psi \rangle + 2 \sum_M \sum_i \frac{\langle \Psi | Q_i e \mathbf{r}_i | M \rangle \langle M | H_{\not P \not T} | \Psi \rangle}{E_0 - E_M} \tag{7.7}$$

where the first term refers to the sum of the intrinsic EDMs of the components and the second term to the contribution of the polarization arising from P, CP-odd interactions. Ψ is the atomic state of interest, unperturbed by P, CP-odd interactions,

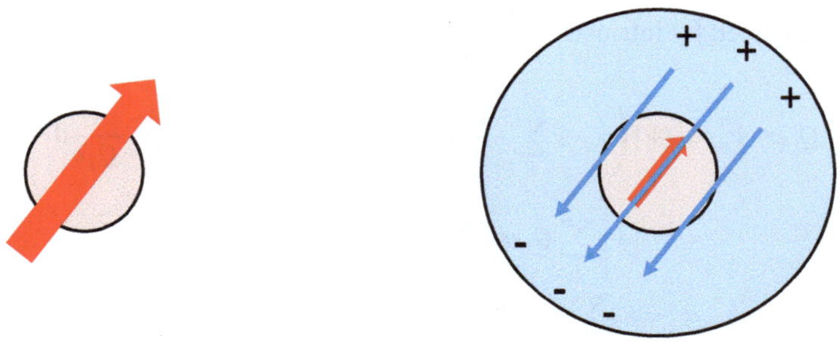

Fig. 7.1 Schematic representations of the bare nuclear EDM (*left side*) and the nuclear EDM screened in atom (*right side*). The effect of nuclear EDM is suppressed by the internal rearrangement of the atomic systems

d_i the intrinsic EDM of the atomic components (electrons and nucleus), Q_i and $\mathbf{r}_i$ are the charges and coordinates of each component, M the intermediate atomic states (with opposite parity against $|\Psi\rangle$) and $H_{\not{P}\not{T}}$ the P, CP-odd interaction. γ_0 and $\boldsymbol{\sigma}_i \equiv \gamma_0\gamma_5\boldsymbol{\gamma}_i$ are Dirac matrices acting independently on each component labeled by i.

The contribution from the first term, i.e. the sum of the EDMs of components can be splitted into the non-relativistic and relativistic parts. Here we write down its hamiltonian:

$$H_{\rm comp} = -\sum_i d_i\gamma_0\boldsymbol{\sigma}_i \cdot \mathbf{E}_{\rm ext} = -\sum_i d_i\boldsymbol{\sigma}_i \cdot \mathbf{E}_{\rm ext} - \sum_i d_i(\gamma_0 - 1)\boldsymbol{\sigma}_i \cdot \mathbf{E}_{\rm ext} \quad (7.8)$$

where $\mathbf{E}_{\rm ext}$ is the external electric field acting on the component EDMs. We will now see that the non-relativistic contribution (first term) will be cancelled by the polarization contribution. The EDMs of the components interact with the internal electric field, thus generating mixing between opposite parity states. The corresponding (non-relativistic) EDM interaction which polarizes the system can be written as

$$H_{\not{P}\not{T}} = -\sum_i \mathbf{d}_i \cdot \mathbf{E}_{\rm int} = \sum_i \frac{1}{Q_i e}\mathbf{d}_i \cdot \nabla_i U(\mathbf{r}) = i\sum_i \frac{1}{Q_i e}[\mathbf{p}_i, H_0] \cdot \mathbf{d}_i \,. \quad (7.9)$$

where $\mathbf{d}_i \equiv d_i\boldsymbol{\sigma}_i$. U and H_0 are respectively the potential energy and the P, CP-even hamiltonian of the atomic system. This equation states that the charge of the system will be polarized proportional to EDMs of components. The contribution from the polarization is then

$$|\Psi'\rangle = |\Psi\rangle + \sum_m \frac{|m\rangle\langle m|H_{\not{P}\not{T}}|\Psi\rangle}{E_0 - E_m} = \left(1 + i\sum_i \frac{1}{Q_i e}\mathbf{d}_i \cdot \mathbf{p}_i\right)|\Psi\rangle \,. \quad (7.10)$$

The EDM induced by intrinsic electron EDM is then

$$\begin{aligned}\langle\Psi'|\sum_i Q_i e\mathbf{r}_i|\Psi'\rangle &= \langle\Psi|\left(1 - i\sum_k \frac{1}{Q_k e}\mathbf{d}_k\cdot\mathbf{p}_k\right)\sum_i Q_i e\mathbf{r}_i\left(1 + i\sum_j \frac{1}{Q_j e}\mathbf{d}_j\cdot\mathbf{p}_j\right)|\Psi\rangle \\ &= \langle\Psi|\, i\left[\sum_l Q_i e\mathbf{r}_i, \sum_k \frac{1}{Q_k e}\mathbf{d}_k\cdot\mathbf{p}_k\right]|\Psi\rangle \\ &= -\langle\Psi|\sum_i \mathbf{d}_i\,|\Psi\rangle\,. \end{aligned} \tag{7.11}$$

We see that the polarization of the system completely cancels the direct contribution given by the sum of EDMs of *non-relativistic* constituents with intrinsic EDM. This cancellation of the non-relativistic EDM contribution by the internal rearrangement is called the Schiff's theorem [5]. This conclusion is very important since the EDM of the nucleus which can be treated as non-relativistic in atomic systems is shielded, so that its effect is largely suppressed. We should also emphasize that the above cancellation applies only for the non-relativistic contribution of the EDM of the constituents. The EDM of relativistic particles actually gives an additional contribution to the polarization of the atomic system (additional effective interaction to $H_{\not P \not T}$), which is not cancelled by the direct contribution. The P, CP-odd electron-nucleon interactions also contribute to the atomic polarization (with no direct contribution), thus giving a non-vanishing EDM to the atom. The effects of relativistic electrons and P, CP-odd electron-nucleon interactions are discussed in the next chapter.

Let us see Schiff's screening phenomenon in the atomic system with a finite size nucleus having an EDM [5, 20]. The hamiltonian of the neutral atomic system with Z electrons is given by

$$H = \sum_{i=1}^{Z}\left[K_i - \int \frac{e^2\rho(\mathbf{r})d^3r}{|\mathbf{R}_i - \mathbf{r}|} - e\mathbf{R}_i\cdot\mathbf{E}_{\text{ext}}\right] + \sum_{i>k}^{Z}\frac{e^2}{|\mathbf{R}_i - \mathbf{R}_k|} - \mathbf{d}_A\cdot\mathbf{E}_{\text{ext}} \tag{7.12}$$

where K_i and R_i are the kinetic energy and the coordinate of the atomic electrons, respectively, and the nuclear charge density $\rho(\mathbf{r})$ is normalized with $\int \rho(\mathbf{r})d^3r = Z$. The nuclear EDM is given by $\mathbf{d}_A \equiv \int e\mathbf{r}\rho(\mathbf{r})d^3r$. Let us now add to this hamiltonian the following additional auxiliary term

$$V_{pol} = \mathbf{d}_A\cdot\mathbf{E}_{\text{ext}} - \frac{1}{eZ}\sum_{i=1}^{Z}\mathbf{d}_A\nabla_i\int\frac{e\rho(\mathbf{r})d^3r}{|\mathbf{R}_i - \mathbf{r}|} \tag{7.13}$$

with $\nabla_i \equiv \partial/\partial R_i$. This new interaction can be expressed by the following commutation relation

$$\frac{i}{Zem_e}\sum_{i=1}^{Z}[\,\mathbf{p}_i, H\,]\cdot\mathbf{d}_A = \mathbf{d}_A\cdot\mathbf{E}_{\text{ext}} - \frac{1}{eZ}\sum_{i=1}^{Z}\mathbf{d}_A\nabla_i\int\frac{e\rho(\mathbf{r})d^3r}{|\mathbf{R}_i - \mathbf{r}|} = V_{pol} \tag{7.14}$$

where we have used the fact that the total electron momentum $\sum_i \mathbf{p}_i$ commutes with the Coulomb interaction between electrons $\sum_{i>k}^{Z} \frac{e^2}{|\mathbf{R}_i-\mathbf{R}_k|}$. The addition of this auxiliary term does not affect the observation in the linear approximation in d_A, since $\langle \Psi | [\, \mathbf{p}_i, H\,] | \Psi \rangle \sim (E_0 - E_0) = 0$. This term is just the effective interaction given by the polarization of the atomic system. We thus obtain the following total hamiltonian shifted by V_{pol}:

$$\tilde{H} = H + V_{pol} = \sum_{i=1}^{Z} [K_i - e\varphi(\mathbf{R}_i) - e\mathbf{R}_i \cdot \mathbf{E}_{\text{ext}}] + \sum_{i>k}^{Z} \frac{e^2}{|\mathbf{R}_i - \mathbf{R}_k|} \tag{7.15}$$

with the electrostatic potential given by

$$\varphi(\mathbf{R}) = \int \frac{e\rho(\mathbf{r})}{|\mathbf{R}-\mathbf{r}|} d^3r + \frac{1}{Z}(\mathbf{d}_A \cdot \nabla) \int \frac{\rho(\mathbf{r})}{|\mathbf{R}-\mathbf{r}|} d^3r \,. \tag{7.16}$$

It is interesting to observe that Eq. (7.15) does not depend on the direct interaction between the nuclear EDM and the external electric field $\mathbf{E}_{\text{ext}}$. This is the screening phenomenon pointed by Schiff [5].

Let us expand the nuclear potential in r/R (multipole expansion). To the first order in r/R, the nuclear electrostatic potential of Eq. (7.16) can be written as

$$-\int e\rho(\mathbf{r}) \left(\mathbf{r} \cdot \nabla \frac{1}{R} \right) d^3r + \frac{1}{Z} (\mathbf{d}_A \cdot \nabla) \frac{1}{R} \int \rho(\mathbf{r}) d^3r = 0 \,. \tag{7.17}$$

This cancellation means that the EDM of the nucleus is totally screened if the nucleus is point-like. This is also the consequence of Schiff's screening phenomenon. The shielding of the nuclear EDM contribution is however not complete if we take the finite size of the nucleus into account. This fact manifests itself through the electrostatic potential of the nucleus by the non-zero contribution of the third order terms in r/R:

$$\varphi^{(3)} = -\frac{1}{6} \int e\rho(\mathbf{r}) r_i r_j r_k d^3r \nabla_i \nabla_j \nabla_k \frac{1}{R} + \frac{1}{2Z} (\mathbf{d}_A \cdot \nabla) \nabla_i \nabla_j \frac{1}{R} \int \rho(\mathbf{r}) r_i r_j d^3r \,. \tag{7.18}$$

Tensors $r_i r_j r_k$ and $r_i r_j$ are reducible, and their decompositions can be written as follows:

$$r_i r_j r_k = \left[r_i r_j r_k - \frac{1}{5} r^2 (r_i \delta_{jk} + r_j \delta_{ik} + r_k \delta_{ij}) \right] + \frac{1}{5} r^2 (r_i \delta_{jk} + r_j \delta_{ik} + r_k \delta_{ij}) \tag{7.19}$$

$$r_i r_j = \left[r_i r_j - \frac{1}{3} r^2 \delta_{ij} \right] + \frac{1}{3} r^2 \delta_{ij}. \tag{7.20}$$

It is thus possible to decompose the third order $\varphi^{(3)}$ to the rank-3 octupole potential [first terms of Eqs. (7.19) and (7.20)] and the rank-1 "Schiff" potential [second terms of Eqs. (7.19) and (7.20)]. The electric octupole moment is a P, CP-odd moment, so it can contribute to the atomic EDM through higher nuclear spin state. In this discussion, the nuclear states in question are spin 1/2 states, so we omit the octupole potential. By subtituting the first term of Eqs. (7.19) and (7.20) to Eq. (7.18), we obtain the following rank-1 Schiff potential:

$$\varphi_{\text{Schiff}} = -\mathbf{S}_A \cdot \nabla \left(\nabla^2 \frac{1}{R} \right) = 4\pi \mathbf{S}_A \cdot \nabla \delta^3(\mathbf{R}) \tag{7.21}$$

where $\mathbf{S}_A$ is defined as

$$\mathbf{S}_A \equiv \frac{1}{10} \left[\int e\rho(\mathbf{r})\mathbf{r}r^2 d^3r - \frac{5}{3}\mathbf{d}_A \frac{1}{Z} \int \rho(\mathbf{r}) r^2 d^3r \right]. \tag{7.22}$$

This is the *nuclear Schiff moment*. The Schiff moment operator is thus written as

$$\hat{\mathbf{S}} = \frac{1}{10} \sum_{p=1}^{Z} e \left(r_p^2 - \frac{5}{3} \langle r^2 \rangle_{\text{ch}} \right) \mathbf{r}_p \tag{7.23}$$

where $\langle r^2 \rangle_{\text{ch}} \equiv \frac{1}{Z} \int r^2 \rho(\mathbf{r}) d^3r$ is the average (squared) charge radius of the nucleus, and $\mathbf{r}_p$ the coordinate operator of the nuclear proton. The above Schiff moment operator satisfies $\mathbf{S}_A = \langle \Psi_A | \hat{\mathbf{S}} | \Psi_A \rangle$ with Ψ_A the nuclear state vector. Note that the Schiff potential (7.21) acts on the atomic electron states, whereas the Schiff operator (7.23) operates in the nuclear space. If we consider also the charge distribution of the nucleons, we have to extend this expression to

$$\hat{\mathbf{S}} = \frac{1}{10} \sum_{N=1}^{A} \sum_{i_N} e_{i_N} \left[(\mathbf{r}_N + \boldsymbol{\rho}_{i_N})^2 - \frac{5}{3} \langle r^2 \rangle_{\text{ch}} \right] (\mathbf{r}_N + \rho_{i_N}) \tag{7.24}$$

where N denotes the index of each nucleon and i_N the index of the charged constituents inside the Nth nucleon. The coordinate of the i_Nth constituent relative to the center of mass of the Nth nucleon $\mathbf{r}_N$ is given by $\boldsymbol{\rho}_{i_N}$, so that the intrinsic EDM of the nucleon N can be written as $\mathbf{d}_N = \sum_{i_N} e_{i_N} \boldsymbol{\rho}_{i_N}$. The charge distribution inside the nucleon is smaller than the size of the nucleus, so taking only up to the first order in ρ_{i_N}, we obtain the following final formula for the nuclear Schiff moment

$$\hat{\mathbf{S}} = \hat{\mathbf{S}}^{\text{ch}} + \hat{\mathbf{S}}^{\text{nucleon}} \tag{7.25}$$

where

$$\hat{\mathbf{S}}^{\text{ch}} = \frac{e}{10}\sum_{p=1}^{Z}\left(r_p^2 - \frac{5}{3}\langle r^2\rangle_{\text{ch}}\right)\mathbf{r}_p \tag{7.26}$$

$$\hat{\mathbf{S}}^{\text{nucleon}} \approx \sum_{N=1}^{A}\left\{\frac{1}{6}\left(r_N^2 - \langle r^2\rangle_{\text{ch}}\right)\mathbf{d}_N + \frac{1}{5}\left((\mathbf{r}_N\cdot\mathbf{d}_N)\mathbf{r}_N - \frac{r_N^2}{3}\mathbf{d}_N\right)\right\}. \tag{7.27}$$

The derivation of the above relations from Eq. (7.24) is given in Appendix F.

7.3 Derivation of P, CP-odd Nuclear Moments in a Simple Model

We will now try to derive the P, CP-odd nuclear moments of heavy nuclei in a simple model to see their qualitative properties. The P, CP-odd nuclear moments of interest are the nuclear EDM, the Schiff moment and the magnetic quadrupole moment, which give the leading contribution to nuclear and atomic P and CP violations. The derivation presented in this section follows the discussion of Ref. [21]. P, CP-odd nuclear moments are generated in the presence of P, CP-odd nucleon-nucleon (N-N) interactions. Let us assume the following P, CP-odd interaction:

$$\begin{aligned} W_{ab} = \frac{G_F}{\sqrt{2}}\frac{1}{2m_N}\{&(\eta_{ab}\boldsymbol{\sigma}_a - \eta_{ba}\boldsymbol{\sigma}_b)\cdot\nabla\delta(\mathbf{r_a}-\mathbf{r_b}) \\ &+\eta'_{ab}(\boldsymbol{\sigma}_a\times\boldsymbol{\sigma}_b)\cdot[(\mathbf{p}_a-\mathbf{p}_b)\delta(\mathbf{r}_a-\mathbf{r}_b)+\delta(\mathbf{r}_a-\mathbf{r}_b)(\mathbf{p}_a-\mathbf{p}_b)]\} \end{aligned} \tag{7.28}$$

where a and b label nucleons in the nuclear system. η_{ab} and η'_{ab} are the P, CP-odd N-N coupling constants. This interaction is the non-relativistic approximation of the general contact P, CP-odd N-N interaction at the lowest order in derivatives. The relation between the scalar-pseudoscalar type P, CP-odd N-N couplings η_{ab} and the P, CP-odd pion-nucleon couplings of Eq. (6.3) is roughly $\eta_{ab} \sim O\left(\frac{\bar{g}_{\pi NN} g_{\pi NN}}{m_\pi^2 G_F/\sqrt{2}}\right) \sim O(10^7 \bar{g}_{\pi NN})$ ($\bar{g}_{\pi NN}$ denotes $\bar{g}^{(0)}_{\pi NN}$, $\bar{g}^{(1)}_{\pi NN}$ and $\bar{g}^{(2)}_{\pi NN}$).

It is known that the nuclear interaction has strong pairing force, which pairs even number of nuclei, so that odd nuclei have one valence nucleon. By assuming the simple shell model (works for stable spherical nuclei with nucleon number $A \geq 20$), the valence nucleon feels the following P, CP-odd nuclear potential:

$$W = \frac{G_F}{\sqrt{2}}\frac{\eta_a}{2m_N}\boldsymbol{\sigma}\cdot\nabla\rho_A(r), \tag{7.29}$$

with $\eta_a \equiv \frac{Z}{A}\eta_{ap} + \frac{N}{A}\eta_{an}$, and $\rho_A(r)$ the density of nucleon inside the nucleus normalized with $\int \rho_A(r)d^3r = A$ (note that the definition of ρ and ρ_A are different!). The tensor type P, CP-odd N-N interaction [second term of Eq. (7.28)] does not contribute to the nucleus with one valence nucleon since nucleons in the core are paired to have zero angular momentum. We see from Eq. (7.29) that in this model the nuclear P, CP-odd effect occurs at the surface, so it grows slower than the total nucleon number A of the system.

It is known that the nuclear potential which is felt by valence nucleon and the nuclear density have the same shape. Their relation is then

$$\rho_A(r) = \frac{\rho_A(0)}{U(0)} U(r) \tag{7.30}$$

where $\rho_A(0)$ and $U(0) \sim -45$ MeV are the nuclear density and potential at the center of the nucleus, respectively. The potential felt by the valence nucleon can therefore be rewritten as

$$\tilde{U}(r) = U(r) + W(r) \approx U(|\mathbf{r} + \xi \boldsymbol{\sigma}|), \tag{7.31}$$

where $\xi \equiv \frac{G_F}{\sqrt{2}} \frac{\eta_a}{2m_N} \frac{\rho_A(0)}{U(0)}$. The wave function in this potential then becomes

$$\tilde{\Psi}(\mathbf{r}) = \Psi(\mathbf{r} + \xi \boldsymbol{\sigma}) = (1 + \xi \boldsymbol{\sigma} \cdot \nabla)\Psi(\mathbf{r}). \tag{7.32}$$

Following Refs. [8, 21], we obtain the next three formulae for the dependence of the nuclear EDM, Schiff moment and magnetic quadrupole moment, respectively, on the P, CP-odd N-N interactions:

$$d_A(\xi) = -e\xi \left(q - \frac{Z}{A} \right) t_j \tag{7.33}$$

$$S_A(\xi) = -\frac{eq}{2} \xi \left[\frac{1}{5} \left(t_j + \frac{1}{j+1} \right) r_{ex}^2 - \frac{1}{3} t_j \langle r^2 \rangle_{\text{ch}} \right] \tag{7.34}$$

$$M_A(\xi) = \frac{e}{m_N} \xi \cdot (\mu - q)(2j - 1)t_j \tag{7.35}$$

where μ is the magnetic moment of the nucleus, $r_{ex}^2 \equiv \int r^2 |\tilde{\Psi}(\mathbf{r})|^2 d^3r$ is the mean square radius of the valence nucleon and $\langle r^2 \rangle_{\text{ch}} \equiv \frac{1}{Z} \int r^2 \rho(\mathbf{r}) d^3r$ the mean square radius of the nuclear charge (with $\rho(\mathbf{r})$ the nuclear charge density). The coefficient q is the charge of the valence nucleon in unit of e ($q = 0$ when the valence nucleon is a neutron, and $q = 1$ for the proton). Here $t_j = 1$ is for a nucleus with $j = l + 1/2$, and $t_j = -\frac{j}{j+1}$ for a $j = l - 1/2$ nucleus. The derivation of the nuclear EDM and the Schiff moment in this simple model is presented in Appendix F.

The above estimation within a simple model gives us important qualitative information on the CP-odd nuclear moments. From the formula (7.34), we see that nuclei with valence neutron (^{129}Xe, ^{199}Hg, ^{211}Rn and ^{225}Ra) have no Schiff moment due to the overall factor $q = 0$. This means that P, CP-odd interactions between the valence neutron and the core cannot directly generate the nuclear Schiff moment. Moreover, in Ref. [8], it is assumed that the mean square radii of the valence nucleon and the nuclear charge are approximately equal ($r_{ex}^2 = \langle r^2 \rangle_{\text{ch}} = \frac{3}{5} A^{2/3} r_0^2$). In this approximation, the nuclear Schiff moment of spin 1/2 s-wave nuclei vanishes. This is the case for ^{205}Tl nucleus. To do more quantitative analysis, more accurate determinations of r_{ex}^2 and $\langle r^2 \rangle_{\rm ch}$ are needed, and these are the subject of the next work. In our discussion, we will take the nuclear Schiff moment of ^{205}Tl to be zero. The magnetic quadrupole moment is not relevant to our discussion, since we are considering nuclei with spin 1/2.

Since we are interested in the EDM of heavy atoms, the EDM of heavy nuclei are also not relevant to our discussion, but here we should add some comments on the nuclear enhancement of P, CP-odd effect compared with the single nucleon EDM. The P, CP-odd N-N coupling generated from isoscalar P, CP-odd pion-nucleon interaction is $\frac{G_F}{\sqrt{2}}\eta_a \sim \frac{g_{\pi NN}\bar{g}^{(0)}_{\pi NN}}{m_\pi^2}$. Comparing with the contribution of the isoscalar P, CP-odd pion-nucleon interaction to the the single nucleon EDM $d_N(\bar{g}^{(0)}_{\pi NN}) \sim \frac{e g_{\pi NN}\bar{g}^{(0)}_{\pi NN}}{4\pi^2 m_N} \ln \frac{m_N}{m_\pi}$ [see Eq. (6.46)], the enhancement of the nuclear EDM against the single nucleon EDM is

$$\frac{d_A}{d_N} \sim \left| \frac{e\xi}{d_N} \right| \sim \left| \frac{3\pi}{2m_\pi^2 r_0^3 U(0) \ln(m_N/m_\pi)} \right| \sim 12 \tag{7.36}$$

where we have assumed $1/\rho_A(0) = \frac{4}{3}\pi r_0^3$ with the internucleon distance $r_0 \sim 1.2$ fm, which is valid for nuclei of interest. We see that the nuclear EDM is sensitive on the P, CP-odd pion-nucleon couplings than the single nucleon EDM by more than one order of magnitude in this simple estimation. The nuclear enhancement is of course dependent on the model of CP violation at the hadron level. For the standard model contribution (see Chap. 9), the enhancement factor can be as large as 60 [24, 25]. We can thus expect a very good improvement of sensitivity against the CP violation of new physics with the progress of experimental studies of the nuclear EDM.

We should also give the dependence of the P, CP-odd nuclear moments on the EDM of the valence nucleon. With the same model assumptions, Ref. [21] (also Ref. [8]) gives the following formulae for the nucleon EDM dependence of the nuclear EDM, Schiff moment and magnetic quadrupole moment:

$$d_A(d_N) = d_N\, t_j \tag{7.37}$$

$$S_A(d_N) = \frac{d_N}{2}\left[\frac{1}{5}\left(t_j + \frac{1}{j+1}\right) r_{ex}^2 - \frac{1}{3} t_j \langle r^2 \rangle_{\rm ch}\right] \tag{7.38}$$

$$M_A(d_N) = \frac{d_N}{m_N}(2j-1)t_j \tag{7.39}$$

where d_N is the EDM of the valence nucleon. The derivation of the nuclear EDM and the Schiff moment is presented in Appendix F. The relations (7.38) and (7.39) have similar form as Eqs. (7.34) and (7.35). This implies again that the nuclear Schiff moment of the spin 1/2 s-wave nuclei is suppressed. This is the case for ^{205}Tl, ^{129}Xe and ^{225}Ra nuclei.

Phenomenologically, the valence nucleon is a superposition of the proton and neutron due to the configuration mixing. The mixing coefficients $\langle\sigma_p\rangle_z$ and $\langle\sigma_n\rangle_z$ can be obtained using the magnetic moment of the nucleus as follows:

$$\begin{cases} \mu_A & = \mu_p \langle\sigma_p\rangle_z + \mu_n \langle\sigma_n\rangle_z \\ \langle\, \Psi : j, j_z = j \,|\, \sigma_z \,|\, \Psi : j, j_z = j \,\rangle & = \langle\sigma_n\rangle_z + \langle\sigma_p\rangle_z \end{cases} \tag{7.40}$$

where μ_A is the nuclear magnetic moment (in unit of the Bohr magneton). Here the matrix element $\langle\, \Psi : j, j_z = j \,|\, \sigma_z \,|\, \Psi : j, j_z = j \,\rangle$ is 1 for $j = l + \frac{1}{2}$ nuclei, and $-\frac{j}{j+1}$ for $j = l - \frac{1}{2}$ nuclei (the same matrix element is needed for the derivation of the nuclear EDM generated by the valence nucleon EDM. The detail is given in Appendix F). The magnetic moment of the proton is $\mu_p = +2.7928$ and that of the neutron is $\mu_n = -1.9130$. The mixing coefficients $\langle\sigma_p\rangle_z$ and $\langle\sigma_n\rangle_z$ are needed to separate the EDM contribution of the valence proton and neutron, but also for the P, CP-odd electron-nucleon interactions, which will be reviewed in the next chapter.

In this section, we have seen that the Schiff moment of nuclei with valence neutron has no dependence on the P, CP-odd N-N interaction in the simple shell model. However, for heavy nuclei such as ^{129}Xe, ^{199}Hg, ^{211}Rn and ^{225}Ra nuclei, the whole nuclear system may be polarized by the P, CP-odd N-N interaction. This effect can be calculated with many-body methods using the *mean-field theory*, which will be reviewed in the next section. We must finally note that we have only *assumed* that the shell model works for odd nuclei in question. We have also assumed that nuclei are spherical. Deformed nuclei can enhance P, CP-odd moments with their close opposite parity levels. In the next section, we will review the calculational methods based on mean-field theory to treat the core polarization of heavy nuclei with valence neutron, and their results.

7.4 Evaluation of the ^{129}Xe, ^{199}Hg, ^{211}Rn and ^{225}Ra Nuclear Schiff Moments within Mean-Field Approach

Let us now present more sophisticated calculations of the nuclear wave functions needed to obtain the nuclear Schiff moment from core polarization. Ideally, the nuclear wave function should be obtained via some *ab initio* calculations, but this task is too difficult as the computational cost for solving the many-body problem increases exponentially with nucleon number. We must therefore introduce some approximations.

In many-body calculations, we often use the *Hartree-Fock method*. In the many-body system, the interaction between particles can be renormalized to a mean-field potential with residual interactions, which will be determined phenomenologically. With this approximation, the many-body calculation is thus reduced to an one-body problem (one-particle interacting with the mean field). The hamiltonian of the N-body system in the Hartree-Fock approximation can be written in the following form

$$H = H_0 + V^{N-1} \tag{7.41}$$

where H_0 is the hamiltonian of the single particle (kinetic terms + mean-field potential), and $V^{N-1} \equiv V_{\text{dir}} + V_{\text{ex}}$ is the Hartree-Fock potential, which satisfies the following relations:

$$\begin{aligned} V_{\text{dir}}(\mathbf{r})\Psi(\mathbf{r}) &= \sum_{n=1}^{N-1} \int \Psi_n^\dagger(\mathbf{r}_1)\Psi_n(\mathbf{r}_1)V(\mathbf{r}_1, \mathbf{r})d^3r_1\, \Psi(\mathbf{r}) \\ V_{\text{ex}}(\mathbf{r})\Psi(\mathbf{r}) &= -\sum_{n=1}^{N-1} \int \Psi_n^\dagger(\mathbf{r}_1)\Psi(\mathbf{r}_1)V(\mathbf{r}_1, \mathbf{r})d^3r_1\, \Psi_n(\mathbf{r})\,. \end{aligned} \tag{7.42}$$

These contributions can be illustrated diagramatically in Fig. 7.2.

Solving the Schrödinger equation for H is equivalent to solving the self-consistent equation depicted in Fig. 7.3.

The physical meaning of the Hartree-Fock potential is the interaction between the single particle and the medium (core) made of the remaining $N - 1$ particles. The

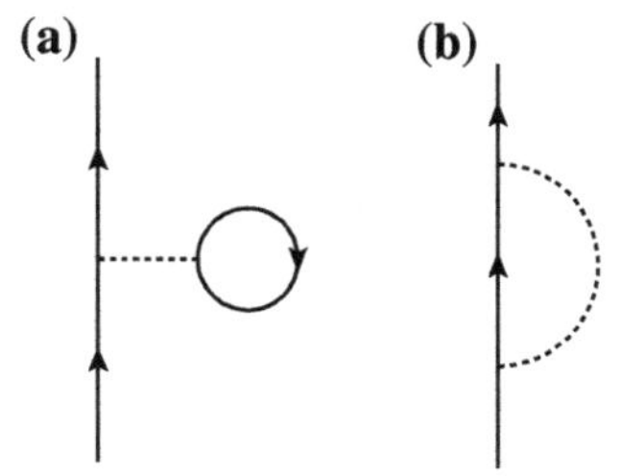

Fig. 7.2 Feynman diagrams representing the Hartree-Fock potential. **a** is the direct contribution (V_{dir}), **b** is the exchange contribution (V_{ex}). Thin lines represent the single particle propagator in the mean-field potential, doted lines represent the inter-particle interaction

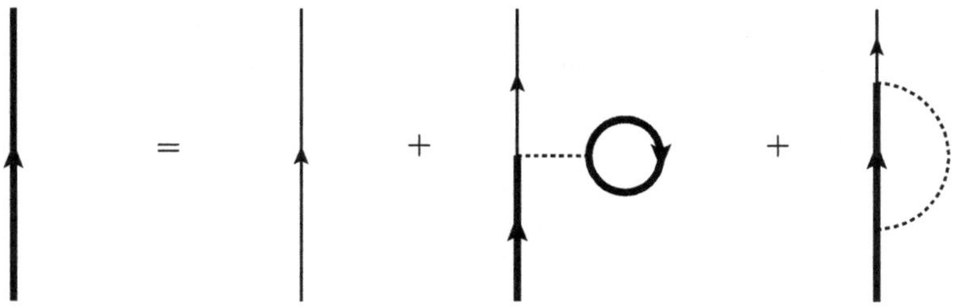

Fig. 7.3 Self-consistent equation for the Hartree-Fock method drawn with Feynman diagrams. Thin lines represent the single particle propagator in the mean-field potential, doted lines represent the interaction. The self-consistent equation will be solved for the thick line

Slater determinant made of Hartree-Fock N particles forms the ground state of the many-body system.

Hartree-Fock method is known to be a very good approximation, but more improvement can be done for the nuclear calculation. In the Hartree-Fock method, the many-body states are Slater determinants formed by the independent single particle states of the mean field. To improve the situation, we should add correlations and introduce the dynamical multi-particle state. We should therefore consider excitations and de-excitations of one-particle states (generated from Hartree-Fock method) through interactions between each other. In this framework, the energy raising operator should be written as

$$X_\eta^\dagger = \sum_{p,h} \left(x_{\eta,ph} a_p^\dagger b_h^\dagger - y_{\eta,ph} b_h a_p \right) \tag{7.43}$$

where a and b are particle and hole annihilating operators, respectively. p and h are the corresponding indices. The determinations of $x_{\eta,ph}$ and $y_{\eta,ph}$ are therefore needed. As $X_\eta^\dagger$'s are energy raising operators, they must satisfy the Heisenberg equation $[H, X_\eta^\dagger] = \hbar\omega_\eta X_\eta^\dagger$. By substituting $X_\eta^\dagger$ and the hamiltonian, we obtain the following equation

$$\begin{aligned}
\hbar\omega_\eta \left(x_{\eta,ph} a_p^\dagger b_h^\dagger - y_{\eta,ph} b_h a_p \right) = & \sum_{p,h} (\epsilon_p - \epsilon_h) x_{\eta,ph} a_p^\dagger b_h^\dagger \\
& + \sum_{p,h,p',h'} (V_{ph'hp'} x_{\eta,p'h'} + V_{pp'hh'} y_{\eta,p'h'} a_p^\dagger b_h^\dagger) \\
& + \sum_{p,h} (\epsilon_p - \epsilon_h) y_{\eta,ph} b_h a_p \\
& + \sum_{p,h,p',h'} (V_{hp'ph'} y_{\eta,p'h'} + V_{hh'pp'} x_{\eta,p'h'} b_h a_p) \\
& + O((a^\dagger b^\dagger, ba)^2)
\end{aligned} \tag{7.44}$$

where V is the particle-hole interaction and higher order terms in $a^\dagger b^\dagger$ and ba are neglected. This self-consistent equation is called the random phase approximation (RPA) equation. The RPA is a dynamical approach: by considering particle-hole interactions, the energy eigenstates of the many-body system become a superposition of particle-hole excitations (see Fig. 7.4).

The Hartree-Fock potential was able to renormalize the interactions between particles into an effective mean-field. The next step is then to include the contribution of the residual interactions. For the nuclear case, the strongest residual interaction is the spin pairing interaction between nucleons. In the presence of attractive pairing interactions, particles can form spin-zero bound-states, which can be effectively seen

Fig. 7.4 Example of diagrammatic representation of 1-particle 1-hole state in RPA. The black dots are particle-hole interactions. Time axis goes from the bottom to the top. The state is a mixing of states with different number of particle-hole pairs

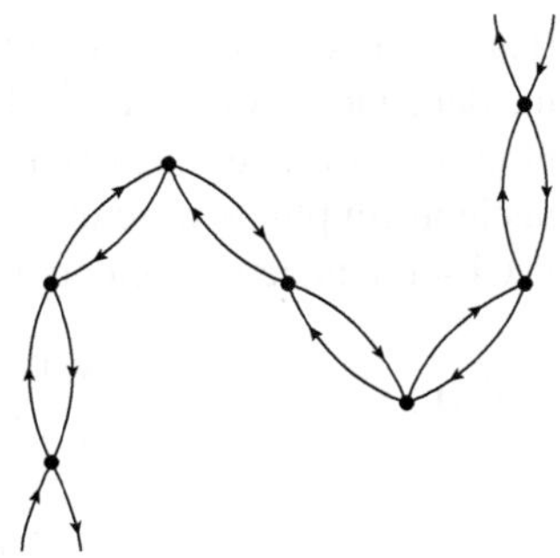

as *mixing between particles and holes*. The mixing between particles and holes is well described in the BCS theory [22, 23]. The hamiltonian of the nuclear system with pairing force can be written as

$$H = \sum_{\alpha}^{N} \epsilon_a c_\alpha^\dagger c_\alpha - \frac{1}{4} G_0 \left(\sum_{\alpha}^{N} (-1)^{j_a - m_\alpha} c_\alpha^\dagger c_{-\alpha}^\dagger \right) \left(\sum_{\beta}^{N} (-1)^{j_b - m_\beta} c_{-\beta} c_\beta \right) \quad (7.45)$$

where c and $c^\dagger$ are annihilation/creation operator of nucleon. The first term in this equation is the Hartree-Fock diagonalized hamiltonian, and the second term is the pairing force. We can see that this hamiltonian is not diagonal, due to the addition of the pairing. To diagonalize this relation, we need to rotate the basis of particle and hole:

$$\begin{aligned} a_\alpha^\dagger &= u_a c_\alpha^\dagger - v_a \tilde{c}_\alpha \\ \tilde{a}_\alpha &= u_a \tilde{c}_\alpha - v_a c_\alpha^\dagger \end{aligned} \quad (7.46)$$

where $\tilde{c}$ and $\tilde{c}^\dagger$ are the annihilation and creation operators of holes, respectively. $\tilde{a}$ and $\tilde{a}^\dagger$ are the corresponding operators for *quasi-particles*, giving the new basis of physical states. This transformation of basis is the Bogoliubov transformation. The determination of u and v will be done by fitting the energy of the system phenomenologically. The extension of the RPA with this formalism can also be done by considering the quasi-particle excitations. The BCS extension of the RPA is called quasi-particle RPA (QRPA).

Once we have explained the basic formalism for calculating nuclear systems, let us now present the calculation of the nuclear Schiff moments. Early calculation of nuclear moments induced by P, CP-odd nuclear forces was done by Flambaum et al. considering the interaction of the valence nucleon with the core within the phenomenological Woods-Saxon potential as a mean-field [24, 25]. RPA calculations

(Hartree-Fock + non-pairing residual interaction) were done by Dmitriev et al., and the dependence of ^{129}Xe, ^{199}Hg, ^{211}Rn and ^{225}Ra Schiff moments on the P, CP-odd pion-nucleon couplings were given in Ref. [26]. The following P, CP-odd pion-nucleon couplings were included at the first order of perturbation, via the following P, CP-odd nucleon-nucleon interaction:

$$W(\mathbf{r}_a - \mathbf{r}_b) = -\frac{g_{\pi NN}}{8\pi m_p}\Big[\Big(\bar{g}^{(0)}_{\pi NN}\tau_a \cdot \tau_b + \bar{g}^{(2)}_{\pi NN}(\tau_a \cdot \tau_b - 3\tau^z_a\tau^z_b)\Big)(\boldsymbol{\sigma}_a - \boldsymbol{\sigma}_b) \\ + \bar{g}^{(1)}_{\pi NN}(\tau^z_a \boldsymbol{\sigma}_a - \tau^z_b \boldsymbol{\sigma}_b)\Big] \cdot \nabla_a \frac{e^{-m_\pi r_{ab}}}{r_{ab}} \tag{7.47}$$

where a and b denote the indices of the two interacting nucleons, and $r_{ab} \equiv |\mathbf{r}_a - \mathbf{r}_b|$. This P, CP-odd interaction was also used in the *ab initio* evaluation of the deuteron and ^{3}He EDMs [see Eq. (7.2)]. The formula for the nuclear Schiff moment is

$$S = \sum_i \frac{\langle \Psi_0 | \hat{S} | \Psi_i \rangle \langle \Psi_i | W | \Psi_0 \rangle}{E_0 - E_i} + \text{c.c.} \tag{7.48}$$

where $\hat{S}$ is the nuclear Schiff moment operator given in Eq. (7.25), and Ψ_0, Ψ_i are nuclear wave functions unperturbed by P, CP-odd N-N interactions W [see Eq. (7.47)] for ground and excited states, respectively.

The result is

$$S_{\mathrm{Hg}} = g_{\pi NN}\left(-0.00004\bar{g}^{(0)}_{\pi NN} - 0.055\bar{g}^{(1)}_{\pi NN} + 0.009\bar{g}^{(2)}_{\pi NN}\right) e\,\mathrm{fm}^3 \tag{7.49}$$

$$S_{\mathrm{Xe}} = g_{\pi NN}\left(0.008\bar{g}^{(0)}_{\pi NN} + 0.006\bar{g}^{(1)}_{\pi NN} - 0.009\bar{g}^{(2)}_{\pi NN}\right) e\,\mathrm{fm}^3 \tag{7.50}$$

$$S_{\mathrm{Rn}} = g_{\pi NN}\left(-0.019\bar{g}^{(0)}_{\pi NN} + 0.061\bar{g}^{(1)}_{\pi NN} + 0.053\bar{g}^{(2)}_{\pi NN}\right) e\,\mathrm{fm}^3 \tag{7.51}$$

$$S_{\mathrm{Ra}} = g_{\pi NN}\left(0.033\bar{g}^{(0)}_{\pi NN} - 0.037\bar{g}^{(1)}_{\pi NN} - 0.043\bar{g}^{(2)}_{\pi NN}\right) e\,\mathrm{fm}^3 \,. \tag{7.52}$$

The small coefficient of $\bar{g}^{(0)}_{\pi NN}$ for the ^{199}Hg Schiff moment is due to an accidental cancellation.

The nucleon EDM dependence of the ^{199}Hg Schiff moment was also calculated in Ref. [27] within the same approach, giving

$$S_{\mathrm{Hg}} = \big((1.895 \pm 0.035)d_n + (0.20 \pm 0.02)d_p\big)\frac{10^{13}}{e\,\mathrm{cm}} e\,\mathrm{fm}^3 \,. \tag{7.53}$$

A more sophisticated calculation including the effects of pairing force with phenomenological Skyrme interactions and nuclear deformation was done for ^{199}Hg [28, 29], ^{211}Rn [29] and for ^{225}Ra [30, 31] Schiff moments. Calculations were made using the computer code HFODD [32] within several models of phenomenological Skyrme interactions: SkO' [33], SkM* [34], SLy4 [35], SV [36] and SIII [36]. SIII

Table 7.1 Coefficients a_i of the dependence of the Schiff moment on P, CP-odd pion-nucleon couplings ($S = g_{\pi NN}(a_0 \bar{g}^{(0)}_{\pi NN} + a_1 \bar{g}^{(1)}_{\pi NN} + a_2 \bar{g}^{(2)}_{\pi NN})$) in unit of e fm^3

Nucleus	Model	$-a_0$	$-a_1$	a_2	$-b$	Ref.
^{199}Hg	SkO'	0.010	0.074	0.018	–	[28]
	SkM* (HFB)	0.041	−0.027	0.069	0.013	
	SLy4 (HFB)	0.013	−0.006	0.024	0.007	[29]
	SLy4 (HF)	0.013	−0.006	0.022	0.003	
	SV (HF)	0.009	−0.0001	0.016	0.002	
	SIII (HF)	0.012	0.005	0.016	0.004	
^{211}Rn	SkM*	0.042	−0.028	0.078	0.015	
	SLy4	0.042	−0.018	0.071	0.016	[29]
	SIII	0.034	−0.0004	0.064	0.015	
^{225}Ra	SkO'	−1.5	6.0	−4.0	–	[31]

The labels HB and HFB stand for calculations in the Hartree-Fock and Hartree-Fock-Bogoliubov approximations, respectively.

may not be as trustworthy as the others: Ref. [28] showed that the interaction was less able to reproduce related observable, the distribution of isoscalar E1 strength in even nuclei. The result is shown in Table 7.1.

The results for ^{211}Rn Schiff moment are almost consistent. The ^{211}Rn nucleus is spherical, so no deformation was considered in the calculations.

In the case of ^{225}Ra Schiff moment, the large enhancement of the P, CP-odd contributions is due to the presence of a nearby parity-doublet states [30]. We have more confidence in the Sko' results, but the uncertainties due to difficulties in treating nuclear deformation in the men-field methods are large. In addition, the optimal Skyrme functional has not yet been identified. These deficiencies render a factor of 2 or 3 of uncertainty [31].

^{199}Hg nucleus has also a small deformation, so it is useful to consider it. The calculation of Ref. [29] took into account the effect of deformation and calculated fully self-consistently, including the P, CP-odd interactions. By comparing with the result of Ref. [28] which considered spherical ^{199}Hg nucleus, we see that the isovector coefficient a_1 significantly decreases with the inclusion of deformation. The results for a_1 vary among Skyrme models and we have no decisive pretext to select one of them.

The calculation of nuclear Schiff moment presents some difficulties in determining the dependence to the P, CP-odd pion-nucleon couplings. This is essentially due to the deficiency in expressing nuclear states of odd nuclei in mean-field theory. To have an accurate description of the nuclei in question, we must wait for new calculational method for odd nuclei. In the subsequent discussion, we will take the average between different calculations shown in Table 7.1 for each coefficient.

7.5 Summary of Nuclear Level Calculation

Deuteron EDM:

$$d_d = -0.015 g_{\pi NN} \bar{g}^{(1)}_{\pi NN} \times 10^{-13} e\,\mathrm{cm} + d_p + d_n \,. \tag{7.54}$$

The deuteron EDM was calculated with realistic N-N potential [6].

^{3}He nucleus EDM:

$$\begin{aligned} d_{\mathrm{He}} =& g_{\pi NN} \left[-0.015 \bar{g}^{(0)}_{\pi NN} - 0.023 \bar{g}^{(1)}_{\pi NN} + 0.036 \bar{g}^{(2)}_{\pi NN} \right] \times 10^{-13} e\,\mathrm{cm} \\ & - 0.04 d_p + 0.90 d_n \end{aligned} \tag{7.55}$$

The ^{3}He nuclear EDM was calculated with the *ab initio* No-core shell model with realistic N-N potential [9].

^{199}Hg nuclear Schiff moment:

$$\begin{aligned} S_{\mathrm{Hg}} =& g_{\pi NN} \Bigg[0.02 \bar{g}^{(0)}_{\pi NN} - 0.007 \bar{g}^{(1)}_{\pi NN} + 0.006 \bar{g}^{(2)}_{\pi NN} \\ & + \frac{0.91 d_n + 0.09 d_p}{e\,\mathrm{cm}} \times 4 \times 10^{12} \Bigg] e\,\mathrm{fm}^3 \,. \end{aligned} \tag{7.56}$$

This is the average of different calculations of mean-field method presented in Ref. [29].

^{129}Xe nuclear Schiff moment:

$$S_{\mathrm{Xe}} = g_{\pi NN} \left[0.008 \bar{g}^{(0)}_{\pi NN} + 0.006 \bar{g}^{(1)}_{\pi NN} - 0.009 \bar{g}^{(2)}_{\pi NN} - \frac{3.2 d_n + 0.06 d_p}{e\,\mathrm{cm}} \times 10^{12} \right] e\,\mathrm{fm}^3. \tag{7.57}$$

The πNN interaction contribution was given by Dmitriev et al. with mean field method [26]. The nucleon EDM contribution was calculated in the refined shell model by Yoshinaga et al. [37].

^{211}Rn nuclear Schiff moment:

$$S_{\mathrm{Rn}} = g_{\pi NN} \left[-0.039 \bar{g}^{(0)}_{\pi NN} + 0.0155 \bar{g}^{(1)}_{\pi NN} + 0.071 \bar{g}^{(2)}_{\pi NN} + \frac{d_n}{e\,\mathrm{cm}} \times 1.5 \times 10^{13} \right] e\,\mathrm{fm}^3. \tag{7.58}$$

This was calculated with mean field method in Ref. [29].

^{225}Ra nuclear Schiff moment:

$$S_{\text{Ra}} = g_{\pi NN}\left[1.5\bar{g}^{(0)}_{\pi NN} - 6.0\bar{g}^{(1)}_{\pi NN} - 4.0\bar{g}^{(2)}_{\pi NN}\right] e\,\text{fm}^3\,. \tag{7.59}$$

This was calculated with mean field method with octupole deformation taken into account [31].

References

1. I.B. Khriplovich, Phys. Lett. B **444**, 98 (1998)
2. F.J.M. Farley et al., Phys. Rev. Lett. **93**, 052001 (2004)
3. Y.K. Semertzidis et al., AIP Conf. Proc. **698**, 200 (2004)
4. Y.F. Orlov, W.M. Morse, Y.K. Semertzidis, Phys. Rev. Lett. **96**, 214802 (2006)
5. L.I. Schiff, Phys. Rev. **132**, 2194 (1963)
6. C.-P. Liu, R.G.E. Timmermans, Phys. Rev. C **70**, 055501 (2004)
7. I.B. Khriplovich, S.K. Lamoreaux, *CP Vioaltion Without Strangeness* (Springer, Berlin, 1997)
8. J.S.M. Ginges, V.V. Flambaum, Phys. Rept. **397**, 63 (2004)
9. I. Stetcu, C.-P. Liu, J.L. Friar, A.C. Hayes, P. Navratil, Phys. Lett. B **665**, 168 (2008)
10. I.B. Khriplovich, R.A. Korkin, Nucl. Phys. A **665**, 365 (2000)
11. R.B. Wiringa, V.G.J. Stoks, R. Schiavilla, Phys. Rev. C **51**, 38 (1995)
12. V.G.J. Stoks, R.A.M. Klomp, C.P.F. Terheggen, J.J. de Swart, Phys. Rev. C **49**, 2950 (1994)
13. P. Navratil, J.P. Vary, B.R. Barrett, Phys. Rev. Lett. **84**, 5728 (2000)
14. P. Navratil, J.P. Vary, B.R. Barrett, Phys. Rev. C **62**, 054311 (2000)
15. R. Machleidt, F. Sammarruca, Y. Song, Phys. Rev. C **53**, 1483 (1996)
16. P. Navratil, Few Body Syst. **41**, 117 (2007)
17. D.R. Entem, R. Machleidt, Phys. Rev. C **68**, 041001(R) (2003)
18. U. van Kolck, Phys. Rev. C **49**, 2932 (1994)
19. E. Epelbaum, A. Nogga, W. Glöckle, H. Kamada, U.-G. Meissner, H. Witala, Phys. Rev. C **66**, 064001 (2002)
20. V. Spevak, N. Auerbach, V.V. Flambaum, Phys. Rev. C **56**, 1357 (1997)
21. O.P. Sushkov, V.V. Flambaum, I.B. Khriplovich, Zh. Eksp. Teor. Fiz. **87**, 1521 (1984) [Sov. Phys. JETP **60**, 873 (1984)]
22. J. Bardeen, L.N. Cooper, J.R. Schrieffer, Phys. Rev. **106**, 162 (1957)
23. J. Bardeen, L.N. Cooper, J.R. Schrieffer, Phys. Rev. **108**, 1175 (1957)
24. V.V. Flambaum, I.B. Khriplovich, O.P. Sushkov, Phys. Lett. B **162**, 213 (1985)
25. V.V. Flambaum, I.B. Khriplovich and O.P. Sushkov, Nucl. Phys. A **449**, 750 (1985)
26. V.F. Dmitriev, R.A. Senkov, N. Auerbach, Phys. Rev. C **71**, 035501 (2005)
27. V.F. Dmitriev, R.A. Senfkov, Phys. Rev. Lett. **91**, 212303 (2003)
28. J.H. de Jesus, J. Engel, Phys. Rev. C **72**, 045503 (2005)
29. S. Ban, J. Dobaczewski, J. Engel, A. Shukla, Phys. Rev. C **82**, 015501 (2010)
30. J. Engel, M. Bender, J. Dobaczewski, J.H. de Jesus, P. Olbratowski, Phys. Rev. C **68**, 025501 (2003)
31. J. Dobaczewski, J. Engel, Phys. Rev. Lett. **94**, 232502 (2005)
32. J. Dobaczewski, W. Satula, B. Carlsson, J. Engel, P. Olbratowski, P. Powalowski, M. Sadziak, N. Schunk, A. Staszczak, M. Stoitsov et al., Comput. Phys. Commun. **180**, 2361 (2009)
33. P.-G. Reinhard, D.J. Dean, W. Nazarewicz, J. Dobaczewski, J.A. Maruhn, M.R. Strayer, Phys. Rev. C **60**, 014316 (1999)
34. J. Bartel, P. Quentin, M. Brack, C. Guet, H.B. Hakansson, Nucl. Phys. A **386**, 70 (1982)
35. E. Chabanat, P. Bonche, P. Haensel, J. Meyer, R. Schaeffer, Nucl. Phys. A **635**, 231 (1998)
36. M. Beiner, H. Flocard, N. Van Giai, P. Quentin, Nucl. Phys. A **238**, 29 (1975)

37. N. Yoshinaga et al., Talk given at 4th International conference on "Fundamental Physics Using Atoms", Osaka University, August 2010 (URL: http://xqw.hep.okayama-u.ac.jp/kakenhi/index.php/fpua2010/fpua2010_top_e/)

Chapter 8
Atomic Level Calculation

The final step of the many-body calculation is the evaluation of the atomic electric dipole moment (EDM). The atomic system can have EDM in the presence of underlying P, CP-odd interactions. This is realized through three leading P, CP-odd processes:

- Enhancement of intrinsic electron EDMs through relativistic effect.
- P, CP-odd electron-nucleon interactions.
- Contribution from the nuclear Schiff moment.

We will first examine the formulation of the atomic EDM due to these P, CP-odd interactions and see the difference between paramagnetic and diamagnetic atoms. We will then present the high quality atomic many-body calculations of the EDMs of the systems of interest (^{205}Tl, ^{129}Xe, ^{199}Hg, ^{211}Rn, ^{225}Ra). We will review the employed mean-field methods and give their results. In the subsequent section, we will derive the contribution to the EDM of diamagnetic atoms using analytic formulae. We finally summarize the dependences of the EDMs of the relevant systems on P, CP-odd processes. Note that the electric charge e is defined as $e = |e| > 0$, the convention of chapter 7.

8.1 Formulation of the Atomic EDM

As we have seen in Eq. (7.7), the EDM of atom is composed of the sum of the intrinsic EDMs of its components and the polarization of the whole system due to P, CP-odd interactions. We have also seen in the previous chapter that the screening phenomenon of Schiff [1] removes any non-relativistic part of the EDM of its components. The formula for the atomic EDM (7.7) can then be rewritten as

$$\mathbf{d}_{\text{atom}} = \sum_i \langle \psi | d_i (\gamma_0 - 1) \boldsymbol{\sigma}_i | \psi \rangle + 2 \sum_M \sum_i \frac{\langle \psi | Q_i e \mathbf{r}_i | M \rangle \langle M | H'_{\not{P} \not{T}} | \psi \rangle}{E_0 - E_M}, \quad (8.1)$$

N. Yamanaka, *Analysis of the Electric Dipole Moment in the R-parity Violating Supersymmetric Standard Model*, Springer Theses,
DOI: 10.1007/978-4-431-54544-6_8, © Springer Japan 2014

where $H'_{\not P \not T}$ is the P, CP-odd hamiltonian, composed of the relativistic contribution of the electron EDM, the P, CP-odd electron-nucleon interactions and the nuclear Schiff moment. Let us present them in detail.

- EDM interaction of the relativistic component of electrons:

$$H_{\text{eEDM}} = \sum_i^Z (\gamma_0 - 1)\boldsymbol{\sigma}_i \cdot \mathbf{E}_{\text{int}} = \sum_i^Z (\gamma_0 - 1)\boldsymbol{\sigma}_i \cdot \boldsymbol{\nabla}_i U(\mathbf{r}), \tag{8.2}$$

  where the sum is taken over all electrons in the atoms. The non-relativistic part has of course been removed, since it does not contribute to the atomic EDM. The relativistic part can however contribute to the EDM of paramagnetic atoms [2–4]. This contribution can be seen as an enhancement of the electron EDM due to relativistic effect inside the atom (it can be expressed as $d_{\text{atom}} = K d_e$), and the coefficient K grows faster than Z^3, where Z is the charge of the nucleus. We have thus for heavy atoms a huge enhancement of the electron EDM. This large amplification can be explained by the strong motional electric field generated from the highly relativistic valence electron in heavy atoms which polarizes the system, and this effect is not cancelled by the bare EDM contribution [first term of Eq. (8.1)]. This is one of the important advantages in the study of paramagnetic systems. The high relativistic enhancement exists also for paramagnetic molecules which are being studied actively.
- P, CP-odd electron-nucleon (e-N) interactions:

$$H_{\text{eN}} = \frac{G_F}{\sqrt{2}} \sum_{N=p,n} \left[C_N^{\text{SP}} \bar{N} N \, \bar{e} i \gamma_5 e + C_N^{\text{PS}} \bar{N} i \gamma_5 N \, \bar{e} e + \frac{1}{2} C_N^{\text{T}} \varepsilon^{\mu\nu\rho\sigma} \bar{N} \sigma_{\mu\nu} N \, \bar{e} \sigma_{\rho\sigma} e \right]. \tag{8.3}$$

  This is the hamiltonian of Eq. (6.4). The generation of atomic EDM within P, CP-odd e-N interactions can naturally be understood, since they polarize the atomic system.
- Nuclear Schiff moment interaction:

$$H_{\text{SM}} = -\frac{3\mathbf{S} \cdot \mathbf{r}}{B} \rho(\mathbf{r}) \,, \tag{8.4}$$

  where $\mathbf{S}$ is the nuclear Schiff moment, $B \equiv \int \rho(r) r^4 dr$. The nuclear Schiff moment is the residual nuclear P, CP-odd effect due to the finite size effect of the nucleus, which could not be shielded with the rearrangement of the atomic system.

The first term of Eq. (8.1), the bare electron EDM contribution, is given by the unpaired valence electrons. The relativistic effect for such electrons should be small, since the valence electron feels potential near atomic radius. Therefore the contribution from the first term of Eq. (8.1) can be neglected. The atomic EDM receives then the leading contribution from the polarization of the system generated by the

P, CP-odd interactions seen above. It can be rewritten as

$$\mathbf{d}_{\text{pol}} = 2\sum_m \frac{\langle\psi\left|-e\sum_i \mathbf{r}_i\right|m\rangle\left\langle m\left|H'_{P/T/}\right|\psi\right\rangle}{E_0 - E_m}. \tag{8.5}$$

The minus sign is due to the negative charge of electrons (note that e is positive!).

Due to the difference in Lorentz structures, the manifestations of the P, CP-odd operators differs for paramagnetic and diamagnetic atoms.

- The paramagnetic atoms have valence electrons, so their EDMs are very sensitive to the electron EDM and scalar-pseudoscalar type P, CP-odd e-N interaction (C_N^{SP}). In our discussion, this is the case for ^{205}Tl atom.
- The diamagnetic atoms have no valence electrons and the electron shell is closed. The contribution from electron EDM and scalar-pseudoscalar type P, CP-odd e-N interaction are suppressed. Instead, the Schiff moment becomes the leading contribution. It should be noted that the effect of electron EDM and scalar-pseudoscalar type P, CP-odd e-N interaction does not vanish, since they can contribute to the EDM via hyperfine interaction. This contribution to diamagnetic atoms will be derived in Sect. 8.3. In our discussion, this is the case for ^{129}Xe, ^{199}Hg, ^{211}Rn and ^{225}Ra atoms.

As for nuclear level calculations, the evaluation of atomic EDMs needs the many-body wave functions. In the atomic level calculation, it is possible to obtain very accurate wave functions. In the next section, we will briefly review the many-body method used and the results of the evaluations of the relevant atomic EDMs.

8.2 Calculational Methods of the Electron Wave Functions

We will now present the atomic level many-body method based on the relativistic mean-field theory. The calculations of the electron wave functions in atoms are essentially the same as those for the nuclear system. *Ideally*, the full hamiltonian we wish to solve is

$$\hat{H} = \sum_{i=1}^{N}\left[\boldsymbol{\alpha}_i \cdot \mathbf{p}_i + (\beta - 1)m - Ze^2/r_i\right] + \sum_{i<j}\frac{e^2}{|\mathbf{r}_i - \mathbf{r}_j|}. \tag{8.6}$$

This is of course not possible, due to the difficulty in treating the huge number of electrons, as seen in the nuclear level calculation, and we must use some approximations.

The first step is to construct the ground state of the atom by using the relativistic Hartree-Fock method. The hamiltonian of the single-electron in a mean field can be written as

$$\hat{h}_0 = \alpha \cdot \mathbf{p} + (\beta - 1)m - Ze^2/r + \hat{V}^{N-1}, \tag{8.7}$$

where $\hat{V}^{N-1}$ is the Hartree-Fock potential which is similar to Eq. (7.42) for the nuclear system. By solving the self-consistent relativistic Hartree-Fock equation, we obtain the single-electron wave functions in a mean-field potential.

The Hartree-Fock method applied to atomic systems is known to be accurate to 10 %, but even more accurate evaluation is possible. For noble gases such as Xe or Rn, the RPA (see Chap. 7) can be used to improve their wave functions. For other atoms (i.e. paramagnetic atoms, Hg, Ra, etc), the many-body perturbation theory (MBPT) with configuration interaction (CI) can be used to improve the accuracy significantly, to the level of 1 % [5]. The essential idea of this approach is to divide the Hilbert space of the electron many-body states into the subspace of valence electron+"frozen core" and its counterpart. The effects on the valence electron state due to the potential of the core then must be treated, since its contribution is dominant in low energy atomic excitations. The effect of virtual core excitation evaluated with the MBPT is included in the correction of the valence electron state. Let us explain this approach in a little more detail. We consider the projection P into the Hilbert subspace in which the core is fixed (only valence electrons can be excited), and Q its counterpart. The hamiltonian and state vectors will be split as

$$H = PHP + PHQ + QHP + QHQ, \tag{8.8}$$

$$\Psi = P\Psi + Q\Psi \equiv \Phi + \chi. \tag{8.9}$$

The Schrödinger equation $H\Psi = E\Psi$ will be rewritten as

$$(PHP)\Phi + (PHQ)\chi = E\Phi, \tag{8.10}$$

$$(QHQ)\chi + (QHP)\Phi = E\chi. \tag{8.11}$$

The first line can be rewritten as

$$(PHP)\Phi + \Sigma(E) = E\Phi, \tag{8.12}$$

where

$$\Sigma(E) \equiv (PHQ)\,[E - (QHQ)]^{-1}\,(QHP). \tag{8.13}$$

The P projected hamiltonian (PHP) is the contribution from the core electron averaged, so it can be expressed as

$$PHP = E_{\rm core} + \sum_{i>N_{\rm core}} h_i^{\rm CI} + \sum_{j>i>N_{\rm core}} \frac{1}{|\mathbf{r}_i - \mathbf{r}_j|}, \tag{8.14}$$

where $E_{\rm core}$ is the kinetic energy of the core electrons + Coulomb interaction (electron-nucleus, electron-electron), $h_{\rm CI}$ is the kinetic energy of the valence electron(s) + Coulomb interaction (valence electron-core, valence electron-nucleus), and the last term the interaction between valence electrons. In this expression, we use the atomic unit ($m_e = 1$, $e = 1$). $\Sigma(E)$ is the contribution from the core excitation (contribution from the Q projected subspace), and can be evaluated using MBPT. In the case of one valence electron system, the PHP is reduced to the relativistic Hartree-Fock hamiltonian, and Σ can be regarded as the self-energy correction. For systems with two valence electrons, there are also corrections from the screening of the Coulomb interaction by the core. The Feynman diagrams for evaluating these corrections can be depicted as shown in Figs. 8.1 and 8.2.

With these methods, the atomic electron wave functions can be calculated accurately for the perturbative evaluation of atomic EDM due to the underlying P, CP-odd interactions. Dzuba et al. have calculated the contribution from P, CP-odd interactions (C_N^{T} and C_N^{PS}), magnetic interaction with the electron EDM, and the nuclear Schiff moment to the EDM of ^{129}Xe, ^{199}Hg, ^{171}Yb, ^{211}Rn and ^{225}Ra atoms [6]. The result is shown in Table 8.1.

The calculation was also performed for the thallium atoms and gave the following result [7, 8]:

$$d_{\mathrm{Tl}}(d_e, C_N^{\mathrm{SP}}) = -582 d_e - 7.0 \times 10^{18} \left(\frac{81}{205} C_p^{\mathrm{SP}} + \frac{124}{205} C_N^{\mathrm{SP}} \right) e\,\mathrm{cm}, \quad (8.15)$$

where the coherency of the non-relativistic nucleon in nucleus was used to derive the ratio of C_N^{SP} contributions (81, 124 and 205 are respectively for the number of protons, neutrons and all nucleons of ^{205}Tl nucleus). The theoretical uncertainty of this calculation is expected to be less than 3 %. As it can be seen, the enhancement of electron EDM in thallium atom is huge. This large amplification can be explained by the strong electric field generated from the highly relativistic valence electron in heavy atoms, and is one of the important advantages in the study of paramagnetic systems.

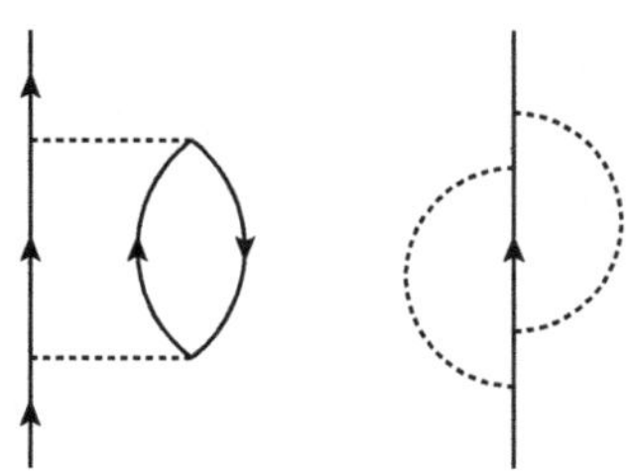

Fig. 8.1 Feynman diagrams representing the self-energy of the valence electron at the second order

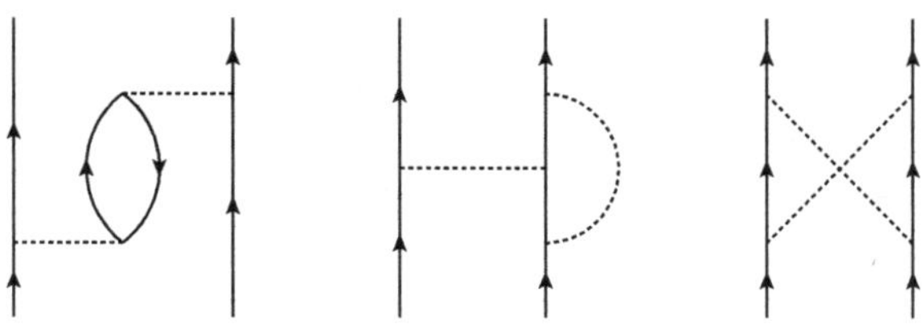

Fig. 8.2 Feynman diagrams representing the screening of the Coulomb potential between valence electrons (at the second order)

Table 8.1 Contributions of the P, CP-odd interactions to the EDM of diamagnetic atoms

	^{129}Xe	^{199}Hg	^{211}Rn	^{225}Ra
$d_{\rm atom}(C_N^{\rm T})$ $(10^{-20} C_N^{\rm T} \langle \sigma_N \rangle_z e\,{\rm cm})$	0.57	−5.1	5.6	−18
$d_{\rm atom}(C_N^{\rm P})$ $(10^{-23} C_N^{\rm P} \langle \sigma_N \rangle_z e\,{\rm cm})$	1.6	−18	21	−64
$d_{\rm atom}(S)$ $(10^{-17}[S/(e\,{\rm fm}^3)] e\,{\rm cm})$	0.38	−2.6	3.3	−8.8
$d_{\rm atom}(d_e$ from $H_B)$ $(10^{-4} d_e)$	1.0	11	−13	−56

8.3 Subleading Contributions to EDMs of Diamagnetic Atoms

For diamagnetic atoms, the electron EDM and the scalar-pseudoscalar type P, CP-odd e-N interaction [term with $C_N^{\rm SP}$ in Eq. (8.3)] cannot contribute at the leading order, due to the closed electron shell. However, they can contribute to the EDM of diamagnetic atoms in the third order of perturbation with the hyperfine interaction [9, 10]

$$h_{\rm hf} = e \frac{\boldsymbol{\alpha} \cdot \mathbf{M} \times \mathbf{r}}{r^3}, \tag{8.16}$$

where $\boldsymbol{\alpha}$ is the Dirac matrix and $\mathbf{M}$ the nuclear magnetic moment. Its contribution to the EDM of diamagnetic atom is given by

$$\mathbf{d}_{\rm hf} = \sum_{m,n} \frac{\langle \psi \left| -e \sum_i \mathbf{r}_i \right| m \rangle \langle m \left| h_{\rm hf} \right| n \rangle \langle n \left| H''_{\not P \not T} \right| \psi \rangle}{(E_0 - E_m)(E_0 - E_n)} + \text{permutations}, \tag{8.17}$$

where $H''_{\not P \not T}$ is the electron EDM interaction or the scalar-pseudoscalar type P, CP-odd e-N interaction ($C_N^{\rm SP}$). The correction of $|\psi\rangle$ by $h_{\rm hf}$ and $H''_{\not P \not T}$ is concentrated in the vicinity of the nucleus, so terms with the matrix element of the operator $-e\mathbf{r}$ at the middle in Eq. (8.17) have negligible contributions. The contribution of these P, CP-odd interactions to atomic EDMs can be given by substituting the P, CP-odd interaction $H'_{\not P \not T}$ of Eq. (8.5) by the following effective P, CP-odd operator

$$H_{{\rm eff}\not P \not T} = \sum_n \frac{\left(H''_{\not P \not T} |n\rangle\langle n| h_{\rm hf} \right) + \left(h_{\rm hf} |n\rangle\langle n| H''_{\not P \not T} \right)}{E_0 - E_n}. \tag{8.18}$$

For the electron EDM contribution to diamagnetic atoms, the interaction between EDM and magnetic field (generated by the nuclear magnetic moment) must also be taken into account. This interaction is written as

$$H_B = -i d_e \gamma \cdot \mathbf{B}, \tag{8.19}$$

where **B** is the magnetic field. The effect of this interaction, although smaller than the hyperfine interaction contribution by about one order [6], cannot be neglected. We should therefore define the following effective P, CP-odd interaction for the electron EDM contribution to the diamagnetic atoms:

$$H_{\text{eEDMeff}} = \sum_n \frac{\left(H''_{\not P \not T}|n\rangle\langle n|h_{\text{hf}}\right) + \left(h_{\text{hf}}|n\rangle\langle n|H''_{\not P \not T}\right)}{E_0 - E_n} + H_B. \tag{8.20}$$

To obtain the contribution of the above effective P, CP-odd operators to the EDM of diamagnetic atoms, we need the analytic formulations of the atomic electron wave functions [10, 11]. The external electron wave function can be written as

$$\psi_{njlm}(\mathbf{r}) = \begin{pmatrix} f_{njl}(r)\Omega_{jlm} \\ g_{njl}(r)\, i(-\boldsymbol{\sigma}\cdot\mathbf{r}/r)\Omega_{jlm} \end{pmatrix}, \tag{8.21}$$

with $\Omega_{jlm} \equiv [Y_{lm'}(\hat{r})\otimes|\, s, s_z\,\rangle\,]_{jm}$ the spherical function with spin s and the spherical harmonics $Y_{lm}(\hat{r})$. n and m are respectively the principal and magnetic quantum numbers, j and l are respectively the total and orbital angular momenta. We must solve the relativistic Dirac equation [$f_{njl}(r)$ and $g_{njl}(r)$] in the unscreened Coulomb field of the nucleus ($r \ll aZ^{1/3}$, with a is the Bohr radius), since the P, CP-odd interactions we want to calculate are all short range interactions. The solution of the Dirac equation of electron in static Coulomb potential is

$$f_{njl}(r) = \frac{c_{njl}}{r}\left[(\gamma+\kappa)J_{2\gamma}(x) - \frac{x}{2}J_{2\gamma-1}(x)\right], \tag{8.22}$$

$$g_{njl}(r) = \frac{c_{njl}}{r} Z\alpha J_{2\gamma}(x), \tag{8.23}$$

where J is the bessel function, $x \equiv \left(\frac{8Zr}{a}\right)^{1/2}$, $\gamma \equiv \sqrt{(j+1/2)^2 - Z^2\alpha^2}$ and $\kappa \equiv (l-j)(2j+1)$. The normalization factor $c_{njl} = \frac{\kappa}{|\kappa|}\left(\frac{1}{Za(n-\sigma_l)^3}\right)^{1/2}$ (with σ_l the quantum defect) was determined phenomenologically to reproduce the non-relativistic solution at long distance [11]. At short distance, this solution can be rewritten as

$$f_{njl}(r) \approx \frac{\kappa}{|\kappa|}(\kappa-\gamma)\left(\frac{Z}{a^3(n-\sigma_l)^3}\right)^{1/2} \frac{2}{\Gamma(2\gamma+1)}\left(\frac{a}{2Zr}\right)^{1-\gamma}, \tag{8.24}$$

$$g_{njl}(r) \approx \frac{\kappa}{|\kappa|} Z\alpha \left(\frac{Z}{a^3(n-\sigma_l)^3}\right)^{1/2} \frac{2}{\Gamma(2\gamma+1)}\left(\frac{a}{2Zr}\right)^{1-\gamma}. \tag{8.25}$$

The factor $\frac{2}{\Gamma(2\gamma+1)}\left(\frac{a}{2Zr}\right)^{1-\gamma}$ goes to 1 as $Z\alpha \to 0$, so it can be seen as a relativistic enhancement factor.

It is now possible to give the analytic formulae for the contribution of the electron EDM and scalar-pseudoscalar type e-N interaction to the EDMs of diamagnetic atoms. We can see that the analytic formulae for the P, CP-odd e-N interactions (C_N^{PS}, C_N^{PS} combined with hyperfine interaction) and the magnetic interaction with the electron EDM (H_B) are proportional to the same matrix elements $\langle s_{1/2}|\boldsymbol{\gamma}\,|p_{1/2}\rangle$. This matrix element is also used in the calculation of the tensor type P, CP-odd e-N interaction (C_N^{T}). Therefore, it is possible to relate the contributions of the electron EDM and scalar-pseudoscalar type P, CP-odd e-N interaction to C_N^{T}.

The following analytic relations are from Refs. [9, 10]:

- Scalar-pseudoscalar type P, CP-odd e-N interaction C_N^{SP} can be written as

$$\left(\frac{Z}{A}C_p^{\mathrm{SP}} + \frac{N}{A}C_n^{\mathrm{SP}}\right)\frac{\mathbf{I}}{I} \leftrightarrow \frac{1.9\times 10^3}{(1+0.3Z^2\alpha^2)A^{2/3}\mu_A}\left\langle C_p^{\mathrm{T}}\sum_p \boldsymbol{\sigma}_p + C_n^{\mathrm{T}}\sum_n \boldsymbol{\sigma}_n\right\rangle, \tag{8.26}$$

  where Z, N and A are respectively the number of protons, neutrons and all nucleons in the nucleus, $\mathbf{I}$ is the spin of the nucleus and μ_A is the magnetic moment of the nucleus in unit of Bohr magneton. The ratios Z/A and N/A are due to the coherence of the non-relativistic nucleons. Terms in squared brackets $\langle\cdots\rangle$ denote the spin average over the nuclear state (the valence nucleon mixing coefficients $\langle\sigma_p\rangle_z$ and $\langle\sigma_n\rangle_z$ are given by solving Eq. (7.40) in the simple shell model). Since ^{129}Xe, ^{199}Hg, ^{211}Rn and ^{225}Ra have only one valence nucleon, it is possible to equate the magnetic moment of the nucleus with the magnetic moment of the proton ($\mu_p = +2.7928$) and of the neutron ($\mu_n = -1.9130$) as in Eq. (7.40). The shell model description works well for ^{129}Xe ($\mu_{\mathrm{Xe}} = -0.7778$) and ^{199}Hg ($\mu_{\mathrm{Hg}} = 0.5059$), so it is possible to give reliable coefficients with this method. For ^{211}Rn ($\mu_{\mathrm{Rn}} = 0.601$) and ^{225}Ra ($\mu_{\mathrm{Ra}} = -0.734$) nuclei however, the shell model is not applicable [6].
- Electron EDM contribution in diamagnetic atoms can be written as

$$d_e\frac{\mathbf{I}}{I} \leftrightarrow \frac{3}{7}\frac{G_F m_p e}{\sqrt{2}\pi\alpha\mu_A}\frac{R}{R-1}\left\langle C_p^{\mathrm{T}}\sum_p \boldsymbol{\sigma}_p + C_n^{\mathrm{T}}\sum_n \boldsymbol{\sigma}_n\right\rangle, \tag{8.27}$$

  where $R = \left(\frac{2}{\Gamma(2\gamma+1)}\left(\frac{a}{2Zr}\right)^{1-\gamma}\right)^2$. We have included the contribution from hyperfine interaction and magnetic interaction of electron EDM.
- Pseudoscalar-scalar type P, CP-odd e-N interaction C_N^{SP} **is**

$$C_N^{\mathrm{PS}} \leftrightarrow \frac{5m_p r_0}{Z\alpha}C_N^{\mathrm{T}} \approx 3.8\times 10^3\frac{A^{1/3}}{Z}C_N^{\mathrm{T}}, \tag{8.28}$$

  where r_0 is the nuclear radius. This formula was derived in Ref. [6] and it can be applied generally to heavy atoms. We will use it to derive the dependence of ^{205}Tl EDM on C_N^{PS}.

The above relations are accurate to $O(Z^2\alpha^2)$. Numerical results are presented in Table 8.2.

Table 8.2 Analytically calculated coefficients of P, CP-odd interactions

	^{129}Xe	^{199}Hg	^{211}Rn	^{225}Ra
$d_{\rm atom}(d_e)$ $(10^{-3} d_e)$	−0.98	−7.9	10.7	42.6
$d(ZC_p^{\rm SP}/A + NC_n^{\rm SP}/A)$ $(10^{-22}(ZC_p^{\rm SP}/A + NC_n^{\rm SP}/A) e\,{\rm cm})$	−0.62	−5.1	7.0	29
$d(C_N^{\rm PS})$ $(10^{-23} C^{\rm PS} \langle \sigma_N \rangle_z e\,{\rm cm})$	1.6	−18	21	−69

We see that the analytic results for pseudoscalar-scalar type e-N interaction ($C_N^{\rm PS}$) is in very good agreement with the mean-field approach (compare with Table 8.1).

Let us also show the result for the ^{205}Tl atom.

$$d_{\rm Tl}(C_N^{\rm T})/C_N^{\rm T} = 0.5 \times 10^{-20} \langle \sigma_N \rangle_z e\,{\rm cm}, \quad (8.29)$$

$$d_{\rm Tl}(C_N^{\rm PS})/C_N^{\rm PS} = 1.8 \times 10^{-23} \langle \sigma_N \rangle_z e\,{\rm cm}, \quad (8.30)$$

where the parenthesis after $d_{\rm Tl}$ means that we take only the dependence of the P, CP-odd interaction in the parenthesis. The valence nucleon mixing coefficients $\langle \sigma_N \rangle_z = \langle \sigma_p \rangle_z, \langle \sigma_n \rangle_z$ are given by solving Eq. (7.40). The contribution of the nuclear Schiff moment to the ^{205}Tl EDM is omitted, since it was neglected at the nuclear level.

8.4 Summary of Atomic Level Calculation

We summarize the results of the calculation of the dependences of the atomic EDM on the underlying P, CP-odd interactions (electron EDM, P, CP-odd electron-nucleon interactions and nuclear Schiff moment). Note that we have omitted the dependences on the tensor type P, CP-odd e-N interaction ($C_N^{\rm T}$), since its effect is not relevant in this thesis.

^{205}Tl atom EDM:

$$d_{\rm Tl} = -582 d_e + \left(-7.0 \times 10^{-18} (0.395 C_p^{\rm SP} + 0.605 C_n^{\rm SP}) + 1.8 \times 10^{-23} C_p^{\rm PS} \right) e\,{\rm cm}\,. \quad (8.31)$$

The first term (electron EDM enhancement) and the coefficients of the P, CP-odd scalar-pseudoscalar e-N interaction ($C_N^{\rm SP}$) are from relativistic Hartree-Fock calculation improved with configuration mixing and many-body perturbation theory [7] [see Eq. (8.15)]. The final term (term with $C_N^{\rm PS}$) is derived from analytic calculation with simple assumptions described in the previous section [10] (see Table 8.2). The Schiff moment contribution was omitted, since its contribution was neglected at the nuclear level [12].

^{199}**Hg atom EDM**:

$$
\begin{aligned}
d_{\mathrm{Hg}} &= -7.9 \times 10^{-3} d_e \\
&\quad + \left(-51(0.40C_p^{\mathrm{SP}} + 0.60C_n^{\mathrm{SP}}) + 6.1(0.09C_p^{\mathrm{PS}} + 0.91C_n^{\mathrm{PS}})\right) \times 10^{-23} e\,\mathrm{cm} \\
&\quad - 2.6 \times 10^{-17} \frac{S_{\mathrm{Hg}}}{e\,\mathrm{fm}^3}\, e\,\mathrm{cm} \qquad (8.32)
\end{aligned}
$$

The first line (electron EDM enhancement) and the coefficients of the P, CP-odd scalar-pseudoscalar e-N interaction (C_N^{SP}) were calculated with simple assumptions specified in the previous section [10] (see Table 8.2). The coefficients of C_N^{PS} and the Schiff moment contribution (third line) were calculated with relativistic Hartree-Fock method improved with configuration mixing and many-body perturbation theory [6] (see Table 8.1).

^{129}**Xe atom EDM**:

$$
\begin{aligned}
d_{\mathrm{Xe}} &= -0.98 \times 10^{-3} d_e \\
&\quad + \left(-6.2(0.42C_p^{\mathrm{SP}} + 0.58C_n^{\mathrm{SP}}) + 1.6(0.24C_p^{\mathrm{PS}} + 0.76C_n^{\mathrm{PS}})\right) \times 10^{-23} e\,\mathrm{cm} \\
&\quad + 0.38 \times 10^{-17} \frac{S_{\mathrm{Xe}}}{e\,\mathrm{fm}^3}\, e\,\mathrm{cm} \qquad (8.33)
\end{aligned}
$$

The first line (electron EDM enhancement) and the coefficients of the P, CP-odd scalar-pseudoscalar e-N interaction (C_N^{SP}) were calculated with the assumptions described in the previous section [10] (see Table 8.2). The coefficients of C_N^{PS} and the Schiff moment contribution (third line) were calculated with RPA [6] (see Table 8.1).

^{211}**Rn atom EDM**:

$$
\begin{aligned}
d_{\mathrm{Rn}} &= 10.7 \times 10^{-3} d_e \\
&\quad + \left(70(0.41C_p^{\mathrm{SP}} + 0.59C_n^{\mathrm{SP}}) - 7.1(0.02C_p^{\mathrm{PS}} + 0.98C_n^{\mathrm{PS}})\right) \times 10^{-23} e\,\mathrm{cm} \\
&\quad + 3.3 \times 10^{-17} \frac{S_{\mathrm{Rn}}}{e\,\mathrm{fm}^3}\, e\,\mathrm{cm} \qquad (8.34)
\end{aligned}
$$

The methods used for the derivation of the coefficients are the same as for the EDM of ^{129}Xe atom.

^{225}**Ra atom EDM**:

$$
\begin{aligned}
d_{\mathrm{Ra}} &= 4.3 \times 10^{-2} d_e \\
&+ \Big(29(0.39C_p^{\mathrm{SP}} + 0.61C_n^{\mathrm{SP}}) - 6.4(0.25C_p^{\mathrm{PS}} + 0.75C_n^{\mathrm{PS}})\Big) \times 10^{-22} e\,\mathrm{cm} \\
&- 8.8 \times 10^{-17} \frac{S_{\mathrm{Ra}}}{e\,\mathrm{fm}^3}\, e\,\mathrm{cm} \qquad (8.35)
\end{aligned}
$$

The methods used for the derivation of the coefficients are the same as for the EDM of ^{199}Hg atom.

References

1. L.I. Schiff, Phys. Rev. **132**, 2194 (1963)
2. P.G.H. Sandars, Phys. Lett. B **14**, 194 (1965)
3. P.G.H. Sandars, Phys. Lett. B **22**, 290 (1966)
4. V.V. Flambaum, Yad. Fiz. **24**, 383 (1976) [Sov. J. Nucl. Phys. 24, 199 (1976)].
5. V.A. Dzuba, V.V. Flambaum, M.G. Kozlov, Phys. Rev. A **54**, 3948 (2002)
6. V.A. Dzuba, V.V. Flambaum, S.G. Porsev, Phys. Rev. A **80**, 032120 (2009)
7. V.A. Dzuba, V.V. Flambaum, Phys. Rev. A **80**, 062509 (2009)
8. S.G. Porsev, M.S. Safronova, M.G. Kozlov, arXiv:1201.5615 [physics.atom-ph].
9. J.S.M. Ginges, V.V. Flambaum, Phys. Rept. **397**, 63 (2004)
10. V.V. Flambaum, I.B. Khriplovich, Zh. Eksp. Teor. Fiz. **89**, 1505 (1985) [Sov. Phys. JETP **62**, 872 (1985)]
11. I.B. Khriplovich, S.K. Lamoreaux, *CP Vioaltion Without Strangeness* (Springer, Berlin, 1997)
12. O.P. Sushkov, V.V. Flambaum, I.B. Khriplovich, Zh. Eksp. Teor. Fiz. 87, 1521 (1984) [Sov. Phys. JETP **60**, 873 (1984)]

Chapter 9
EDM in the Standard Model

9.1 P, CP-odd Processes Generated by the CKM Phase

Fermion EDMs

The fermion EDM in the SM is generated for both quarks and charged leptons by CKM matrix elements [1]. For the quark EDM, the first non-vanishing contribution appears at the three-loop level [2–4]. The one-loop contribution does not exist trivially, since the quark EDM is a process without flavor change, and the CKM matrix elements contributing at the one-loop level appear only with their complex conjugates. At the two-loop level, CP violating diagrams can be drawn, but their total sum cancels and does not contribute to the quark EDM [5]. This cancellation at the two-loop level seems to be rather accidental. The complete calculation of the three-loop level u and d quark EDMs was done in Ref. [4], with a careful analysis of the top quark contribution. The leading diagrams are those with two W bosons and one gluon, an example is shown in Fig. 9.1.

The values for u and d quark EDMs are

$$\begin{aligned} d_d &= -0.7 \times 10^{-32} \frac{m_d}{\text{GeV}} e \text{ cm} \approx -3.5 \times 10^{-35} e \text{ cm}, \\ d_u &= -0.3 \times 10^{-32} \frac{m_u}{\text{GeV}} e \text{ cm} \approx -0.8 \times 10^{-35} e \text{ cm}, \end{aligned} \tag{9.1}$$

with $m_d = 5.0\,\text{MeV}$, $m_u = 2.5\,\text{MeV}$. The dependence between quark EDM and neutron EDM is $O(1)$ (for example, the non-relativistic quark model predicts $d_n = \frac{4}{3}d_d - \frac{1}{3}d_u$), so the neutron EDM given from the CKM matrix elements via quark EDM is $d_n(d_{q\text{CKM}}) \sim 10^{-35} e\,\text{cm}$.

For the electron EDM, the first non-vanishing contribution appears at the four-loop order [6]. To provide the CKM phase, the electron must be attached to a quark loop via W boson, so the EDM contribution must start from the one-loop. In this case again, the lowest order quark loop diagram must have a CKM matrix element

N. Yamanaka, *Analysis of the Electric Dipole Moment in the R-parity Violating Supersymmetric Standard Model*, Springer Theses,
DOI: 10.1007/978-4-431-54544-6_9,

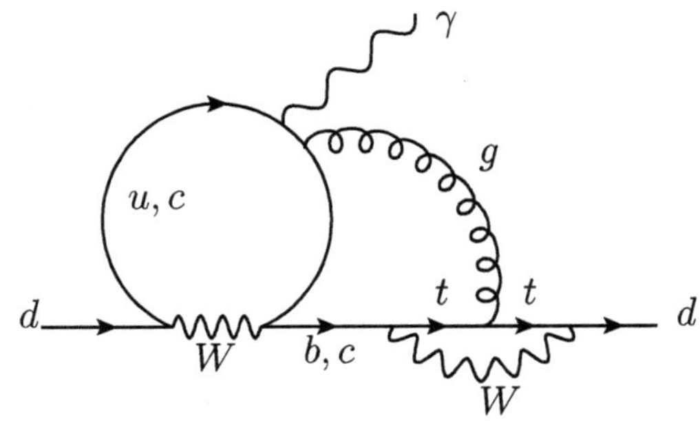

Fig. 9.1 Example of 3-loop diagram contributing to quark EDM

accompanied by its complex conjugate, so that the two-loop contribution cancels. The three-loop contribution cancels also accidentally [6], and the leading EDM diagrams appear only at the four-loop order. The electron EDM in the SM was estimated roughly to be $d_{e\text{CKM}} \sim 10^{-41} e\,\text{cm}$, adding QCD loop corrections to the three-loop diagrams [7]. The muon and τ lepton EDMs arise with the same mechanism, and the corresponding values are $d_{\mu\text{CKM}} = \frac{m_\mu}{m_e} d_{e\text{CKM}} \sim 10^{-38} e\,\text{cm}$ and $d_{\tau\text{CKM}} = \frac{m_\tau}{m_e} d_{e\text{CKM}} \sim 10^{-37} e\,\text{cm}$. We add finally that in the presence of the Majorana neutrino with see-saw model, the lepton EDM receives from the lepton sector CP violation contribution of $O(em_l m_\nu^2 G_F^2) \sim 10^{-43} e\,\text{cm}$ [8], which is smaller than the CKM contribution.

Gluonic P, CP-odd Interactions from CKM Phase

The P, CP-odd gluonic operators such as the θ-term or the Weinberg operator [9] can also be generated with the CP phase of the CKM matrix.

The non-vanishing θ-term contribution from the CKM phase appears at the 4-loop order, with two W boson exchange and QCD radiative correction [10, 11], as shown in Fig. 9.2.

The θ-term generated by the CKM matrix is [11]

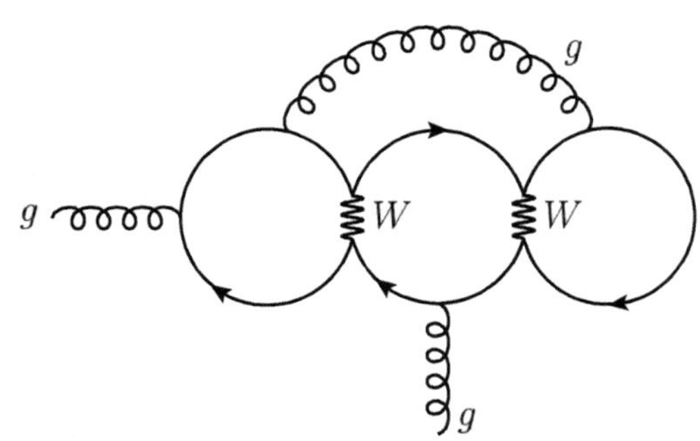

Fig. 9.2 Example of diagram contributing to θ-term in the SM. The *solid lines* indicate quark propagators

$$\theta_{\text{CKM}} \sim 10^{-19}. \tag{9.2}$$

which gives the following contribution to the neutron EDM

$$d_n(\theta_{\text{CKM}}) \sim 10^{-35} e\,\text{cm}. \tag{9.3}$$

The Weinberg operator within CKM phase was calculated in Refs. [12–14]. It receives contribution from the three-loop level diagram with W boson exchange, as shown in Fig. 9.3.

By using the hadron matrix element calculated with QCD sum rules [15], we obtain

$$d_n(w_{\text{CKM}}) \sim 10^{-42} e\,\text{cm}. \tag{9.4}$$

We see that the CP violating contribution of the CKM matrix has small contributions to the gluonic P, CP-odd operators.

Penguin Diagram Contribution

We have seen previously that the contribution of the quark EDM from the CKM matrix element gives $d_n(d_{q\text{CKM}}) \sim 10^{-35} e$ cm. However, the neutron EDM receives larger contribution from the long distance effect generated by pion-loop diagrams with strangeness non-conserving penguin diagram interactions as shown in Fig. 9.4 [16].

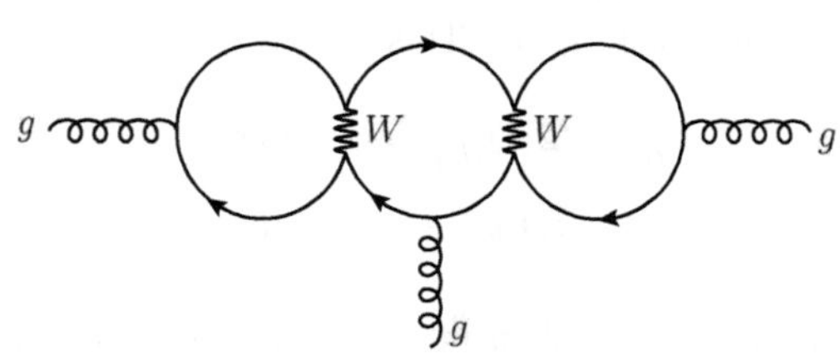

Fig. 9.3 Diagram contributing to the Weinberg operator in the SM. The *solid lines* indicate quark propagators

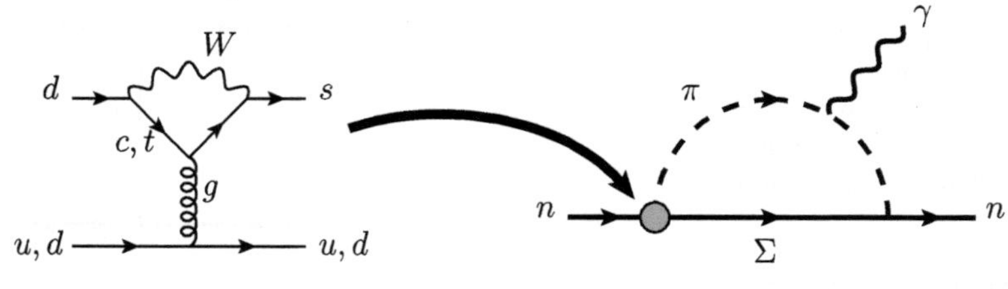

Fig. 9.4 Example of SM contribution to the neutron EDM. Penguin diagrams give rise to P, CP-odd $|\Delta S| = 1$ meson-baryon interaction (*grey blob*)

This pion-loop processes generates a large contribution to the neutron EDM[1] with

$$d_n^{(\text{penguin})} \sim 10^{-32} e\,\text{cm}. \tag{9.5}$$

This gives actually larger contribution than the quark EDM ($\sim 10^{-35} e$ cm). We must note that the quark EDM induced by weak diquark interaction contributes also to the neutron EDM, but this contribution is also estimated to be small ($\sim 10^{-34} e$ cm) [18–21].

For many-nucleon systems, their EDM receives the leading contribution from the P, CP-odd nucleon-nucleon (N-N) interactions, which are also generated by penguin diagrams [22]. The P, CP-odd N-N interactions in the SM are given by P, CP-odd $|\Delta S| = 1$ process given in Fig. 9.5. It is then possible to generate contributions to nuclear EDMs or Schiff moments in the SM [22–24].
The P, CP-odd N-N interaction

$$H_{N-N} = -\eta_{NN'} \frac{G_f}{\sqrt{2}} \bar{N} i \gamma_5 N \, \bar{N}' N' \tag{9.6}$$

contributes to the Schiff moments of ^{129}Xe and ^{199}Hg as follows:

$$S_{\text{Xe}} = 1.75 \eta_{np} \times 10^{-8} e\,\text{fm}^3, \tag{9.7}$$

$$S_{\text{Hg}} = -1.4 \eta_{np} \times 10^{-8} e\,\text{fm}^3. \tag{9.8}$$

These numbers were obtained from the calculation using the Woods-Saxon potential with spin-orbit correction [25, 26]. We can obtain the expressions for ^{211}Rn and ^{225}Ra Schiff moment by multiplying the ratio of $A^{2/3}$ since the nuclear Schiff moment is a surface effect of the nucleus [25, 26]. This yields

$$S_{\text{Rn}} = -1.4 \eta_{np} \times 10^{-8} e\,\text{fm}^3, \tag{9.9}$$

$$S_{\text{Ra}} = 2.4 \eta_{np} \times 10^{-8} e\,\text{fm}^3. \tag{9.10}$$

This simple estimation has not taken into account the deformation of the ^{225}Ra nucleus, so the value for S_{Ra} may be more enhanced. The P, CP-odd coupling

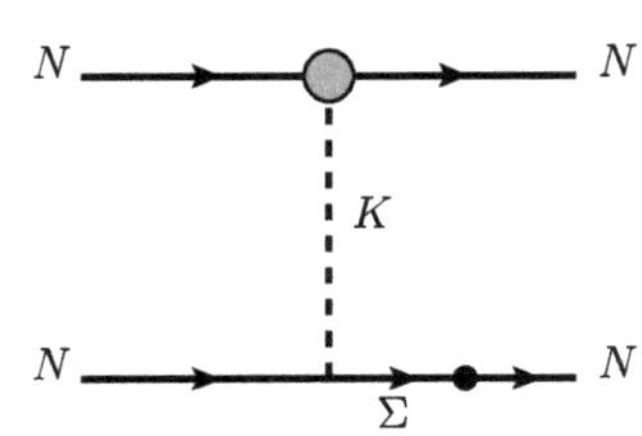

Fig. 9.5 Example of SM contribution to the P, CP-odd N-N interaction. Penguin diagrams give rise to P, CP-odd $|\Delta S| = 1$ meson-baryon interaction (*grey blob*)

[1] Recently, it was pointed that tree level bound state effect at the second order in the weak interaction can generate nucleon EDM of $O(10^{-31})e$ cm [17].

$\eta \sim 10^{-9}$ was given in Refs. [22–24]. By accounting for the dependences of the nuclear Schiff moments on the atomic EDM given in Table 8.1, we obtain

$$d_{\mathrm{Xe}}^{(\mathrm{SM})} \sim 10^{-35} e\ \mathrm{cm}, \tag{9.11}$$

$$d_{\mathrm{Hg}}^{(\mathrm{SM})} \sim 10^{-34} e\ \mathrm{cm}, \tag{9.12}$$

$$d_{\mathrm{Rn}}^{(\mathrm{SM})} \sim 10^{-34} e\ \mathrm{cm}, \tag{9.13}$$

$$d_{\mathrm{Ra}}^{(\mathrm{SM})} \sim 10^{-33} e\ \mathrm{cm}. \tag{9.14}$$

We must note that the P, CP-odd electron-nucleon (e-N) interactions contribute also to the atomic EDMs. The contribution is estimated to be small of order $C \sim 10^{-16} \frac{G_F}{\sqrt{2}}$ [27]. This can generate EDMs of diamagnetic atoms of order $d < 10^{-37} e\,\mathrm{cm}$, so it cannot be the leading contribution. The P, CP-odd e-N interaction gives however for the paramagnetic atoms (^{205}Tl) the leading contribution in the SM because of the large enhancement of the scalar-pseudoscalar type interaction C_N^{SP}. The value can be estimated to be $d_{\mathrm{Tl}}^{(\mathrm{SM})} \sim 10^{-34} e\,\mathrm{cm}$. These values are of course well beyond the current experimental sensitivity.

For the estimation of the EDMs of the deuteron and ^{3}He nucleus in the SM, we will relate the P, CP-odd N-N coupling with the P, CP-odd pion-nucleon coupling: $\eta \leftrightarrow \frac{g_{\pi NN} \bar{g}_{\pi NN}^{(1)} \sqrt{2}}{G_F m_\pi^2}$. By using the dependence of the deuteron and ^{3}He EDM on the P, CP-odd pion-nucleon coupling, we obtain

$$d_d^{(\mathrm{SM})} \sim 10^{-31} e\,\mathrm{cm}, \tag{9.15}$$

$$d_{\mathrm{He}}^{(\mathrm{SM})} \sim 10^{-31} e\ \mathrm{cm}, \tag{9.16}$$

with closer value with the proton or neutron EDMs.

9.2 Summary of the SM Contribution to EDM Observables

We have finally the following results for the SM contribution to the EDMs of interest (Table 9.1):

We can see that the present experimental data are well below the SM predictions. This makes clear that the EDMs are very good observables to probe the new physics.

Table 9.1 EDM in the SM. Units are in e cm

System	SM contribution	Present exp. limit	Prospected exp. sensitivity
e^-	$\sim 10^{-41}$	($<1.05 \times 10^{-27}$) [28]	($\sim 10^{-31}$) [29]
μ	$\sim 10^{-38}$	$<1.8 \times 10^{-19}$ [30]	$\sim 10^{-25}$ [31–34]
n	$\sim 10^{-32}$	$<2.9 \times 10^{-26}$ [35]	$\sim 10^{-28}$ [29, 36]
p	$\sim 10^{-32}$	Not available	$\sim 10^{-27}$ [31–34]
Deuteron	$\sim 10^{-31}$	Not available	$\sim 10^{-27}$ [31–34]
^{3}He nucleus	$\sim 10^{-31}$	Not available	–
^{205}Tl atom	$\sim 10^{-34}$	$<9 \times 10^{-25}$ [37]	–
^{199}Hg atom	$\sim 10^{-34}$	$<3.1 \times 10^{-29}$ [38]	$\sim 10^{-30}$
^{129}Xe atom	$\sim 10^{-35}$	$<4.1 \times 10^{-27}$ [39]	$\sim 10^{-30}$ [29]
^{211}Rn atom	$\sim 10^{-34}$	Not available	$\sim 10^{-29}$
^{225}Ra atom	$\sim 10^{-33}$	Not available	$\sim 10^{-28}$ [29]

Data for the electron EDM were enclosed with parenthesis since they were given from the experimental data of the YbF molecules [28]

References

1. M. Kobayashi, T. Maskawa, Prog. Theor. Phys. **49**, 652 (1973)
2. E.P. Shabalin, Yad. Fiz. **31**, 1665 (1980) [Sov. J. Nucl. Phys. **31**, 864 (1980)]
3. I.B. Khriplovich, Yad. Fiz. **44**, 1019 (1986) [Sov. J. Nucl. Phys. **44**, 659 (1986)]
4. A. Czarnecki, B. Krause, Phys. Rev. Lett. **78**, 4339 (1997)
5. E.P. Shabalin, Yad. Fiz. **28**, 151 (1978) [Sov. J. Nucl. Phys. **28**, 75 (1978)]
6. M.E. Pospelov, I.B. Khriplovich, Yad. Fiz. **53**, 1030 (1991) [Sov. J. Nucl. Phys. **53**, 638 (1991)]
7. M.J. Booth, hep-ph/9301293
8. J.P. Archambault, A. Czarnecki, M. Pospelov, Phys. Rev. D **70**, 073006 (2004)
9. S. Weinberg, Phys. Rev. Lett. **63**, 2333 (1989)
10. J. Ellis, M.K. Gaillard, Nucl. Phys. B **150**, 141 (1979)
11. I.B. Khriplovich, Phys. Lett. B **173**, 193 (1986)
12. I. Bigi, N.G. Uraltsev, Sov. Phys. JETP **100**, 198 (1991)
13. I. Bigi, N.G. Uraltsev, Nucl. Phys. B **353**, 321 (1991)
14. M.E. Pospelov, Phys. Lett. B **328**, 441 (1994)
15. D. Demir, M. Pospelov, A. Ritz, Phys. Rev. D **67**, 015007 (2003)
16. I.B. Khriplovich, A.R. Zhitnitsky, Phys. Lett. B **109**, 490 (1982)
17. T. Mannel, N. Uraltsev, Phys. Rev. D **85**, 096002 (2012)
18. B.F. Morel, Nucl. Phys. B **157**, 23 (1979)
19. E.P. Shabalin, Yad. Fiz. **32**, 443 (1980) [Sov. J. Nucl. Phys. **32**, 228 (1980)]
20. D.V. Nanopoulos, A. Yildiz, P.H. Cox, Ann. Phys. **127**, 126 (1980)
21. M.B. Gavela, A. Le Yaouanc, L. Oliver, O. Pene, J.-C. Raynal, T.N. Pham, Phys. Lett. B **109**, 215 (1982)
22. X.-G. He, B. McKellar, Phys. Rev. D **46**, 2131 (1992)
23. O.P. Sushkov, V.V. Flambaum, I.B. Khriplovich, Zh. Eksp. Teor. Fiz. **87**, 1521 (1984) [Sov. Phys. JETP **60**, 873 (1984)]
24. J.F. Donoghue, B.R. Holstein, M.J. Musolf, Phys. Lett. B **196**, 196 (1987)
25. V.V. Flambaum, I.B. Khriplovich, O.P. Sushkov, Phys. Lett. B **162**, 213 (1985)
26. V.V. Flambaum, I.B. Khriplovich, O.P. Sushkov, Nucl. Phys. A **449**, 750 (1985)
27. X.-G. He, B.H.J. McKellar, S. Pakvasa, Phys. Lett. B **283**, 348 (1992)
28. J.J. Hudson et al., Nature **473**, 493 (2011)
29. P. Mueller et al., in *5th International conference on Fundamental Physics Using Atoms*, Okayama University, October 2011 (URL: http://xqw.hep.okayama-u.ac.jp/kakenhi/index.php/fpua2011_home/fpua2011_top_e/)

30. G.W. Bennett et al., (Muon $(g-2)$ Collaboration), Phys. Rev. D **80**, 052008 (2009)
31. I.B. Khriplovich, Phys. Lett. B **444**, 98 (1998)
32. F.J.M. Farley et al., Phys. Rev. Lett. **93**, 052001 (2004)
33. Y.K. Semertzidis et al., AIP Conf. Proc. **698**, 200 (2004)
34. Y.F. Orlov, W.M. Morse, Y.K. Semertzidis, Phys. Rev. Lett. **96**, 214802 (2006)
35. C.A. Baker et al., Phys. Rev. Lett. **97**, 131801 (2006)
36. M. Burghoff et al., arXiv:1110.1505 [nucl-ex]
37. B.C. Regan et al., Phys. Rev. Lett. **88**, 071805 (2002)
38. W.C. Griffith et al., Phys. Rev. Lett. **102**, 101601 (2009)
39. M.A. Rosenberry et al., Phys. Rev. Lett. **86**, 22 (2001)

Chapter 10
Constraints on Supersymmetric Models from Electric Dipole Moments

10.1 Constraints to MSSM via Fermion EDM

The leading (R-parity conserving) supersymmetric contribution to the fermion EDMs arises at the one-loop level, and was analyzed by many authors. The analyses of the supersymmetric contribution to the neutron EDM were made in the context of the phenomenological search of CP violation [1–21]. The leading one-loop contribution is generated by the supersymmetric fermion-sfermion-gaugino interactions (see Fig. 10.1).

The expressions for the one-loop contribution to the electron EDM, quark EDM and chromo-EDM are [15–21]

$$
\begin{aligned}
d_e =& \frac{em_e}{16\pi^2 M_{\rm SUSY}^2}\left(\frac{g_1^2}{12}\sin\theta_A + \frac{5g_2^2+g_1^2}{24}\sin\theta_\mu \tan\beta\right),\\
d_q =& \frac{em_q}{16\pi^2 M_{\rm SUSY}^2}\frac{2g_3^2}{9}\left(\sin\theta_\mu[\tan\beta]^\pm - \sin\theta_A\right) + O(g_2^2, g_1^2),\\
d_q^c =& \frac{m_q}{16\pi^2 M_{\rm SUSY}^2}\frac{5g_3^2}{18}\left(\sin\theta_\mu[\tan\beta]^\pm - \sin\theta_A\right) + O(g_2^2, g_1^2),
\end{aligned}
\tag{10.1}
$$

where θ_A is the CP phase of the trilinear soft breaking interactions, and θ_μ the phase of the μ-term (mixing of the up and down type Higgs). The notation $[\tan\beta]^\pm$ means that we take $\tan\beta$ for down type quarks, and $1/\tan\beta$ for u quarks. The detail of the one-loop EDM calculation is collected in Appendix D. In deriving these expressions, we have taken the general assumption that the trilinear soft breaking terms are proportional to the corresponding Yukawa matrix (which gives mass to fermions) with the same CP phase θ_A, to simplify the analysis. Masses of the sparticles were also taken to be the same, $M_{\rm SUSY}$. By comparing Eq. (10.1) with the EDM experimental data [22–24], we obtain stringent constraints on the CP phases of the soft-breaking terms θ_μ and θ_A, suppressed by about 2 orders in magnitudes,

N. Yamanaka, *Analysis of the Electric Dipole Moment in the R-parity Violating Supersymmetric Standard Model*, Springer Theses,
DOI: 10.1007/978-4-431-54544-6_10, © Springer Japan 2014

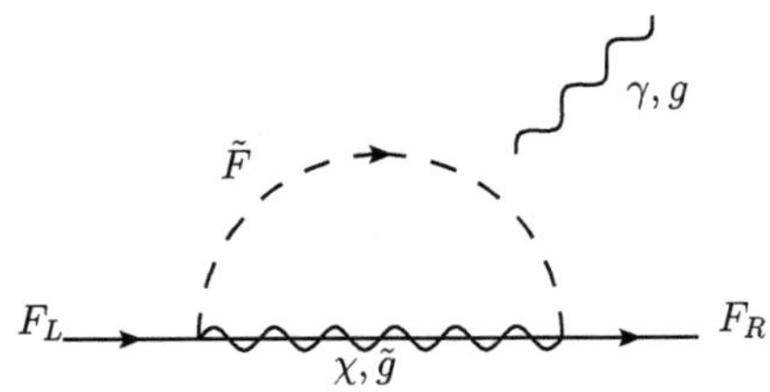

Fig. 10.1 One-loop supersymmetric contribution to the fermion EDM (or chromo-EDM). The external gauge boson (photon or gluon) line is attached to one of the sparticles in the loop. Diagrams with external gluon and internal gluino lines are only allowed when external line corresponds to quarks. When the external gauge boson is a gluon, the amplitude contributes to the chromo-EDM

for $M_{\text{SUSY}} \sim$ TeV. The above analysis was also extended to the flavor changing sector of sfermions, where strong constraints can be obtained with processes involving the third generation [25–32].

We must note that there are also similar diagrams contributing to the θ-term as shown in Fig. 10.2. As we have seen in Sect. 6.5, the CP-odd quark mass $-\eta\bar{q}i\gamma_5 q$ contributes to the θ-term via chiral rotation and gives a large correction to the tree level θ ($\theta_{\text{eff}} \sim \theta_{\text{tree}} + 10^{-2}\delta_{\text{CP}}$, where δ_{CP} is the linear combination of SUSY breaking CP phases θ_A and θ_μ with $O(1)$ coefficients). The θ-term is strongly constrained by neutron EDM experiment, so that the corrected θ-term with SUSY CP phases are also tightly bound (at the order of $\delta_{\text{CP}} < 10^{-8}$ if $\theta_{\text{tree}} = 0$!). However, if the Peccei-Quinn symmetry "unphysicalize" the θ-term (see Sect. 6.5), the contribution from the one-loop diagrams of Fig. 10.2 can be completely omitted. Under this condition, the leading constraints on supersymmetric CP phases are given by the one-loop fermion EDM seen above.

The above one-loop level analysis yields very strong upper limits on the CP violation of the first and second generation sfermions. These stringent constraints lead us to think of "natural" supersymmetric models with no CP phases in the first and second generations. This put also a significant constraint on the CP violation at the Grand unification scale via the renormalization group analysis [33–39].

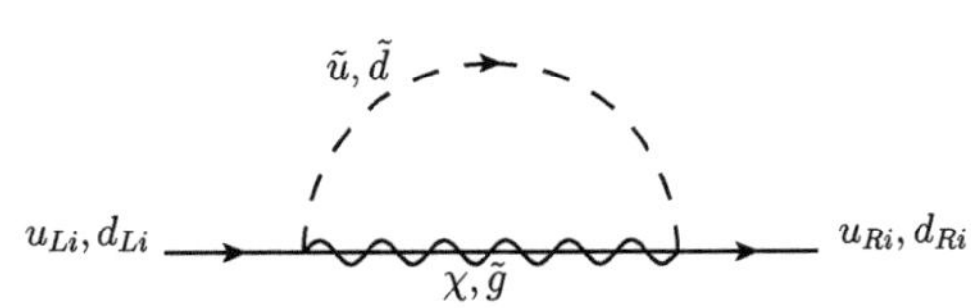

Fig. 10.2 One-loop supersymmetric contribution to the θ-term. The "CP-odd mass" of quarks is generated at the one-loop level

10.2 Subleading Supersymmetric Contributions to EDMs

The subleading supersymmetric contributions known so far are the two-loop fermion EDMs, the Weinberg operator, the P, CP-odd 4-fermion interactions and the fermion-gluon interaction.

Two-Loop Level Fermion EDMs: Barr-Zee Type Contribution

Among two-loop level fermion EDM contributions, the leading effect is given by the Barr-Zee type diagrams [40–54], depicted in Fig. 10.3. These Barr-Zee type graphs can contribute significantly to the fermion EDM when the third generation particles (top, bottom quarks and τ lepton) run in the loop especially when $\tan\beta$ becomes very large. At very large $\tan\beta$, the one-loop level θ_A contribution is relatively suppressed [see Eq. (10.1)]. Actually, the two-loop level Barr-Zee type diagrams with sfermion loop have contribution of order $\sim \frac{m_F}{16\pi^2 M_{\rm SUSY}^2}\alpha Y_f \sin(\theta_A+\theta_\mu)\tan\beta$, and θ_A contributes significantly. In the situation where sfermion masses are very heavy (split SUSY scenario [55], for example), the one-loop and fermionic Barr-Zee type contributions are suppressed, so that the Barr-Zee type graph with chargino loop becomes dominant.

4-Fermion Interaction

The P, CP-odd 4-fermion interactions can have an important contribution to the atomic EDM [59–61]. Despite the suppression due to Yukawa couplings, this

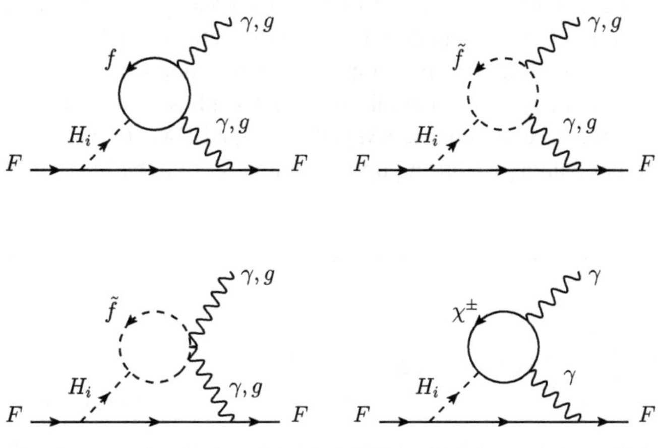

Fig. 10.3 Leading two-loop supersymmetric contribution to the fermion EDM. H_i denotes all possible neutral Higgs bosons. When the external gauge boson is a gluon, the diagram contributes to the chromo-EDM

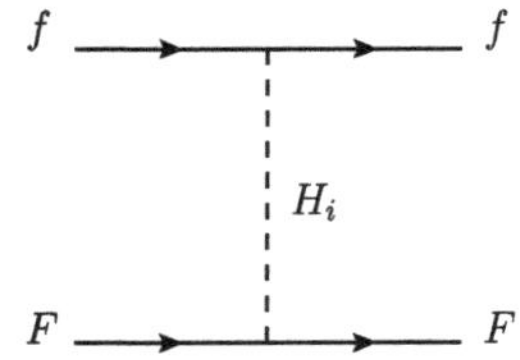

Fig. 10.4 4-fermion interaction with intermediate Higgs boson. H_i denotes all possible neutral Higgs bosons

contribution is proportional to $(\tan\beta)^3$, so it can be greatly enhanced, even more than Barr-Zee diagrams. An example of this contribution is depicted in Fig. 10.4.

One-Loop Electron-Gluon Interaction

Atomic EDM is also generated via the electron-gluon interaction at the one-loop level. The contributing processes are depicted in Fig. 10.5. These mechanisms can induce the P, CP-odd electron-nucleon interaction [54] through the relation $\frac{\alpha_s}{8\pi}\langle N|G^a_{\mu\nu}G^{\mu\nu,a}|N\rangle = -(0.1\,\text{GeV})\bar{N}N$. This contribution is comparable in size for top/bottom loop contribution with the P, CP-odd electron-quark interaction, and is also expected to have similar behavior under large $\tan\beta$.

Weinberg Operator

The Weinberg operator[55–58] can give higher order contributions to the hadronic EDM. Its leading contribution is depicted in Fig. 10.6. The gluino loop can contribute significantly to the neutron EDM when heavy fermion is involved, and have comparable sensitivity as the one-loop contribution in the case where top quark run in the loop. The quantitative comparison is unfortunately difficult due to the large theoretical uncertainty of the relation between the neutron EDM and the Weinberg operator. We must also add that this contribution does not grow with $\tan\beta$ [59–61].

It is now clear that, with "natural size" of CP phases of soft supersymmetry breaking terms, the supersymmetric SM yields a large EDM observables. The current EDM experimental data give strong upper bounds, so that the SUSY CP phases are

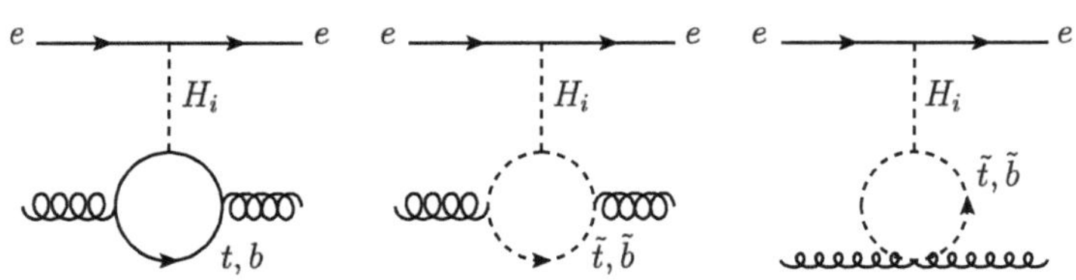

Fig. 10.5 Supersymmetric contribution to the P, CP-odd electron-nucleon interaction generated from gluon loop. H_i denotes all possible neutral Higgs bosons

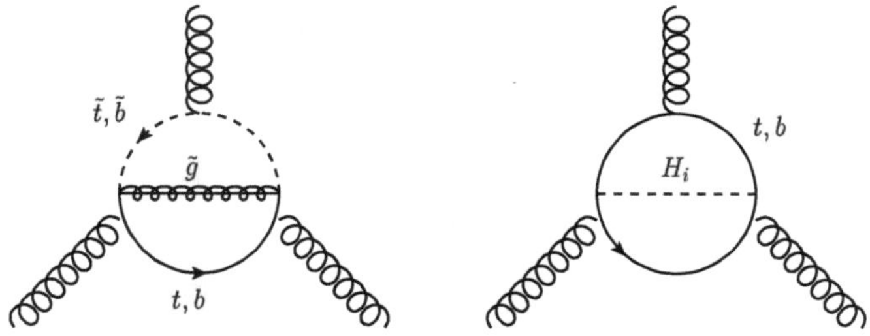

Fig. 10.6 Supersymmetric contribution to the Weinberg operator. H_i denotes all possible neutral Higgs bosons

constrained by about 2 or 3 orders in magnitudes. This is the *SUSY CP problem*. The possible explanation of the SUSY CP problem is as follows [16–21]:

- Heavy superpartners: The masses of sparticles are all above 100 TeV, such that the contributions discussed above decouple.
- Small phases: The CP symmetry may be an exact or approximate symmetry. This assumption depends on the model. There is also a possibility for scenarios in which CP violation is only provided via flavor off-diagonal components [25–32], as in the standard model.
- Accidental cancellation: This scenario allows every CP phases to be $O(1)$. Then the cancellation due to phases could happen, since the EDM constraints (^{205}Tl, ^{199}Hg, neutron, etc) are combinations of different CP violating sources. To find such combinations, sophisticated methods for scanning the SUSY parameter space are required. This topic will be reviewed in the next section.
- No electroweak scale SUSY: In this scenario, we give up the explanation of the hierarchy problem within the supersymmetry. This gives way to other mechanisms which may solve the fine-tuning problem.

All the contributions cited above were carefully rederived in detail in the paper of Ellis et al. [66].

10.3 Geometric Approach to SUSY CP Violation

In the previous analysis, the limits on the supersymmetric phases were set by comparing upper bounds given by each EDM experimental data (^{205}Tl [22], ^{199}Hg [24] and neutron [23]), then taking the tightest one. However, as pointed in Refs. [16–21, 62–64], the possibility of $O(1)$ SUSY CP phases remains, if there is accidental cancellation among CP violations in the EDM data used. To analyze the remaining $O(1)$ regions of the SUSY CP phases, J. Ellis et al. developed a new approach based on the differential-geometry to find the optimal value of some observables within the constraints of the existing experimental data [63–65]. They analyzed the EDMs

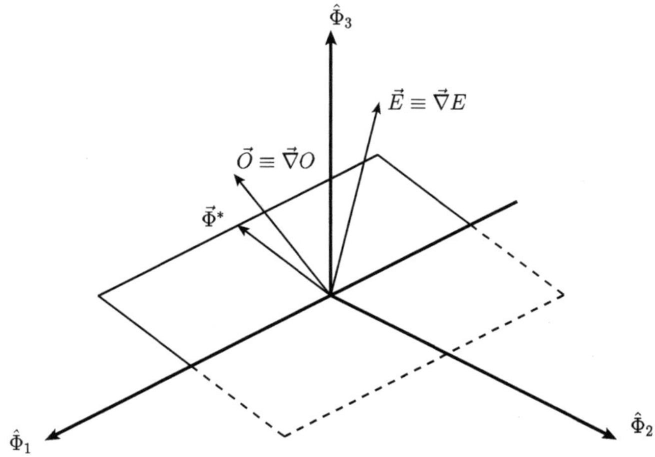

Fig. 10.7 3D example of the geometric analysis [63–65]. The (hyper)plane orthogonal to **E** is the EDM-constrained parameter space. **O** is the true optimal direction for the observable O, and ϕ^* is its EDM-constrained optimal direction, which can be obtained by projection onto the EDM-constrained hyperplane

in the minimally flavor violating maximally CP violating (MFVMCP) models of supersymmetric SM and have shown the prospect of measuring some CP violating observables. The MFVMCP SUSY is the supersymmetric SM with 6 CP phases, composed of three phases from the gaugino masses Φ_1, Φ_2, Φ_3, and three phases from the trilinear soft breaking interactions Φ_{A_u}, Φ_{A_d} and Φ_{A_e} (the CP phase of the μ parameter θ_μ was absorbed in the phase of the gluino mass).

The geometric approach is as follows. First consider the Taylor expansion of the EDM experimental constraints $E_i = 0$ and the observable O we want to maximize. In the small phase approximation, it suffices to take the first order to write

$$O = \phi \cdot \mathbf{O}, \qquad\qquad E = \phi \cdot \mathbf{E}, \tag{10.2}$$

where $\phi \equiv (\Phi_1, \Phi_2, \Phi_3, \ldots)$ are the set of parameters of the new physics (in this case, $\phi = (\Phi_1, \Phi_2, \Phi_3, \Phi_{A_u}, \Phi_{A_d}, \Phi_{A_e})$). The gradients are defined as $\mathbf{O} \equiv \nabla O$ and $\mathbf{E} \equiv \nabla E$. In this case, there are 3 constraints $E^{\rm Hg} = E^{\rm Tl} = E^n = 0$, corresponding respectively to the zero value of the ^{199}Hg EDM, ^{205}Tl EDM and neutron EDM, since those EDM experimental data give very strict constraints. We can now express the "EDM-constrained" parameter space in terms of these gradients. The corresponding equation can be expressed with the exterior product

$$A_{\alpha\beta\gamma} = E^{\rm Hg}_{[\alpha} E^{\rm Tl}_{\beta} E^{b}_{\gamma]} \,, \tag{10.3}$$

where the square brackets means that we antisymmetrize the indices. The EDM-constraint (hyperplane) is represented schematically in Fig. 10.7.

We need now to find the optimal direction of the observable O (which is given by the gradient ∇O) within the EDM-constrained hyperplane. We can do so by projecting orthogonally the optimal direction of O to the EDM-constrained space. The easiest way is to find the plane which contains both $\mathbf{E}^i$ and $\mathbf{O}$, and take the intersection with the EDM constrained hyperplane, where i labels the system giving EDM-constraints (^{199}Hg, ^{205}Tl atoms and neutron). The plane which contains $\mathbf{E}^i$ and $\mathbf{O}$ is expressed with

$$B_{\mu\nu} = \varepsilon_{\mu\nu\lambda\rho\sigma\tau} O_\lambda E_\rho^{\mathrm{Hg}} E_\sigma^{\mathrm{Tl}} E_\tau^n, \tag{10.4}$$

where the epsilon tensor is totally antisymmetric. The components of the optimal EDM-constrained direction are then

$$\phi_\alpha^* = \mathcal{N} \varepsilon_{\alpha\beta\gamma\delta\mu\nu} A_{\beta\gamma\delta} B_{\mu\nu}, \tag{10.5}$$

where $\mathcal{N}$ is the normalization constant which must be determined to keep the linear approximation safe. In fact, the above equation determines the optimal direction, but not the size of the vector ϕ^*. The size of ϕ^* can be determined by extending the overall factor until the EDM-constraints are broken. The formula above neither determines the sign of the optimal ϕ^*, but this problem can be resolved by analyzing higher derivatives which prefer some signs.

In this case, the EDM-constraints are ^{199}Hg, ^{205}Tl and neutron EDMs. The optimal observables we want to obtain are those of the deuteron, muon, ^{225}Ra EDMs and the CP violating asymmetry in $b \to s\gamma$ decay [63–65]. It was concluded that the optimal deuteron EDM in the MFVMCP SUSY is one order of magnitude larger than the prospective experimental sensitivity. The analysis was also performed for ^{225}Ra atom EDM, which shows also a large optimal value. On the other hand, the muon EDM and the branching ratio of $b \to s\gamma$ are too small to be observed in planned experiments. These analyses were the first to treat analytically several CP violating parameters on the same footing, and the results given may be significantly larger than the analysis with single phase. Moreover, these analyses are more efficient than the naive scan of all parameter space, and the computational cost is greatly reduced. This method has thus a significant impact on the search for new physics with large parameter space.

References

1. J.R. Ellis, S. Ferrera, D.V. Nanopoulos, Phys. Lett. B **114**, 231 (1982)
2. W. Buchmüller, D. Wyler, Phys. Lett. B **121**, 321 (1983)
3. J. Polchinski, M.B. Wise, Phys. Lett. B **125**, 393 (1983)
4. F. del Aguila, M.B. Gavela, J.A. Grifols, A. Mendez, Phys. Lett. B **126**, 71 (1983)
5. D.V. Nanopoulos, M. Srednicki, Phys. Lett. B **128**, 61 (1983)
6. M. Dugan, B. Grinstein, L.J. Hall, Nucl. Phys. B **255**, 413 (1985)
7. P. Nath, Phys. Rev. Lett. **66**, 2565 (1991)

8. Y. Kizukuri, N. Oshimo, Phys. Rev. D **45**, 1806 (1992)
9. Y. Kizukuri, N. Oshimo, Phys. Rev. D **46**, 3025 (1992)
10. W. Fischler, S. Paban, S.D. Thomas, Phys. Lett. B **289**, 373 (1992)
11. T. Inui, Y. Mimura, N. Sakai, T. Sasaki, Nucl. Phys. B **449**, 49 (1995)
12. S. Pokorski, J. Rosiek, C.A. Savoy, Nucl. Phys. B **570**, 81 (2000)
13. S.Y. Ayazi, Y. Farzan, Phys. Rev. D **74**, 055008 (2006)
14. S.Y. Ayazi, Y. Farzan, JHEP **0706**, 013 (2007)
15. M. Pospelov, A. Ritz, Ann. Phys. (For review) **318**, 119 (2005)
16. T. Ibrahim, P. Nath, Phys. Rev. D **57**, 478 (1998)
17. T. Ibrahim, P. Nath, Phys. Lett. B **418**, 98 (1998)
18. T. Ibrahim, P. Nath, Phys. Rev. D **58**, 111301 (1998)
19. M. Brhlik, G.J. Good, G.L. Kane, Phys. Rev. D **59**, 115004 (1999)
20. S. Abel, S. Khalil, O. Lebedev, Nucl. Phys. B **606**, 151 (2001)
21. Y. Li, S. Profumo, M.J. Ramsey-Musolf, JHEP **1008**, 062 (2010)
22. B.C. Regan et al., Phys. Rev. Lett. **88**, 071805 (2002)
23. C.A. Baker et al., Phys. Rev. Lett. **97**, 131801 (2006)
24. W.C. Griffith et al., Phys. Rev. Lett. **102**, 101601 (2009)
25. J. Hisano, Y. Shimizu, Phys. Lett. B **581**, 224 (2004)
26. J. Hisano, Y. Shimizu, Phys. Rev. D **70**, 093001 (2004)
27. M. Endo, M. Kakizaki, M. Yamaguchi, Phys. Lett. B **583**, 186 (2004)
28. G.-C. Cho, N. Haba, M. Honda, Mod. Phys. Lett. A **20**, 2969 (2005)
29. J. Hisano, M. Nagai, P. Paradisi, Phys. Lett. B **642**, 510 (2006)
30. J. Hisano, M. Nagai, P. Paradisi, Phys. Rev. D **78**, 075019 (2008)
31. J. Hisano, M. Nagai, P. Paradisi, Phys. Rev. D **80**, 095014 (2009)
32. W. Altmannshofer, A.J. Buras, P. Paradisi, Phys. Lett. B **688**, 202 (2010)
33. S.A. Abel, W.N. Cottingham, I.B. Whittingham, Phys. Lett. B **370**, 106 (1996)
34. R. Gariato, J.D. Wells, Phys. Rev. D **55**, 1611 (1997)
35. A. Romanino, A. Strumia, Nucl. Phys. B **490**, 3 (1997)
36. A. Bartl, T. Gajdosik, W. Porod, P. Stöckinger, H. Stremnitzer, Phys. Rev. D **60**, 073003 (1999)
37. M. Brhlik, L.L. Everett, G.L. Kane, J.D. Lykken, Phys. Rev. Lett. **83**, 2124 (1999)
38. E. Accomando, R.L. Arnowitt, B. Dutta, Phys. Rev. D **61**, 115003 (2000)
39. K.A. Olive, M. Pospelov, A. Ritz, Y. Santoso, Phys. Rev. D **72**, 075001 (2005)
40. S.M. Barr, A. Zee, Phys. Rev. Lett. **65**, 21 (1990)
41. T.H. West, Phys. Rev. D **50**, 7025 (1994)
42. T. Kadoyoshi, N. Oshimo, Phys. Rev. D **55**, 1481 (1997)
43. D. Chang, W.-Y. Keung, A. Pilaftsis, Phys. Rev. Lett. **82**, 900 (1999)
44. A. Pilaftsis, Phys. Lett. B **471**, 174 (1999)
45. D. Chang, W.-F. Chang, W.-Y. Keung, Phys. Lett. B **478**, 239 (2000)
46. A. Pilaftsis, Phys. Rev. D **62**, 016007 (2000)
47. D. Chang, W.-F. Chang, W.-Y. Keung, Phys. Rev. D **66**, 116008 (2002)
48. T.-F. Feng, X.-Q. Li, L. Lin, J. Maalampi, X.-M. Zhang, Phys. Rev. D **71**, 056005 (2005)
49. N. Arkani-Hamed, S. Dimopoulos, G.F. Giudice, A. Romanino, Nucl. Phys. B **709**, 3 (2005)
50. D. Chang, W.-F. Chang, W.-Y. Keung, Phys. Rev. D **71**, 076006 (2005)
51. G.F. Giudice, A. Romanino, Phys. Lett. B **634**, 307 (2006)
52. T.-F. Feng, X.-Q. Li, L. Lin, J. Maalampi, H.-S. Song, Phys. Rev. D **73**, 116001 (2006)
53. Y. Li, S. Profumo, M.J. Ramsey-Musolf, Phys. Rev. D **78**, 075009 (2008)
54. A. Pilaftsis, Nucl. Phys. B **644**, 263 (2002)
55. N. Arkani-Hamed, S. Dimopoulos, JHEP **0506**, 073 (2005)
56. T. Falk, K.A. Olive, M. Pospelov, R. Roiban, Nucl. Phys. B **560**, 3 (1999)
57. O. Lebedev, M. Pospelov, Phys. Rev. Lett. **89**, 101801 (2002)
58. D. Demir, O. Lebedev, K.A. Olive, M. Pospelov, A. Ritz, Nucl. Phys. B **680**, 339 (2004)
59. J. Dai, H. Dykstra, R.G. Leigh, S. Paban, D. Dicus, Phys. Lett. B **237**, 216 (1990)
60. R. Arnowitt, J.L. Lopez, D.V. Nanopoulos, Phys. Rev. D **42**, 2423 (1990)
61. R. Arnowitt, M.J. Duff, K.S. Stelle, Phys. Rev. D **43**, 3085 (1991)
62. O. Lebedev, K.A. Olive, M. Pospelov, A. Ritz, Phys. Rev. D **70**, 016003 (2004)
63. J. Ellis, J.S. Lee, A. Pilaftsis, JHEP **0810**, 049 (2008)
64. J. Ellis, J.S. Lee, A. Pilaftsis, JHEP **1010**, 049 (2010)
65. J. Ellis, J.S. Lee, A. Pilaftsis, JHEP **1102**, 045 (2011)
66. J. Ellis, J.S. Lee and A. Pilaftsis, arXiv:1009.1151 [math.OC]

Part III
Analysis of the Electric Dipole Moment in the R-parity Violating Supersymmetric Model

Chapter 11
Leading RPV Contributions to the EDM Observables

We have previously seen that the general renormalizable supersymmetric extension of the SM allows also baryon and lepton number violating interactions, the R-parity violating (RPV) interactions. The assumption of the R-parity is *ad hoc*, and RPV interactions must be investigated phenomenologically. The RPV interactions are generated by the following superpotential:

$$W_{\not R} = \mu'_i \varepsilon_{ab} \hat{L}_i^a \hat{H}_u^b + \frac{1}{2}\lambda_{ijk}\varepsilon_{ab}\hat{L}_i^a \hat{L}_j^b (\hat{E}^c)_k + \lambda'_{ijk}\varepsilon_{ab}\hat{L}_i^a \hat{Q}_j^b (\hat{D}^c)_k + \frac{1}{2}\lambda''_{ijk}(\hat{U}^c)_i (\hat{D}^c)_j (\hat{D}^c)_k, \tag{11.1}$$

with $i, j, k = 1, 2, 3$ indicating the generation, $a, b = 1, 2$ the $SU(2)_L$ indices. For the baryon number violating interactions (terms with λ''), the $SU(3)_c$ indices have been omitted. The lepton left-chiral superfields $\hat{L}$ and $\hat{E}^c$ are respectively $SU(2)_L$ doublet and singlet. The quark superfields $\hat{Q}$, $\hat{U}^c$ and $\hat{D}^c$ denote respectively the quark $SU(2)_L$ doublet, up quark singlet and down quark singlet left-chiral superfields, and $\hat{H}_u$ the up type Higgs left-chiral superfield.

The trilinear R-parity violating (RPV) interactions contribute to the EDM observables via two leading processes: the Barr–Zee type 2-loop fermion EDM and the P, CP-odd 4-fermion interaction. In deriving the Barr–Zee type two-loop contribution to the fermion EDM, our result did not agree with that of the previous works [1, 2]. We have found that the values from our correct formula [3] are one order in magnitudes smaller than the results from previous works. This has an important consequences on phenomenological analysis, since its contribution relative to the other leading contributions (such as that from 4-fermion interaction) changes.

In this chapter, the detail of the derivation of the RPV contribution to the EDM is carefully explained. We first explain the absence of one-loop level fermion EDM diagram within trilinear RPV supersymmetry. We will then present the two leading P, CP-odd contributions, the Barr–Zee type two-loop level fermion EDM and the tree level P, CP-odd 4-fermion interactions.

N. Yamanaka, *Analysis of the Electric Dipole Moment in the R-parity Violating Supersymmetric Standard Model*, Springer Theses,
DOI: 10.1007/978-4-431-54544-6_11,

11.1 Absence of One-Loop Level RPV Contribution to Fermion EDM

The leading contributions of the R-parity violating interactions to the quark and electron EDMs are from the 2-loop level diagrams. It was shown by Godbole et al. that there is no one-loop contribution to the fermion EDM generated from trilinear R-parity violating couplings [4]. Here we briefly explain its mechanism. The analysis is based on examining the helicity flip processes. From the Lorentz structure of the EDM operator, diagrams contributing to the EDMs must flip the helicity. Since the helicity flip by mass insertions to the external line can be removed by the equation of motion (see Fig. 11.1), the helicity flip must be generated from the loop itself, i.e. the external left-handed (Q_i and L_i) and right-handed (D_i, U_i and E_i) chiral fermion lines must be directly attached to the vertices of the loop. For the case of charged lepton EDM, the diagrams must be proportional to $\lambda\lambda^*$ or $\lambda'\lambda'^*$. For $\lambda\lambda^*$, it is impossible to generate a one-loop graph with left-handed lepton (chiral super-field L) and right-handed lepton (chiral super-field E^c) attached directly to the loop (Fig. 11.2a). For $\lambda'\lambda'^*$, we cannot give EDM contribution since there is no vertex to give E^c (Fig. 11.2b). Similar discussion holds for the case of down type quark. In this case, the diagrams must be proportional to $\lambda'\lambda'^*$ or $\lambda''\lambda''^*$. As for the lepton EDM, it is not possible to draw such one-loop diagrams (see Fig. 11.2c and d). For the up type quark, it is also not possible to generate one-loop diagram since RPV interactions do not involve $SU(2)_L$ singlet U_i field. We must note that it is possible to generate one-loop EDM graphs if we consider bilinear R-parity violation [5–8], but we do not consider the bilinear R-parity violation in this work.

11.2 Two-Loop Level Barr–Zee Type Fermion EDM

The analysis of the two-loop quark EDM contribution was done by Chang et al. [1]. They enumerated all possible graphs contributing to the down-type quark EDM, and found that the Barr–Zee type graphs with strong or electromagnetic gauge bosons give

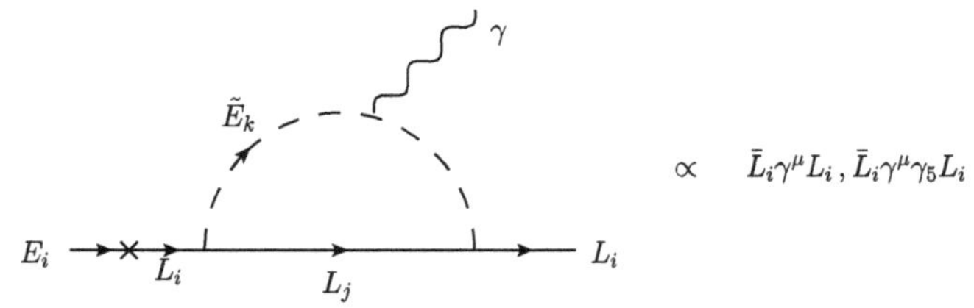

Fig. 11.1 The helicity flip in the external line cannot generate EDM operator: it becomes the renormalization of the vector and axial vector currents. The helicity flip must come directly from the loop

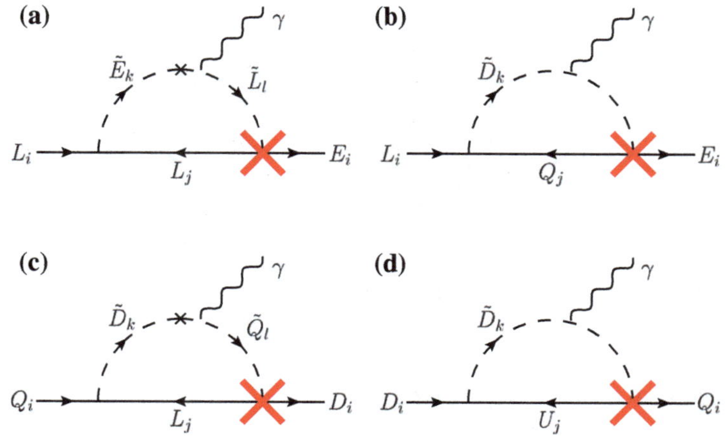

Fig. 11.2 Attempt to draw one-loop level EDM diagrams within R-parity violating supersymmetry. We cannot construct one-loop diagram within RPV and supersymmetric interactions. There are no interactions which give the Yukawa vertices with the big cross. The lepton EDM is given by diagrams (**a**) and (**b**), where (**a**) is the contribution from the lepton loop and (**b**) from the quark loop. The quark EDM is given by diagrams (**c**) and (**d**), where (**c**) is the contribution involving neutrino in the loop and (**d**) is that generated by baryon number violating interactions. Fields Q_i and L_j refer to the down type quark ($SU(2)_L$ doublet) and the neutrino, respectively

the leading contributions. The essential point was that the other two-loop diagrams must have a Higgs boson which give a suppression of at least one power of light quark mass plus some CKM mixing angle.[1]

We now derive the formula of the Barr–Zee type fermion EDM. The EDM d_F of the fermion F is defined as follows:

$$\mathscr{L}_{\text{EDM}} = -i\frac{d_F}{2}\bar{\psi}_F\gamma_5\sigma^{\mu\nu}\psi_F F_{\mu\nu}\,, \tag{11.2}$$

where $F_{\mu\nu}$ is the electromagnetic field strength. With the RPV lagrangian (4.2), the sneutrino exchange Barr–Zee type diagrams shown in Fig. 11.3 contribute to the EDM. Here the emission (absorption) of the sneutrino from fermion is accompanied by $P_R \equiv \frac{1}{2}(1+\gamma_5)$ ($P_L \equiv \frac{1}{2}(1-\gamma_5)$) projection operator as is apparent from Eq. (4.2).

There are also additional diagrams with the inverted flow of the inner loop fermion f_j, and also diagrams in which the incident fermion F_k interacts first with the gauge boson and then to the sneutrino before exiting. For them, the calculation is essentially the same.

[1] This section is based on the article of N. Yamanaka, T. Sato and T. Kubota, Phys. Rev. D **85**, 117701 (2012), Copyright © American Physical Society All Rights Reserved.

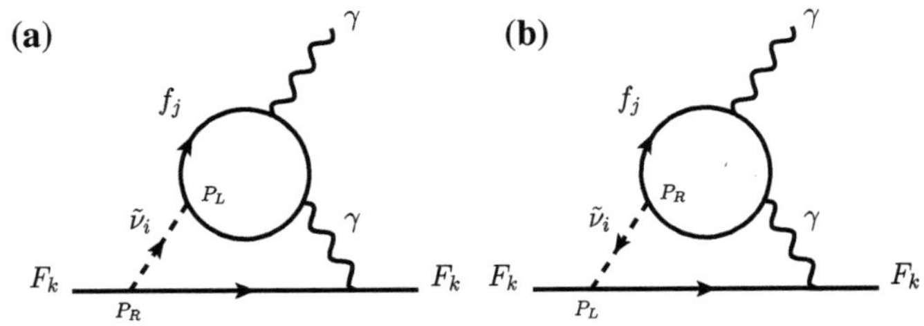

Fig. 11.3 Examples of Barr–Zee type two-loop contributions to the fermion EDM within RPV interactions. The projections of the chirality (P_L and P_R) were explicitly given for the RPV vertices

Effective One-Loop $\tilde{\nu}\gamma\gamma$ *Vertex*

The first step of its computation is to obtain the formula for the effective one-loop $\tilde{\nu}\gamma\gamma$ vertex. The $\tilde{\nu}\gamma\gamma$ vertex has two contributions as shown in Fig. 11.4.

The amplitude of the effective $\tilde{\nu}\gamma\gamma$ vertex is

$$\begin{aligned} i\mathcal{M}_{\tilde{\nu}\gamma\gamma} = & -\hat{\lambda}_{ijj} n_c (Q_f e)^2 \varepsilon^*_\mu(q_1)\varepsilon^*_\nu(q_2) \\ & \times \int \frac{d^4k}{(2\pi)^4} \frac{\mathrm{Tr}\left[(\not{k}+\not{q}_1+m_{f_j})\gamma^\mu(\not{k}+m_{f_j})\gamma^\nu(\not{k}-\not{q}_2+m_{f_j})(1-\gamma_5)\right]}{\left[(k+q_1)^2-m_{f_j}^2\right]\left[k^2-m_{f_j}^2\right]\left[(k-q_2)^2-m_{f_j}^2\right]}, \end{aligned} \tag{11.3}$$

where i and j denote the flavor indices of $\tilde{\nu}$ and loop fermion, respectively. The coupling $\hat{\lambda}$ is the R-parity violating coupling, $\hat{\lambda} = \lambda$ when charged lepton runs in the loop, and $\hat{\lambda} = \lambda'$ in the case of down type quark. Here n_c is the number of colors of the fermion in the loop, i.e. $n_c = 1$ if leptons run in the loop, and $n_c = 3$ when quarks run. The mass and the charge in unit of e of the loop fermion are denoted respectively by m_{f_j} and Q_f. We have to pay attention on the minus sign provided by the R-parity violating coupling, the trace of the fermion loop and the i's from vertices and propagators.

We now explain the steps to calculate Eq. (11.3):

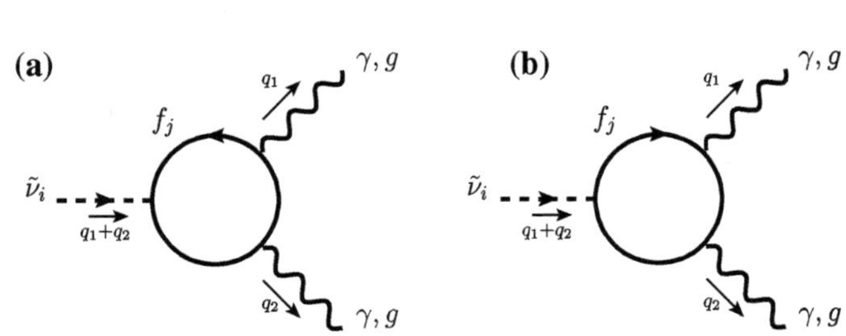

Fig. 11.4 One-loop diagrams contributing to the effective $\tilde{\nu}\gamma\gamma$ (or $\tilde{\nu}gg$) vertex

1. **Feynman parameter calculation of propagator**

$$
\begin{aligned}
&\frac{1}{\left[(k+q_1)^2-m_{f_j}^2\right]\left[k^2-m_{f_j}^2\right]\left[(k-q_2)^2-m_{f_j}^2\right]} \\
&=\int_0^1 \frac{2!\delta(1-x-y-z)\,dxdydz}{\left[x(k+q_1)^2+yk^2+z(k-q_2)^2-m_{f_j}^2\right]^3} \\
&=\int_0^1 \frac{2\delta(1-x-y-z)\,dxdydz}{\left[k^2+2x(k\cdot q_1)-2z(k\cdot q_2)+xq_1^2+zq_2^2-m_{f_j}^2\right]^3} \\
&=\int_0^1 \frac{2\delta(1-x-y-z)\,dxdydz}{\left[(k+xq_1-zq_2)^2-(xq_1-zq_2)^2+xq_1^2+zq_2^2-m_{f_j}^2\right]^3} \\
(k'\equiv k+xq_1-zq_2)&=\int_0^1 \frac{2\delta(1-x-y-z)\,dxdydz}{\left[k'^2+x(1-x)q_1^2+z(1-z)q_2^2+2xz(q_1\cdot q_2)-m_{f_j}^2\right]^3}
\end{aligned}
\tag{11.4}
$$

2. **Trace calculation**

$$
\begin{aligned}
T &= \mathrm{Tr}\left[(\not k+\not q_1+m_{f_j})\gamma^\mu(\not k+m_{f_j})\gamma^\nu(\not k-\not q_2+m_{f_j})(1-\gamma_5)\right] \\
&= m_{f_j}\,\mathrm{Tr}\Big[(\not k+\not q_1)\gamma^\mu\not k\gamma^\nu(1-\gamma_5)+(\not k+\not q_1)\gamma^\mu\gamma^\nu(\not k-\not q_2)(1-\gamma_5) \\
&\quad+\gamma^\mu\not k\gamma^\nu(\not k-\not q_2)(1-\gamma_5)+m_{f_j}^2\gamma^\mu\gamma^\nu\Big] \\
&= 4m_{f_j}\Big[4k^\mu k^\nu+2q_1^\mu k^\nu-2k^\mu q_2^\nu+[m_{f_j}^2-k^2-(q_1\cdot q_2)]g^{\mu\nu} \\
&\quad-q_1^\mu q_2^\nu+q_2^\mu q_1^\nu-i\varepsilon^{\mu\nu\alpha\beta}(q_1)_\alpha(q_2)_\beta\Big] \\
&= 4m_{f_j}\Big[4(k'^\mu-xq_1^\mu+zq_2^\mu)(k'^\nu-xq_1^\nu+zq_2^\nu)+2q_1^\mu(k'^\nu-xq_1^\nu+zq_2^\nu) \\
&\quad-2(k'^\mu-xq_1^\mu+zq_2^\mu)q_2^\nu+[m_{f_j}^2-(k'-xq_1+zq_2)^2-(q_1\cdot q_2)]g^{\mu\nu} \\
&\quad-q_1^\mu q_2^\nu+q_2^\mu q_1^\nu-i\varepsilon^{\mu\nu\alpha\beta}(q_1)_\alpha(q_2)_\beta\Big] \\
&= 4m_{f_j}\Big[4k'^\mu k'^\nu+4(xq_1^\mu-zq_2^\mu)(xq_1^\nu-zq_2^\nu)-2q_1^\mu(xq_1^\nu-zq_2^\nu) \\
&\quad+2(xq_1^\mu-zq_2^\mu)q_2^\nu+[m_{f_j}^2-k'^2-(xq_1-zq_2)^2-(q_1\cdot q_2)]g^{\mu\nu} \\
&\quad-q_1^\mu q_2^\nu+q_2^\mu q_1^\nu-i\varepsilon^{\mu\nu\alpha\beta}(q_1)_\alpha(q_2)_\beta\Big] \\
&= 4m_{f_j}\Big[\frac{4-d}{d}k'^2g^{\mu\nu}+[m_{f_j}^2-x^2q_1^2-z^2q_2^2+(2xz-1)(q_1\cdot q_2)]g^{\mu\nu} \\
&\quad+(1-4xz)q_2^\mu q_1^\nu+(4x^2-2x)q_1^\mu q_1^\nu+(-4xz+2z+2x-1)q_1^\mu q_2^\nu \\
&\quad+(4z^2-2z)q_2^\mu q_2^\nu-i\varepsilon^{\mu\nu\alpha\beta}(q_1)_\alpha(q_2)_\beta\Big].
\end{aligned}
\tag{11.5}
$$

In the fifth line, we have omitted terms linear in p' since they cancel by symmetry when integrated. The convention we used for ε and γ_5 is

$$\varepsilon^{0123} \equiv +1, \tag{11.6}$$

$$\gamma_5 \equiv i\gamma^0\gamma^1\gamma^2\gamma^3. \tag{11.7}$$

3. **Momentum integration**
Diergent part: needed for obtaining the first term of the last line of Eq. (11.5).

$$\begin{aligned}
I_{\text{div}} &= \int \frac{d^d k'}{(2\pi)^d} \frac{k'^2}{\left[k'^2 - \Delta\right]^3} \frac{4-d}{d} \\
&= \int \frac{i d^d k'_E}{(2\pi)^d} \frac{-k'^2_E}{\left[-k'^2_E - \Delta\right]^3} \frac{4-d}{d} \\
&= i\frac{4-d}{d} \cdot \frac{1}{(4\pi)^{d/2}} \frac{d}{2} \frac{\Gamma(2-d/2)}{\Gamma(3)} \left(\frac{1}{\Delta}\right)^{2-\frac{d}{2}} \\
(d \to 4) &\to i\frac{1}{(4\pi)^{2-\frac{\varepsilon}{2}}} \frac{\varepsilon}{4} \Gamma(\varepsilon/2) \left(\frac{1}{\Delta}\right)^{\frac{\varepsilon}{2}} \qquad (\varepsilon \equiv 4-d) \\
&= i\frac{1}{(4\pi)^2} \frac{\varepsilon}{4} \left(\frac{2}{\varepsilon} - \gamma + O(\varepsilon)\right) e^{-\frac{\varepsilon}{2}\log\Delta} e^{\frac{\varepsilon}{2}\log 4\pi} \\
&= \frac{i}{2(4\pi)^2}.
\end{aligned} \tag{11.8}$$

We must note here that the regularization method used is the dimensional reduction [9], to keep the supersymmetry (in this situation, the manipulation is the same as for the dimensional regularization).

Convergent part:

$$I_{\text{conv}} = \int \frac{d^d k'}{(2\pi)^d} \frac{1}{\left[k'^2 - \Delta\right]^3} = \frac{-i}{2(4\pi)^2} \frac{1}{\Delta},$$

where $\Delta = -x(1-x)q_1^2 - z(1-z)q_2^2 - 2xz(q_1 \cdot q_2) + m_{f_j}^2$.

With the above ingredients, Eq. (11.3) then gives

$$\begin{aligned}
i\mathcal{M}_{\tilde{\nu}\gamma\gamma} =& -4m_{f_j}\hat{\lambda}_{ijj} n_c (Q_f e)^2 \varepsilon^*_\mu(q_1)\varepsilon^*_\nu(q_2) \int_0^1 2\delta(1-x-y-z)dxdydz \\
&\times \Big\{ I_{\text{div}} g^{\mu\nu} \\
&+ I_{\text{conv}} \Big[(m_{f_j}^2 - x^2 q_1^2 - z^2 q_2^2 + [2xz-1][q_1 \cdot q_2]) g^{\mu\nu} \\
&+ (1-4xz) q_2^\mu q_1^\nu
\end{aligned}$$

$$
\begin{aligned}
&+ (4x^2 - 2x) q_1^\mu q_1^\nu + (-4xz + 2z + 2x - 1) q_1^\mu q_2^\nu \\
&+ (4z^2 - 2z) q_2^\mu q_2^\nu - i\varepsilon^{\mu\nu\alpha\beta}(q_1)_\alpha (q_2)_\beta \Big]\Big\} \\
= & -\frac{4i}{(4\pi)^2} m_{f_j} \hat{\lambda}_{ijj} n_c (Q_f e)^2 \varepsilon_\mu^*(q_1) \varepsilon_\nu^*(q_2) \\
&\times \int_0^1 \frac{\delta(1-x-y-z)\, dxdydz}{-x(1-x)q_1^2 - z(1-z)q_2^2 - 2xz(q_1 \cdot q_2) + m_{f_j}^2} \\
&\times \Big\{ \Big[(2x^2 - x) q_1^2 + (2z^2 - z) q_2^2 + (1 - 4xz)(q_1 \cdot q_2) \Big] g^{\mu\nu} \\
&- (1 - 4xz) q_2^\mu q_1^\nu \\
&- (4x^2 - 2x) q_1^\mu q_1^\nu - (-4xz + 2z + 2x - 1) q_1^\mu q_2^\nu \\
&- (4z^2 - 2z) q_2^\mu q_2^\nu + i\varepsilon^{\mu\nu\alpha\beta}(q_1)_\alpha (q_2)_\beta \Big\} \\
= & \frac{4i}{(4\pi)^2} m_{f_j} \hat{\lambda}_{ijj} n_c (Q_f e)^2 \varepsilon_\mu^*(q_1) \varepsilon_\nu^*(q_2) \\
&\times \int_0^1 dx \int_0^{1-x} dz \frac{(1-4xz)\left[q_2^\mu q_1^\nu - (q_1 \cdot q_2) g^{\mu\nu} \right] - i\varepsilon^{\mu\nu\alpha\beta}(q_1)_\alpha (q_2)_\beta}{m_{f_j}^2 - x(1-x)q_1^2 - z(1-z)q_2^2 - 2xz(q_1 \cdot q_2)} \\
&- \frac{4i}{(4\pi)^2} m_{f_j} \hat{\lambda}_{ijj} n_c (Q_f e)^2 \varepsilon_\mu^*(q_1) \varepsilon_\nu^*(q_2) \\
&\times \left[\frac{1}{2}(B_1 + B_2) q_1^2 q_2^2 g^{\mu\nu} - B_1 q_2^2 q_1^\mu q_1^\nu - B_2 q_1^2 q_2^\mu q_2^\nu + B_3 (q_1 \cdot q_2) q_1^\mu q_2^\nu \right],
\end{aligned}
\tag{11.9}
$$

where B_1, B_2 and B_3 are defined as

$$
\begin{aligned}
B_1 &= \frac{1}{q_2^2} \int_0^1 dx \int_0^{1-x} dz \frac{2x(1-2x)}{x(1-x)q_1^2 + z(1-z)q_2^2 + 2xz(q_1 \cdot q_2) - m_{f_j}^2}, \\
B_2 &= \frac{1}{q_1^2} \int_0^1 dx \int_0^{1-x} dz \frac{2z(1-2z)}{x(1-x)q_1^2 + z(1-z)q_2^2 + 2xz(q_1 \cdot q_2) - m_{f_j}^2}, \\
B_3 &= -\frac{1}{q_1 \cdot q_2} \int_0^1 dx \int_0^{1-x} dz \frac{(1-2x)(1-2z)}{x(1-x)q_1^2 + z(1-z)q_2^2 + 2xz(q_1 \cdot q_2) - m_{f_j}^2}.
\end{aligned}
\tag{11.10}
$$

We now note that the gauge invariance requires $B_1 = B_2 = B_3$. The equality $B_1 = B_3$ is shown as follows:

$$
\begin{aligned}
q_2^2 (q_1 \cdot q_2)(B_1 - B_3) &= \int_0^1 dx \int_0^{1-x} dz \frac{(1-2x)\left[2x(q_1 \cdot q_2) + q_2^2(1-2z) \right]}{F(x,z)} \\
&= \int_0^1 dx\, (1-2x) \int_0^\infty dw \int_0^{1-x} dz\, F(x,z) \left(-\frac{\partial}{\partial z} \right) e^{-wF(x,z)}
\end{aligned}
$$

$$
\begin{aligned}
&= \int_0^1 dx\,(1-2x)\int_0^\infty dw\Bigg\{-F(x,z)e^{-wF(x,z)}\Big|_0^{1-x} \\
&\quad + \int_0^{1-x} dz\, e^{-wF(x,z)}\frac{\partial}{\partial z}F(x,z)\Bigg\} \\
&= \int_0^1 dx\,(1-2x)\int_0^\infty dw\left[-F(x,z)e^{-wF(x,z)} - \frac{1}{w}e^{-wF(x,z)}\right]_0^{1-x},
\end{aligned}
\tag{11.11}
$$

where $F(x,z) \equiv x(1-x)q_1^2 + z(1-z)q_2^2 + 2xz(q_1\cdot q_2) - m_{f_j}^2$. The dw integral is a function of $x(1-x)$. By converting $x' = x - \frac{1}{2}$, the above integral cancels (the integral is odd in x'). We then have $B_1 = B_3$. Similar derivation is also possible for B_2.

$\Rightarrow$ We have finally $B_1 = B_2 = B_3 \equiv B$.

We shall show that B is not divergent as $q_1 \to 0$. By Taylor expanding B_3 in $(q_1\cdot q_2)$,

$$
\begin{aligned}
B_3 &= -\frac{1}{q_1\cdot q_2}\int_0^1 dx\int_0^{1-x} dz\frac{(1-2x)(1-2z)}{x(1-x)q_1^2 + z(1-z)q_2^2 + 2xz(q_1\cdot q_2) - m_{f_j}^2} \\
&\approx -\frac{1}{q_1\cdot q_2}\int_0^1 dx\int_0^{1-x} dz\frac{(1-2x)(1-2z)}{z(1-z)q_2^2 - m_{f_j}^2} + O(1) \\
&= -\frac{1}{q_1\cdot q_2}\int_0^1 dx\,(1-2x)\frac{1}{q_2^2}\int_0^{1-x} dz\frac{\frac{\partial}{\partial z}\left[z(1-z)q_2^2 - m_{f_j}^2\right]}{z(1-z)q_2^2 - m_{f_j}^2} + O(1) \\
&= -\frac{1}{(q_1\cdot q_2)q_2^2}\int_0^1 dx\,(1-2x)\ln\left(\frac{m_{f_j}^2 - x(1-x)q_2^2}{m_{f_j}^2}\right) + O(1) \\
&= 0 + O(1).
\end{aligned}
\tag{11.12}
$$

The first divergent term vanishes since it is an odd function in $x' = x - \frac{1}{2}$. Thus we have shown that B is not divergent.

In Eq. (11.9), the term with B is higher order in q_1 and q_2. The EDM is the first order coefficient of the external momentum, so we are not interested in it. Thus the final result for the $\tilde{\nu}\gamma\gamma$ loop of our interest here is the following leading gauge invariant amplitude in q_2:

$$
\begin{aligned}
i\mathcal{M}_{\tilde{\nu}\gamma\gamma} &\approx \frac{4i}{(4\pi)^2}m_{f_j}\hat{\lambda}_{ijj}n_c(Q_f e)^2\varepsilon_\mu^*(q_1)\varepsilon_\nu^*(q_2) \\
&\quad\times\int_0^1 dx\int_0^{1-x} dz\frac{(1-4xz)\left[q_2^\mu q_1^\nu - (q_1\cdot q_2)g^{\mu\nu}\right] - i\varepsilon^{\mu\nu\alpha\beta}(q_1)_\alpha(q_2)_\beta}{m_{f_j}^2 - x(1-x)q_1^2}.
\end{aligned}
\tag{11.13}
$$

Since

$$\int_0^1 dx \int_0^{1-x} dz \frac{(1-4xz)}{m_{f_j}^2 - x(1-x)q_1^2} = \int_0^1 dx \frac{(1-x)[1-2x(1-x)]}{m_{f_j}^2 - x(1-x)q_1^2}$$
$$= \frac{1}{2}\int_0^1 dx \frac{1-2x(1-x)}{m_{f_j}^2 - x(1-x)q_1^2},$$
$$\int_0^1 dx \int_0^{1-x} dz \frac{1}{m_{f_j}^2 - x(1-x)q_1^2} = \frac{1}{2}\int_0^1 dx \frac{1}{m_{f_j}^2 - x(1-x)q_1^2}, \tag{11.14}$$

we obtain finally

$$i\mathcal{M}_{\tilde{\nu}\gamma\gamma} = \frac{2i}{(4\pi)^2} m_{f_j} \hat{\lambda}_{ijj} n_c (Q_f e)^2 \varepsilon_\mu^*(q_1)\varepsilon_\nu^*(q_2)$$
$$\times \int_0^1 dx \frac{[1-2x(1-x)]\left[q_2^\mu q_1^\nu - (q_1\cdot q_2)g^{\mu\nu}\right] - i\varepsilon^{\mu\nu\alpha\beta}(q_1)_\alpha(q_2)_\beta}{m_{f_j}^2 - x(1-x)q_1^2}. \tag{11.15}$$

For the gluon contribution ($\mathcal{M}_{\tilde{\nu}gg}$), we just have to replace $(Q_f e)^2$ by g_s^2, the polarization vectors of photon by those of gluon, and n_c by $\frac{1}{2}$ which is the constant originating from the color trace of the quark loop.

Note that the amplitude and the lagrangian are related with

$$F_{\mu\nu}F^{\mu\nu} = (\partial_\mu A_\nu^{(1)} - \partial_\nu A_\mu^{(1)})(\partial^\mu A^{(2)\nu} - \partial^\nu A^{(2)\mu})$$
$$\leftrightarrow\ 2\varepsilon_\mu^*(q_1)\varepsilon_\nu^*(q_2)\left[q_2^\mu q_1^\nu - (q_1\cdot q_2)g^{\mu\nu}\right]. \tag{11.16}$$

Second Loop

We first calculate the amplitude of the diagram in Fig. 11.3.

$$i\mathcal{M}_{\rm BZ} = \frac{2}{(4\pi)^2} m_{f_j} \hat{\lambda}_{ijj} \tilde{\lambda}_{ikk}^* n_c (Q_f e)^2 \frac{Q_F e}{2} \varepsilon_\nu^*(q) \int \frac{d^4k}{(2\pi)^4}$$
$$\times \left[\frac{\bar{u}(p-q)\gamma_\mu(\not{p} - \not{q} - \not{k} + m_{F_k})\gamma_5 u(p)\cdot[(q\cdot k)g^{\mu\nu} - q^\mu k^\nu]}{k^2\left[(q+k)^2 - m_{\tilde{\nu}_i}^2\right]\left[(p-q-k)^2 - m_{F_k}^2\right]}\right.$$
$$\times \int_0^1 dx \frac{1-2x(1-x)}{m_{f_j}^2 - x(1-x)k^2}$$
$$+ \frac{\bar{u}(p-q)\gamma_\mu(\not{p} - \not{q} - \not{k} + m_{F_k})u(p)\cdot i\varepsilon^{\mu\nu\alpha\beta}k_\alpha q_\beta}{k^2\left[(q+k)^2 - m_{\tilde{\nu}_i}^2\right]\left[(p-q-k)^2 - m_{F_k}^2\right]}$$

$$\times \int_0^1 dx \frac{1}{m_{f_j}^2 - x(1-x)k^2}\Bigg], \tag{11.17}$$

where $\tilde{\lambda} = \lambda$ for lepton EDM contribution and $\tilde{\lambda} = \lambda'$ for quark EDM contribution. We have taken only the EDM contribution to the amplitude $i\mathscr{M}_{\rm BZ}$.

First term of Eq. (11.23):

$$\begin{aligned}
I_1 &= \int \frac{d^4k}{(2\pi)^4} \frac{\bar{u}(p-q)\gamma_\mu(\not{p} - \not{q} - \not{k} + m_{F_k})\gamma_5 u(p)\cdot[(q\cdot k)g^{\mu\nu} - q^\mu k^\nu]}{k^2\left[(q+k)^2 - m_{\tilde{\nu}_i}^2\right]\left[(p-q-k)^2 - m_{F_k}^2\right]} \\
&\quad \times \int_0^1 dx \frac{1-2x(1-x)}{m_{f_j}^2 - x(1-x)k^2} \\
&\approx \int \frac{d^4k}{(2\pi)^4} \frac{\bar{u}\,[\not{q}\, k^\nu - (q\cdot k)\gamma^\nu]\not{k}\,\gamma_5 u}{k^2(k^2 - m_{\tilde{\nu}_i}^2)k^2} \int_0^1 dx \frac{1-2x(1-x)}{m_{f_j}^2 - x(1-x)k^2} \\
&= \frac{1}{4}\int \frac{d^4k}{(2\pi)^4} \frac{\bar{u}\,(\not{q}\,\gamma^\nu - \gamma^\nu\not{q}\,)\,\gamma_5 u\cdot k^2}{k^2(k^2 - m_{\tilde{\nu}_i}^2)k^2} \int_0^1 dx \frac{1-2x(1-x)}{m_{f_j}^2 - x(1-x)k^2} \\
&= \frac{-i}{2}\int \frac{d^4k}{(2\pi)^4} \frac{\bar{u}\sigma^{\mu\nu}q_\mu\gamma_5 u}{k^2(k^2 - m_{\tilde{\nu}_i}^2)} \int_0^1 dx \frac{1-2x(1-x)}{m_{f_j}^2 - x(1-x)k^2} \\
&= \frac{-i}{2}\int_0^1 dx \int_0^\infty \frac{i\pi^2 k_E^2 d(k_E^2)}{(2\pi)^4}\cdot\frac{\bar{u}\sigma^{\mu\nu}q_\mu\gamma_5 u}{k_E^2(k_E^2 + m_{\tilde{\nu}_i}^2)}\cdot\frac{1-2x(1-x)}{m_{f_j}^2 + x(1-x)k_E^2}.
\end{aligned} \tag{11.18}$$

Second term of Eq. (11.23):

$$\begin{aligned}
I_2 &= \int \frac{d^4k}{(2\pi)^4} \frac{\bar{u}(p-q)\gamma_\mu(\not{p} - \not{q} - \not{k} + m_{F_k})u(p)\cdot i\varepsilon^{\mu\nu\alpha\beta}k_\alpha q_\beta}{k^2\left[(q+k)^2 - m_{\tilde{\nu}_i}^2\right]\left[(p-q-k)^2 - m_{F_k}^2\right]} \\
&\quad \times \int_0^1 dx \frac{1}{m_{f_j}^2 - x(1-x)k^2} \\
&= \int \frac{d^4k}{(2\pi)^4} \frac{\bar{u}\gamma_\mu(-\not{k}\,)u\cdot i\varepsilon^{\mu\nu\alpha\beta}k_\alpha q_\beta}{k^2(k^2 - m_{\tilde{\nu}_i}^2)k^2} \int_0^1 dx \frac{1}{m_{f_j}^2 - x(1-x)k^2} \\
&= \frac{-1}{4}\int \frac{d^4k}{(2\pi)^4} \frac{\bar{u}\gamma_\mu\gamma_\alpha u\cdot i\varepsilon^{\mu\nu\alpha\beta}q_\beta}{k^2(k^2 - m_{\tilde{\nu}_i}^2)} \int_0^1 dx \frac{1}{m_{f_j}^2 - x(1-x)k^2} \\
&= \frac{-1}{2}\int \frac{d^4k}{(2\pi)^4} \frac{\bar{u}i\gamma_5\sigma^{\nu\beta}q_\beta u}{k^2(k^2 - m_{\tilde{\nu}_i}^2)} \int_0^1 dx \frac{1}{m_{f_j}^2 - x(1-x)k^2} \\
&= \frac{i}{2}\int_0^1 dx \int_0^\infty \frac{i\pi^2 k_E^2 d(k_E^2)}{(2\pi)^4}\cdot\frac{\bar{u}\gamma_5\sigma^{\mu\nu}q_\mu u}{k_E^2(k_E^2 + m_{\tilde{\nu}_i}^2)}\cdot\frac{1}{m_{f_j}^2 + x(1-x)k_E^2}.
\end{aligned} \tag{11.19}$$

We have used

$$\frac{1}{2}\varepsilon^{\mu\nu\rho\sigma}\sigma_{\rho\sigma} = -i\gamma_5\sigma^{\mu\nu}. \tag{11.20}$$

We then obtain

$$\begin{aligned}
I_1 + I_2 &= \frac{-i}{2}\int_0^1 dx \int_0^\infty \frac{i\pi^2 k_E^2 d(k_E^2)}{(2\pi)^4} \cdot \frac{\bar{u}\sigma^{\mu\nu}q_\mu\gamma_5 u}{k_E^2(k_E^2+m_{\tilde{\nu}_i}^2)} \cdot \frac{-2x(1-x)}{m_{f_j}^2 + x(1-x)k_E^2} \\
&= \frac{-1}{(4\pi)^2}\bar{u}\sigma^{\mu\nu}q_\mu\gamma_5 u \int_0^1 dx\, x(1-x) \int_0^\infty dr \frac{1}{r+m_{\tilde{\nu}_i}^2} \cdot \frac{1}{m_{f_j}^2 + x(1-x)r} \\
&= \frac{-1}{(4\pi)^2}\bar{u}\sigma^{\mu\nu}q_\mu\gamma_5 u \int_0^1 dx \int_0^\infty dr \frac{1}{r+m_{\tilde{\nu}_i}^2} \cdot \frac{1}{r + \frac{m_{f_j}^2}{x(1-x)}} \\
&= \frac{-1}{(4\pi)^2}\bar{u}\sigma^{\mu\nu}q_\mu\gamma_5 u \int_0^1 dx \frac{x(1-x)}{m_{f_j}^2 - m_{\tilde{\nu}_i}^2 x(1-x)} \\
&\quad \times \int_0^\infty dr \left\{ \frac{1}{r+m_{\tilde{\nu}_i}^2} - \frac{1}{r+\frac{m_{f_j}^2}{x(1-x)}} \right\} \\
(\tau \equiv m_{f_j}^2/m_{\tilde{\nu}_i}^2) &= \frac{-1}{(4\pi)^2}\frac{1}{m_{\tilde{\nu}_i}^2}\bar{u}\sigma^{\mu\nu}q_\mu\gamma_5 u \int_0^1 dx \frac{x(1-x)}{x(1-x)-\tau}\ln\left(\frac{x(1-x)}{\tau}\right).
\end{aligned} \tag{11.21}$$

We have finally

$$\begin{aligned}
i\mathcal{M}_{\mathrm{BZ}} = &- \hat{\lambda}_{ijj}\tilde{\lambda}_{ikk}^* \frac{\alpha_{\mathrm{em}}}{(4\pi)^3} n_c Q_f^2 Q_F e \frac{m_{f_j}}{m_{\tilde{\nu}_i}^2}\int_0^1 dx \frac{x(1-x)}{x(1-x)-\tau}\ln\left(\frac{x(1-x)}{\tau}\right) \\
&\times \varepsilon_\nu^*(q)\bar{u}\sigma^{\mu\nu}q_\mu\gamma_5 u.
\end{aligned} \tag{11.22}$$

We have obtained the amplitude corresponding to the diagram of Fig. 11.3. We must now add to them the remaining diagrams. The diagram with inverted flow is just the complex conjugate of $\mathcal{M}_{\mathrm{BZ}}$ (except the gluon polarization vector). Diagrams with sneutrino and gauge boson between the external fermion (F) line and the inner loop replaced have just the same contribution as the original diagrams. This can be easily shown by replacing the gamma matrices and inverting the sign of the loop momentum in Eqs. (11.18) and (11.19). The total Barr–Zee contribution is then

$$\begin{aligned}
i\mathcal{M}_{\mathrm{BZtot}} \approx &- 4i\,\mathrm{Im}(\hat{\lambda}_{ijj}\tilde{\lambda}_{ikk}^*)\frac{\alpha_{\mathrm{em}}}{(4\pi)^3} n_c Q_f^2 Q_F e \frac{m_{f_j}}{m_{\tilde{\nu}_i}^2}\int_0^1 dx \frac{x(1-x)}{x(1-x)-\tau}\ln\left(\frac{x(1-x)}{\tau}\right) \\
&\times \varepsilon_\nu^*(q)\bar{u}\sigma^{\mu\nu}q_\mu\gamma_5 u \\
= &\, 4i\mathrm{Im}(\hat{\lambda}_{ijj}\tilde{\lambda}_{ikk}^*)\frac{\alpha_{\mathrm{em}}}{(4\pi)^3} n_c Q_f^2 Q_F e \frac{1}{m_{f_j}} \left\{ f\left(\frac{m_{f_j}^2}{m_{\tilde{\nu}_i}^2}\right) - g\left(\frac{m_{f_j}^2}{m_{\tilde{\nu}_i}^2}\right) \right\}
\end{aligned}$$

$$\times\, \varepsilon_\nu^*(q)\bar{u}\sigma^{\mu\nu}q_\mu\gamma_5 u, \tag{11.23}$$

where the functions f and g are defined as

$$f(z) = \frac{z}{2}\int_0^1 dx \frac{1-2x(1-x)}{x(1-x)-z}\ln\left(\frac{x(1-x)}{z}\right), \tag{11.24}$$

$$g(z) = \frac{z}{2}\int_0^1 dx \frac{1}{x(1-x)-z}\ln\left(\frac{x(1-x)}{z}\right), \tag{11.25}$$

in the notation of the original notation of Barr and Zee [10]. For small z, we have $f(z) \approx \frac{z}{2}\left(\frac{\pi^2}{3}+4+2\ln z+(\ln z)^2\right)$, and $g(z) \approx \frac{z}{2}\left(\frac{\pi^2}{3}+(\ln z)^2\right)$. In the last line of Eq. (11.23), we have taken only the part of $i\mathcal{M}_{\rm BZ}$ which contributes to the EDM, disregarding $\mathrm{Re}(\hat{\lambda}_{ijj}\tilde{\lambda}^*_{ikk})$. The total EDM of the fermion F from the Barr–Zee type diagrams with R-parity violating interaction is then

$$\begin{aligned} d_{F_k} &= \mathrm{Im}(\hat{\lambda}_{ijj}\tilde{\lambda}^*_{ikk})\frac{\alpha_{\rm em} n_c Q_f^2 Q_F e}{16\pi^3 m_{f_j}} \cdot \{f(\tau) - g(\tau)\} \\ &\approx \mathrm{Im}(\hat{\lambda}_{ijj}\tilde{\lambda}^*_{ikk})\frac{\alpha_{\rm em} n_c Q_f^2 Q_F e}{16\pi^3 m_{f_j}} \cdot \tau(2+\ln\tau+\cdots), \end{aligned} \tag{11.26}$$

where $\tau = m_{f_j}^2/m_{\tilde{\nu}_i}^2$. The flavor index, electric charge and number of color of the fermion F are denoted respectively by k, $Q_F e$ and n_c ($n_c = 3$ if the inner loop fermion is a quark, otherwise $n_c = 1$), and j and $Q_f e$ are respectively the flavor index and the electric charge of the inner loop fermion f. The second line of the above equation is the approximated expression for small τ. Note also that the Barr–Zee type diagram gives EDM contribution only to down-type quarks, and the same property holds also for the chromo-EDM (cEDM) seen below.

We see from Eq. (11.17) (see also Fig. 11.3) that the chirality structure of the scalar exchange between internal loop and external line (RPV vertices with sneutrino exchange) has the form $P_L \otimes P_R$ and $P_R \otimes P_L$, which is a consequence of the lepton number conservation of the whole EDM process. This gives as a result the structure

$$f(\tau) - g(\tau) \approx \tau(2+\ln\tau), \tag{11.27}$$

in the final formula (11.26). This is consistent with the result obtained in the analysis of the Barr–Zee type diagram analogues with the exchange of Higgs bosons in the two Higgs doublet model, done originally by Barr and Zee [10] (see also [11]). In the two Higgs doublet model, there are also additional contributions with the structures $P_L \otimes P_L$ and $P_R \otimes P_R$ which yield contribution proportional to

$$f(\tau) + g(\tau) \approx \tau\left(\frac{\pi^2}{3}+2+\ln\tau+(\ln\tau)^2\right), \tag{11.28}$$

which is absent in the RPV supersymmetric models.

The small τ behaviour in Eq. (11.26) is in contradiction with the result presented in Refs. [1, 2, 12, 13], where the RPV Barr–Zee type diagrams receive the leading contribution proportional to $\tau(\ln\tau)^2$. The difference between these two results is large. For example, if we consider the Barr–Zee type diagram with the bottom quark in the loop and with sneutrino mass = 1 TeV, we have $[f(\tau)-g(\tau)]/[f(\tau)+g(\tau)] \approx -0.08$. By using our correct formula, the experimental upper bounds on RPV interactions given from the RPV Barr–Zee type contribution are therefore loosened by one order of magnitude.

We can also evaluate Barr–Zee type diagrams which contribute to the quark cEDM. The lagrangian of the cEDM interaction is given by

$$\mathscr{L}_{\mathrm{cEDM}} = -i\frac{d_Q^c}{2}\bar{\psi}_Q\gamma_5\sigma^{\mu\nu}T_a\psi_Q G^a_{\mu\nu}, \tag{11.29}$$

where $G^a_{\mu\nu}$ is the gluon field strength. The Barr–Zee type contribution of the down-type quark q_k is then

$$\begin{aligned} d^c_{Q_k} &= -\,\mathrm{Im}(\lambda'_{ijj}\lambda'^*_{ikk})\frac{\alpha_s g_s}{32\pi^3}\frac{m_{q_j}}{m^2_{\tilde{\nu}_i}}\int_0^1 dx\frac{x(1-x)}{x(1-x)-\tau}\ln\left(\frac{x(1-x)}{\tau}\right) \\ &= \mathrm{Im}(\lambda'_{ijj}\lambda'^*_{ikk})\frac{\alpha_s g_s}{32\pi^3 m_{q_j}}\cdot\{f(\tau)-g(\tau)\}, \end{aligned} \tag{11.30}$$

where $\tau = m^2_{q_j}/m^2_{\tilde{\nu}_i}$. The flavor indices of the external quark Q and the quark of the inner loop q are denoted respectively by k and j.

11.3 P, CP-Odd 4-Fermion Interactions

The EDM and the chromo-EDM of fermions seen above are P, CP-odd processes of a single fermion. For the case of composite systems such as hadrons, nuclei and atoms, their EDM can receive contribution from the P, CP-odd many-body interactions at the elementary level, in addition to the EDM of their constituents. The leading P, CP-odd many-body interactions are the *P, CP-odd 4-fermion interactions*. Trilinear R-parity violating interactions contribute to the P, CP-odd 4-fermion interaction at the tree level as shown in Fig. 11.5.

The importance of the contribution of the P, CP-odd electron-quark interaction from R-parity violation was first pointed by Herczeg [2]. The analysis of P, CP-odd 4-quark interaction within R-parity violation was made by Faessler et al. [12, 13]. In both cases, it was possible to constrain R-parity violating couplings contributing to P, CP-odd 4-fermion interactions from the current experimental data.

The 4-fermion interaction can be written as follows:

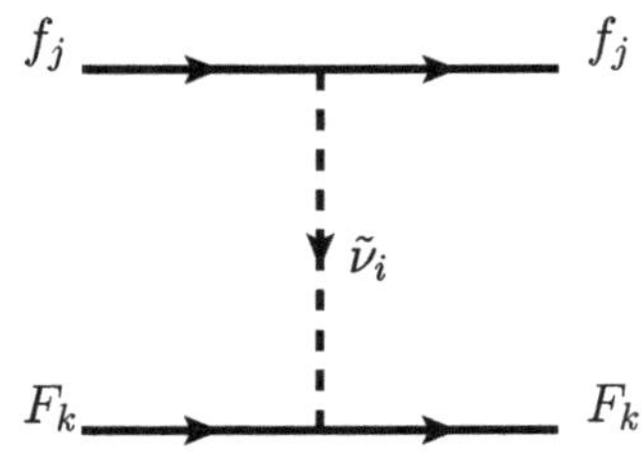

Fig. 11.5 Diagram contributing to 4-fermion interaction within R-parity violation. We must also consider its complex conjugate

$$\mathscr{L}_{4-\mathrm{f}} = \frac{\mathrm{Im}(\hat{\lambda}_{ijj}\tilde{\lambda}^*_{ikk})}{2m^2_{\tilde{\nu}_i}} \left[\bar{f}_j f_j \cdot \bar{F}_k i\gamma_5 F_k - \bar{f}_j i\gamma_5 f_j \cdot \bar{F}_k F_k \right]$$
$$- \frac{\mathrm{Re}(\hat{\lambda}_{ijj}\tilde{\lambda}^*_{ikk})}{2m^2_{\tilde{\nu}_i}} \left[\bar{f}_j f_j \cdot \bar{F}_k F_k + \bar{f}_j i\gamma_5 f_j \cdot \bar{F}_k i\gamma_5 F_k \right], \tag{11.31}$$

where $\hat{\lambda} = \lambda, \lambda'$ when the fermion f is a lepton or quark, respectively. Also, $\tilde{\lambda} = \lambda, \lambda'$ when the fermion F is a lepton or quark, respectively. The second line is the CP-even part of the 4-fermion interaction, and is of no interest in this discussion, so we omit them from now.

The tree level RPV contribution to the P, CP-odd 4-quark interaction can then be written in terms of Eq. (6.8) as follows:

$$C_{jk} = \frac{\mathrm{Im}(\lambda'_{ijj}\lambda'^*_{ikk})}{\sqrt{2}m^2_{\tilde{\nu}_i} G_F}, \tag{11.32}$$

where j and k are the quark flavor. Note that there are no tensor type P, CP-odd 4-quark interaction ($C'_{qq'}$) in R-parity violation at the tree level.

The tree level RPV contribution to the P, CP-odd electron-quark interaction can be written in terms of Eq. (6.9) as follows:

$$C^{\mathrm{SP}}_{ej} = \frac{\mathrm{Im}(\lambda_{i11}\lambda'^*_{ijj})}{\sqrt{2}m^2_{\tilde{\nu}_i} G_F}, \tag{11.33}$$

$$C^{\mathrm{PS}}_{ej} = -\frac{\mathrm{Im}(\lambda_{i11}\lambda'^*_{ijj})}{\sqrt{2}m^2_{\tilde{\nu}_i} G_F}. \tag{11.34}$$

We should be careful on the minus sign in the definition of Eq. (6.9), which was introduced to be consistent with many references [14, 15]. In this case also, we do not have tensor type P, CP-odd electron-quark interaction (C^{T}_{eq}) at the tree level within R-parity violation.

References

1. D. Chang, W.-F. Chang, M. Frank, W.-Y. Keung, Phys. Rev. D **62**, 095002 (2000)
2. P. Herczeg, Phys. Rev. D **61**, 095010 (2000)
3. N. Yamanaka, T. Sato, T. Kubota, Phys. Rev. D **85**, 117701 (2012)
4. R.M. Godbole, S. Pakvasa, S.D. Rindani, X. Tata, Phys. Rev. D **61**, 113003 (2000)
5. K. Choi, E.J. Chun, K. Hwang, Phys. Rev. D **63**, 013002 (2000)
6. Y.Y. Keum, O.C.W. Kong, Phys. Rev. Lett. **86**, 393 (2001)
7. Y.Y. Keum, O.C.W. Kong, Phys. Rev. D **63**, 113012 (2001)
8. C.-C. Chiou, O.C.W. Kong, R.D. Vaidya, Phys. Rev. D **76**, 013003 (2007)
9. W. Siegel, Phys. Lett. B **84**, 193 (1979)
10. S.M. Barr, A. Zee, Phys. Rev. Lett. **65**, 21 (1990)
11. V. Barger, A. Das, C. Kao, Phys. Rev. D **55**, 7099 (1997)
12. A. Faessler, T. Gutsche, S. Kovalenko, V.E. Lyubovitskij, Phys. Rev. D **73**, 114023 (2006)
13. A. Faessler, T. Gutsche, S. Kovalenko, V.E. Lyubovitskij, Phys. Rev. D **74**, 074013 (2006)
14. I.B. Khriplovich, S.K. Lamoreaux, *CP Vioaltion Without Strangeness* (Springer, Berlin, 1997)
15. J.S.M. Ginges, V.V. Flambaum, Phys. Rept. **397**, 63 (2004)

Chapter 12
Classification of RPV Couplings and RPV Dependence to EDM Observables

12.1 Classification of RPV Contribution to the EDMs

Let us now discuss the general properties and the classification of the RPV contributions to the EDM observables. The P, CP-odd processes involved in the EDM do not change flavors and fermion numbers, so the relevant RPV interactions at the tree level should be combined so as to create and annihilate sneutrino in intermediate states, as we can see in Eqs. (11.26), (11.30) and (11.31). The relevant RPV interactions are then λ_{ijj} and λ'_{ikk}, where the first index indicates the flavor of the virtual sneutrino, and the subsequent two are those of the incident and exiting fermions, which have to be the same. Of course, two complex conjugate RPV interactions appear in each process, to avoid the fermion number violation of the total process. The possible RPV combinations are then *leptonic bilinears* ($\lambda_{ijj}\lambda^*_{ikk}$), *semi-leptonic bilinears* ($\lambda_{ijj}\lambda'^*_{ikk}$), and *hadronic bilinears* ($\lambda'_{ijj}\lambda^*_{ikk}$). In Table 12.1, we have written all possible RPV combinations relevant at the tree level, and the P, CP-odd processes sensitive to them.

Let us give a more phenomenologically refined classification. We have summarized in Table 12.2 the RPV bilinears classified using the sensitivity of the available EDM observables on them.

We see that each fermion EDMs and P, CP-odd 4-fermion interactions has its own dependence on the imaginary parts of RPV bilinears. When one RPV bilinear contributes to two or more P, CP-odd operators, the relative size of its effect to the final EDM observables has to be investigated. This depends of course on the final observables, but here we give a rough comparison. For RPV bilinears contributing simultaneously to the electron EDM and to the P, CP-odd electron-quark interaction, ($\mathrm{Im}(\lambda_{i11}\lambda'^*_{ijj})$ with $i = 2, 3$ and $j = 1, 2, 3$), the P, CP-odd electron-quark interaction has generally larger effect to the EDM of atoms. For hadronic RPV interactions ($\mathrm{Im}(\lambda'_{i11}\lambda'^*_{i22})$, $\mathrm{Im}(\lambda'_{i11}\lambda'^*_{i22})$ and $\mathrm{Im}(\lambda'_{i11}\lambda'^*_{i22})$ with $i = 1, 2, 3$), the quark chromo-EDM contribution to the hadronic P, CP violation (nucleon EDMs, nuclear Schiff moments) generally dominates over the d-quark EDM or the P, CP-odd 4-quark interactions. The dominance of the chromo-EDM over the EDM is due to the suppression

N. Yamanaka, *Analysis of the Electric Dipole Moment in the R-parity Violating Supersymmetric Standard Model*, Springer Theses,
DOI: 10.1007/978-4-431-54544-6_12, © Springer Japan 2014

Table 12.1 All possible combinations of RPV couplings which generate the fermion EDM, the chromo-EDM and the P, CP-odd 4-fermion interactions at the leading order, and P, CP-odd processes sensitive to them

	$j \backslash k$	λ^*_{ikk} 2	3
λ_{ijj}	1	$d_e, d_\mu, V_{e\mu}$	$d_e, d_\tau, V_{e\tau}$
	2	–	$d_\mu, d_\tau, V_{\mu\tau}$

	$j \backslash k$	λ'^*_{ikk} 1	2	3
λ_{ijj}	1	d_e, d_d, V_{ed}	d_e, d_s, V_{es}	d_e, d_b, V_{eb}
	2	$d_\mu, d_d, V_{\mu d}$	$d_\mu, d_s, V_{\mu s}$	$d_\mu, d_b, V_{\mu b}$
	3	$d_\tau, d_d, V_{\tau d}$	$d_\tau, d_s, V_{\tau s}$	$d_\tau, d_b, V_{\tau b}$

	$j \backslash k$	λ'^*_{ikk} 2	3
λ'_{ijj}	1	$d_d, d_s, d^c_d, d^c_s, V_{ds}$	$d_d, d_b, d^c_d, d^c_b, V_{db}$
	2	–	$d_s, d_b, d^c_s, d^c_b, V_{sb}$

d_f denotes the EDM of fermion f, d^c_q the chromo-EDM of the quark q, and $V_{ff'}$ the P, CP-odd 4-fermion interactions between fermions f and f'. The upper, middle and lower tables treat respectively the leptonic, semi-leptonic and hadronic RPV combinations

of the Barr-Zee type EDM contribution by the electromagnetic coupling $\alpha_{\rm em}$ and also by the electric charge of the d-type quarks. The dominance of the chromo-EDM over the P, CP-odd 4-quark interactions is however strongly model dependent and involves a large theoretical uncertainty. We must also note that the Barr-Zee type diagrams with electron or d-quark in the inner loop have a small contribution so we neglect them (these are denoted by "small" in Table 12.2).

With these criteria, we can classify the RPV bilinears contributing to the fermion EDMs and tree level P, CP-odd 4-fermion interactions into several types, depending on how they contribute to the P, CP-odd processes. Our classification is as follow:

- Type 1: Leptonic bilinears which contribute only to the electron EDM via the Barr-Zee diagram [$\mathrm{Im}(\lambda_{311}\lambda^*_{322})$ and $\mathrm{Im}(\lambda_{211}\lambda^*_{233})$]. The EDM of paramagnetic atoms and molecules are very sensitive to them.
- Type 2: Semi-leptonic bilinears involving electron which contribute both to the electron EDM and the P, CP-odd electron-nucleon (e-N) interactions [$\mathrm{Im}(\lambda_{i11}\lambda'^*_{i11})$, $\mathrm{Im}(\lambda_{i11}\lambda'^*_{i22})$ and $\mathrm{Im}(\lambda_{i11}\lambda'^*_{i33})$ $(i = 2, 3)$]. Atomic EDMs (paramagnetic and diamagnetic) are very sensitive to them.
- Type 3: Semi-leptonic bilinears involving d-quark and heavy leptons. These can be only constrained via the nucleon EDM [$\mathrm{Im}(\lambda_{i22}\lambda'^*_{i11})$ $(i = 1, 3)$ and $\mathrm{Im}(\lambda_{j33}\lambda'^*_{j11})$ $(j = 1, 2)$].
- Type 4: Hadronic bilinears. They contribute to the d-quark EDM, the chromo-EDM and to the P, CP-odd 4-quark interactions [$\mathrm{Im}(\lambda'_{i11}\lambda'^*_{i22})$, $\mathrm{Im}(\lambda'_{i11}\lambda'^*_{i33})$ and $\mathrm{Im}(\lambda'_{i22}\lambda'^*_{i33})$ $(i = 1, 2, 3)$]. Purely hadronic EDMs (nucleon EDM, bare nuclear EDM) are highly sensitive to them.

Table 12.2 Dependences of bilinears of RPV couplings to the EDM subprocesses (electron EDM denoted by d_e, muon EDM by d_μ, d-quark EDM by d_d, quark chromo-EDM by d_q^c, P, CP-odd electron-quark interaction by C_{eq}, P, CP-odd 4-quark interaction with d-quark by C_{qd})

	d_e	d_μ	d_d	d_q^c	C_{eq}	C_{qd}	Classification
$\mathrm{Im}(\lambda_{311}\lambda^*_{322})$	◯	small	–	–	–	–	Type 1
$\mathrm{Im}(\lambda_{211}\lambda^*_{233})$	◯	–	–	–	–	–	Type 1
$\mathrm{Im}(\lambda_{122}\lambda^*_{133})$	–	◯	–	–	–	–	Type 5
$\mathrm{Im}(\lambda_{i11}\lambda'^*_{i11})$ $(i=2,3)$	small	–	small	–	◯	–	Type 2
$\mathrm{Im}(\lambda_{i11}\lambda'^*_{i22})$ $(i=2,3)$	◯	–	–	–	◯	–	Type 2
$\mathrm{Im}(\lambda_{i11}\lambda'^*_{i33})$ $(i=2,3)$	◯	–	–	–	◯	–	Type 2
$\mathrm{Im}(\lambda_{i22}\lambda'^*_{i11})$ $(i=1,3)$	–	small	◯	–	–	–	Type 3
$\mathrm{Im}(\lambda_{i22}\lambda'^*_{i22})$ $(i=1,3)$	–	◯	–	–	–	–	Type 5
$\mathrm{Im}(\lambda_{i22}\lambda'^*_{i33})$ $(i=1,3)$	–	◯	–	–	–	–	Type 5
$\mathrm{Im}(\lambda_{i33}\lambda'^*_{i11})$ $(i=1,2)$	–	–	◯	–	–	–	Type 3
$\mathrm{Im}(\lambda_{i33}\lambda'^*_{i22})$ $(i=1,2)$	–	–	–	–	–	–	Type 6
$\mathrm{Im}(\lambda_{i33}\lambda'^*_{i33})$ $(i=1,2)$	–	–	–	–	–	–	Type 6
$\mathrm{Im}(\lambda'_{i11}\lambda'^*_{i22})$ $(i=1,2,3)$	–	–	◯	◯	–	◯	Type 4
$\mathrm{Im}(\lambda'_{i11}\lambda'^*_{i33})$ $(i=1,2,3)$	–	–	◯	◯	–	◯	Type 4
$\mathrm{Im}(\lambda'_{i22}\lambda'^*_{i33})$ $(i=1,2,3)$	–	–	–	◯	–	–	Type 4

All combinations of RPV couplings relevant in the Barr-Zee type diagrams and P, CP-odd 4-fermion interactions at the tree level were listed. Circles (◯) denote sizable sensitivity and minus signs (−) denote the null sensitivity of the RPV couplings to the P, CP-odd operators (fermion EDM or P, CP-odd 4-fermion interactions). The classification of RPV bilinears are also shown in the last column

- Type 5: Bilinears which contribute only to muon EDM [$\mathrm{Im}(\lambda_{122}\lambda^*_{133})$, $\mathrm{Im}(\lambda_{i22}\lambda'^*_{i22})$ and $\mathrm{Im}(\lambda_{i22}\lambda'^*_{i33})$ $(i=1,3)$].
- Type 6: Remaining RPV bilinears which cannot be constrained in this analysis [$\mathrm{Im}(\lambda_{i33}\lambda'^*_{i22})$ and $\mathrm{Im}(\lambda_{i33}\lambda'^*_{i33})$ $(i=1,2)$]. They are expected to contribute to the EDMs of τ, or s and b quarks, but we have not considered them in our discussion, since these couplings are not actually accessible experimentally.

The classification of RPV bilinears is also shown in Table 12.2.

12.2 Dependences of EDM Observables on RPV Couplings

Let us now review the dependences of the EDM observables on RPV couplings relevant to our analysis. In our analysis there are 10 relevant RPV variables. For their choice, it is adequate to choose the combinations of RPV bilinears as follows

$$x_1 = \mathrm{Im}(\lambda_{311}\lambda^*_{322}),$$
$$x_2 = \mathrm{Im}(\lambda_{211}\lambda^*_{233}),$$

$$
\begin{aligned}
x_3 &= \sum_{i=2,3} \mathrm{Im}(\lambda_{i11}\lambda'^{*}_{i11}),\\
x_4 &= \sum_{i=2,3} \mathrm{Im}(\lambda_{i11}\lambda'^{*}_{i22}),\\
x_5 &= \sum_{i=2,3} \mathrm{Im}(\lambda_{i11}\lambda'^{*}_{i33}),\\
x_6 &= \sum_{i=1,3} \mathrm{Im}(\lambda_{i22}\lambda'^{*}_{i11}),\\
x_7 &= \sum_{i=1,2} \mathrm{Im}(\lambda_{i33}\lambda'^{*}_{i11}),\\
x_8 &= \sum_{i=1,2,3} \mathrm{Im}(\lambda'_{i11}\lambda'^{*}_{i22}),\\
x_9 &= \sum_{i=1,2,3} \mathrm{Im}(\lambda'_{i11}\lambda'^{*}_{i33}),\\
x_{10} &= \sum_{i=1,2,3} \mathrm{Im}(\lambda'_{i22}\lambda'^{*}_{i33}).
\end{aligned}
\tag{12.1}
$$

Explicitly, x_1 and x_2 belong to the type 1, x_3, x_4 and x_5 to the type 2, x_6 and x_7 to the type 3, and finally x_8, x_9 and x_{10} to the type 4. The type 5 and 6 can be omitted from the linear programming, since they are not related at all with the other observables. These x_i's are unknown parameters to be determined in our analysis, and this notation will be used hereafter.

Let us now present the gradients of EDM-observables. In this analysis, the P, CP-odd observables (EDMs of the ^{205}Tl, ^{199}Hg, ^{129}Xe atoms, neutron, proton, deuteron, ^{3}He nucleus, ^{211}Rn, ^{225}Ra atoms, muon, and the R-correlation of the neutron beta decay) were taken up to the first order in RPV bilinears, so it is possible to express the observables as

$$
d_a = \sum_i c_{ai} x_i, \tag{12.2}
$$

where a is the label of the system (for example, He for ^{3}He nucleus). The coefficients c_{ai} are determined from the elementary, hadronic, nuclear and atomic calculations reviewed in this thesis. The components of the gradient vector of each P, CP-odd observables are the coefficients c_{ai}'s. The relevant coefficient in this analysis are shown in Tables 12.3 and 12.4. Large coefficient c_{ai} means that the EDM-observable a is sensitive to the corresponding RPV variable x_i.

The values shown in Tables 12.3 and 12.4 can be derived by following the discussion of Part II, but here we should review their characteristics briefly.

Table 12.3 Coefficients of proportionality $c_{ai} \equiv \frac{\partial d_a}{\partial x_i}$ for the EDMs of electron, ^{205}Tl, ^{199}Hg, ^{129}Xe atoms and neutron

Obs.	d_e	d_{Tl}	d_{Hg}	d_n	d_{Xe}
c_{a1}	4.9×10^{-25}	2.9×10^{-22}	3.9×10^{-27}	0	4.8×10^{-28}
c_{a2}	5.5×10^{-24}	3.2×10^{-21}	4.3×10^{-26}	0	5.4×10^{-27}
c_{a3}	0	-3.1×10^{-18}	-1.0×10^{-21}	0	-1.8×10^{-22}
c_{a4}	1.6×10^{-25}	-4.2×10^{-20}	6.2×10^{-24}	0	2.1×10^{-24}
c_{a5}	3.6×10^{-24}	-2.1×10^{-21}	-1.7×10^{-25}	0	-5.1×10^{-27}
c_{a6}	0	0	2.9×10^{-29}	2.2×10^{-25}	-2.7×10^{-30}
c_{a7}	0	0	3.2×10^{-28}	2.4×10^{-24}	-2.9×10^{-29}
c_{a8}	0	0	9.4×10^{-26}	9.3×10^{-23}	-9.5×10^{-27}
c_{a9}	0	0	2.7×10^{-24}	2.1×10^{-21}	-2.8×10^{-25}
c_{a10}	0	0	3.1×10^{-25}	2.4×10^{-21}	-2.6×10^{-26}

The sneutrino mass was taken to be 1 TeV. The unit is in e cm

Table 12.4 Coefficients of proportionality $c_{ai} \equiv \frac{\partial d_a}{\partial x_i}$ for the EDMs of proton, deuteron, ^{3}He nucleus, ^{211}Rn, ^{225}Ra atoms and the R-correlation of the neutron beta decay

Obs.	d_p	d_d	d_{He}	d_{Rn}	d_{Ra}	R
c_{a1}	0	0	0	-5.3×10^{-27}	-2.1×10^{-26}	0
c_{a2}	0	0	0	-5.9×10^{-26}	-2.3×10^{-25}	0
c_{a3}	0	0	0	1.3×10^{-21}	7.8×10^{-21}	-1.1×10^{-2}
c_{a4}	0	0	0	-6.5×10^{-24}	-8.6×10^{-23}	0
c_{a5}	0	0	0	2.6×10^{-25}	3.5×10^{-25}	0
c_{a6}	-5.4×10^{-26}	1.6×10^{-25}	2.0×10^{-25}	-1.1×10^{-28}	0	0
c_{a7}	-6.0×10^{-25}	1.8×10^{-24}	1.3×10^{-24}	-1.2×10^{-27}	0	0
c_{a8}	-9.4×10^{-23}	6.1×10^{-22}	9.8×10^{-22}	-2.9×10^{-25}	-2.2×10^{-22}	0
c_{a9}	-2.1×10^{-21}	1.8×10^{-20}	2.9×10^{-20}	-8.1×10^{-24}	-6.4×10^{-21}	0
c_{a10}	-1.5×10^{-21}	9.3×10^{-22}	2.2×10^{-21}	-1.2×10^{-24}	0	0

The sneutrino mass was taken to be 1 TeV. The unit for EDM observables is in e cm

- General comment on atomic EDMs:
  Atomic EDMs are very sensitive to the P, CP-odd e-N interaction contributions (c_3, c_4 and c_5), and in particular to c_3 which issues from the P, CP-odd electron-d-quark interaction.
- Electron EDM:
  The electron EDM is not a direct observable, but we have included it in the list because its upper limit derived from the EDM experiment of YbF molecule was reported in Ref. [1].
- ^{205}Tl EDM:
  Coefficients c_1, c_2, c_3, c_4 and c_5 are very large, since ^{205}Tl EDM is highly sensitive to the electron EDM (c_1 and c_2) and to the P, CP-odd e-N interactions (c_3, c_4 and c_5). The ^{205}Tl EDM should also receive contribution from the hadronic RPV bilinears (c_6, c_7, c_8, c_9 and c_{10}) via the nuclear Schiff moment. These coefficients

however receive cancellation from two approximately equal terms [see Eqs. (7.34) and (7.38)]. In our analysis we just omit the effect of the ^{205}Tl Schiff moment.

- Nucleon EDM:
  The nucleon EDM does not receive contribution from the purely leptonic RPV bilinears x_1 and x_2. Here x_3, x_4 and x_5 also do not contribute to the nucleon EDM, since we have neglected the Barr-Zee type contribution with electron in the inner loop.
- Deuteron and ^{3}He EDM:
  As for the nucleon EDM, the nuclear EDM does not receive contributions from x_1 and x_2, x_3, x_4 and x_5. Note also that for the deuteron EDM, c_{10} is rather suppressed compared to the nucleon EDM. This is due to the fact that c_{10} is generated from the chromo-EDM of the strange quark with smaller isoscalar contribution. This makes the gradient of the deuteron EDM less aligned compared with the nucleon or ^{3}He EDMs. The EDM of ^{3}He has larger sensitivity to the RPV bilinears x_6, x_7, x_8, x_9 and x_{10} than nucleon and deuteron EDMs.
- Leptonic contribution to the EDM of diamagnetic atoms (^{129}Xe, ^{199}Hg, ^{211}Rn and ^{225}Ra):
  The diamagnetic atoms has closed electron shell, so the electron EDM cannot contribute to the atomic EDM directly, and consequently the coefficients c_1 and c_2 are suppressed. It should also be noted that the sensitivity to P, CP-odd e-N interactions (c_3, c_4 and c_5) are also more suppressed than that of the paramagnetic atom (^{205}Tl), for the same reason.
- Nuclear Schiff moment of diamagnetic atoms (^{129}Xe, ^{199}Hg, ^{211}Rn and ^{225}Ra):
  The diamagnetic atoms relevant to this analysis have all odd neutron number, so hadronic RPV bilinears (x_6, x_7, x_8, x_9 and x_{10}) can contribute to the nuclear Schiff moment. We should be careful for c_6, c_7 and c_{10} which are known from the EDM of the valence nucleon. These coefficients receive contribution from two approximately equal terms with opposite sign for spin one half s-wave nuclei [see Eqs. (7.34) and (7.38)]. Therefore, these coefficients for ^{129}Xe and ^{225}Ra should be vanishing. For ^{129}Xe, we have used the result of the shell model [2] which predicts non-zero c_6, c_7 and c_{10}. For ^{225}Ra, the shell model picture is not applicable due to the deformation of the nucleus, and no results were available, so we have set c_6, c_7 and c_{10} to zero. It should also be noted that the ^{225}Ra has particularly large c_8, c_9 compared to other diamagnetic atoms. This is due to the octupole deformation (which induces close levels of opposite parity) which enhances the nuclear Schiff moment by about 1000 times [3].
- R-correlation:
  The R-correlation of the neutron beta decay is the decay asymmetry (see Appendix G for detail) which is odd under P and CP [4, 5], and the SM contribution is known to be very small [6], like the EDM. It is then also a good probe of new physics [7], so we have included it in our analysis. The R-correlation is sensitive to the semi-leptonic combination $\mathrm{Im}(\lambda_{i11}\lambda'^*_{i11})$ $(i = 2, 3)$.

References

1. J.J. Hudson et al., Nature **473**, 493 (2011)
2. N. Yoshinaga et al., inTalk Given at 4th International Conference on Fundamental Physics Using Atoms, Osaka University, Aug 2010 (http://xqw.hep.okayama-u.ac.jp/kakenhi/index.php/fpua2010/fpua2010_top_e/)
3. J. Dobaczewski, J. Engel, Phys. Rev. Lett. **94**, 232502 (2005)
4. J. Jackson, S. Treiman, H. Wyld, Phys. Rev. **106**, 517 (1957)
5. J. Jackson, S. Treiman, H. Wyld, Nucl. Phys. **4**, 206 (1957)
6. P. Herczeg, I.B. Khriplovich, Phys. Rev. D **56**, 80 (1997)
7. P. Herczeg, Prog. Part. Nucl. Phys. **46**, 413 (2001)

Chapter 13
Reappraisal of Constraints on R-parity Violation from EDM at the Leading Order

13.1 Analysis and Results

To obtain upper limits on RPV couplings, it suffices to divide the experimental upper bounds on the EDMs of electron, neutron, ^{205}Tl and ^{199}Hg atoms by the coefficients of Table 12.3. If several EDM experimental data give upper limits for one RPV bilinears, we take the tightest limit. We will see that in some cases, the upper limits are changed from previous analyses, due to the use of our correct formula for the Barr-Zee type contribution derived in Chap. 11.

Type 1

The first type is the leptonic bilinears $\lambda_{311}\lambda^*_{322}$ and $\lambda_{211}\lambda^*_{233}$ which contribute exclusively to the electron EDM via the Barr-Zee type diagram [see Eq. (11.26)]. The electron EDM is actually strongly constrained by the result of the EDM experiment using YbF molecules and the current limit is $d_e < 1.05 \times 10^{-27} e\,\mathrm{cm}$ [1]. From the calculation of the RPV contribution above, we can give the upper limits shown in Table 13.1 [we have also shown the limits given to the semi-leptonic RPV bilinears which contribute to the electron EDM (type 2)]. The limits, which are from the EDM of ^{205}Tl, a paramagnetic atom sensitive to the electron EDM, are also displayed for comparison.

Type 2

The next type is the semi-leptonic RPV bilinears (which involve the electron) of the form $\lambda_{i11}\lambda'^*_{ikk}$ ($i = 2, 3$ and $k = 1, 2, 3$). They can be constrained by the P, CP-odd e-N interactions, as was first pointed by Herczeg [4]. These interactions are constrained via the electron Barr-Zee type graph, and also by the P, CP-odd electron-nucleon (e-N) interactions. The P, CP-odd e-N interactions [see Eq. (8.3)]

N. Yamanaka, *Analysis of the Electric Dipole Moment in the R-parity Violating Supersymmetric Standard Model*, Springer Theses,
DOI: 10.1007/978-4-431-54544-6_13, © Springer Japan 2014

Table 13.1 Upper limits on the absolute value of combinations of RPV couplings ($i = 2, 3$) via the electron EDM

	$\lvert\mathrm{Im}(\lambda_{322}\lambda^*_{311})\rvert$	$\lvert\mathrm{Im}(\lambda_{233}\lambda^*_{211})\rvert$	$\lvert\mathrm{Im}(\lambda'_{i22}\lambda^*_{i11})\rvert$	$\lvert\mathrm{Im}(\lambda'_{i33}\lambda^*_{i11})\rvert$
Limits to RPV from YbF molecule				
Upper limits ($m_{\tilde{\nu}} = 1$ TeV)	2.1×10^{-3}	1.9×10^{-4}	6.6×10^{-3}	2.9×10^{-4}
Upper limits ($m_{\tilde{\nu}} = 5$ TeV)	4.5×10^{-2}	3.7×10^{-3}	1.4×10^{-1}	5.4×10^{-3}
Upper limits ($m_{\tilde{\nu}} = 100$ GeV)	2.9×10^{-5}	3.3×10^{-6}	9.2×10^{-4}	5.9×10^{-6}
Formulae of Ref. [2] ($m_{\tilde{\nu}} = 100$ GeV)	1.9×10^{-6}	3.3×10^{-7}	6.0×10^{-6}	6.6×10^{-7}
Limits to RPV from ^{205}Tl EDM				
Upper limits ($m_{\tilde{\nu}} = 1$ TeV)	3.1×10^{-3}	2.8×10^{-4}	9.7×10^{-3}	4.2×10^{-4}
Upper limits ($m_{\tilde{\nu}} = 5$ TeV)	6.6×10^{-2}	5.5×10^{-3}	2.1×10^{-1}	8.2×10^{-3}
Upper limits ($m_{\tilde{\nu}} = 100$ GeV)	4.4×10^{-5}	4.9×10^{-6}	1.4×10^{-4}	8.6×10^{-6}
Formulae of Ref. [2] ($m_{\tilde{\nu}} = 100$ GeV)	2.8×10^{-6}	4.8×10^{-7}	8.7×10^{-6}	9.7×10^{-7}

The upper table shows limits given from the experimental analysis of the YbF molecule. The lower table shows limits deduced from the EDM of ^{205}Tl atom [3] using the relation (8.31)

can be given by multiplying the P, CP-odd electron-quark interaction of Eq. (11.31) with the quark contents of nucleon. In this work, we use the scalar and pseudoscalar contents of the nucleon derived in Sect. 6.1 [see Eqs. (6.12), (6.13), (6.17), (6.21) and (6.23)] [5–7]. The EDM of atoms is a very sensitive probe of the P, CP-odd e-N interactions. The type 2 RPV bilinears receive the tightest upper bounds from the EDM of ^{199}Hg atom [8]. The dependence of the EDM of ^{199}Hg atom on the P, CP-odd e-N interactions are given in Eq. (8.32) [9, 10]. The upper limits are shown in Table 13.2.

We see that the EDM of the ^{199}Hg atom gives a very tight constraints on the CP phases between RPV couplings. We must be careful for $\mathrm{Im}(\lambda_{i11}\lambda'^*_{i33})$, since the Barr-Zee graph contributing to electron EDM can give also tight constraints on these same RPV couplings. The previous analysis pointed that the Barr-Zee graph can give tighter constraints [4], but the formula used in this analysis was not correct [11]. By using Eq. (11.26), we see that the EDM of ^{199}Hg atom can give through the P, CP-odd electron-nucleon tighter constraints than that from the electron EDM using the YbF molecule experiment.

Type 3

For the type 3, RPV bilinears can be constrained by the Barr-Zee type diagram of the d-quark EDM with heavy fermions in the loop. The d-quark EDM contributes to the nucleon EDM. The strongest limit on the type 3 RPV bilinears is given by the

Table 13.2 Upper limits on the absolute values of combinations of RPV couplings from the EDM of ^{199}Hg atom via the P, CP-odd electron-nucleon interactions

RPV couplings	$\lvert\mathrm{Im}(\lambda_{i11}\lambda'^*_{i11})\rvert$	$\lvert\mathrm{Im}(\lambda_{i11}\lambda'^*_{i22})\rvert$	$\lvert\mathrm{Im}(\lambda_{i11}\lambda'^*_{i33})\rvert$
Upper limits	$7.9 \times 10^{-10}[m_{\tilde{\nu}}]^2$	$3.6 \times 10^{-8}[m_{\tilde{\nu}}]^2$	$1.8 \times 10^{-6}[m_{\tilde{\nu}}]^2$

$i = 2, 3$. $[m_{\tilde{\nu}}]$ is the mass of sneutrino in unit of 100 GeV. Upper limits were given independently

Table 13.3 Upper limits on the absolute value of combinations of RPV couplings from neutron EDM via the Barr-Zee type graph of the d quark. Here $i = 2, 3$

RPV couplings	$\lvert\mathrm{Im}(\lambda_{i22}\lambda'^{*}_{i11})\rvert$	$\lvert\mathrm{Im}(\lambda_{i33}\lambda'^{*}_{i11})\rvert$	$\lvert\mathrm{Im}(\lambda'_{i22}\lambda'^{*}_{i11})\rvert$	$\lvert\mathrm{Im}(\lambda'_{i33}\lambda'^{*}_{i11})\rvert$
Upper limits ($m_{\tilde{\nu}} = 1\,\mathrm{TeV}$)	1.3×10^{-1}	1.2×10^{-2}	4.2×10^{-1}	1.8×10^{-2}
Upper limits ($m_{\tilde{\nu}} = 5\,\mathrm{TeV}$)	2.8	2.3×10^{-1}	8.7	3.3×10^{-1}
Upper limits ($m_{\tilde{\nu}} = 100\,\mathrm{GeV}$)	1.7×10^{-3}	2.0×10^{-4}	5.5×10^{-3}	3.5×10^{-4}
Formulae of Ref. [2] ($m_{\tilde{\nu}} = 100\,\mathrm{GeV}$)	1.1×10^{-4}	1.9×10^{-5}	3.5×10^{-4}	3.9×10^{-5}

neutron EDM experimental data [12]. The corresponding upper bounds are shown in Table 13.3 (we have also shown hadronic type bilinears which contribute also to the electromagnetic Barr-Zee type graphs of the d-quark). We have used the relation (6.83) to obtain the dependence of the neutron EDM on the d-quark EDM.

The limits given here by the single neutron EDM are rather moderate. One of the simple explanation is the absence of some large enhancement mechanism like the electron EDM contribution to the paramagnetic atoms. Nevertheless, these semi-leptonic RPV bilinears can only be constrained efficiently with the neutron EDM. They can also be constrained from the atomic EDM, but the nucleon EDM contribution suffers from Schiff's screening which weakens their effect.

Type 4

The fourth type is the hadronic RPV bilinears ($\lambda'_{ijj}\lambda'^{*}_{ikk}$, $i, j, k = 1, 2, 3$ with $j \neq k$). These can be strongly constrained by the purely hadronic EDMs and the Schiff moments of diamagnetic atoms. The elementary level processes involve the Barr-Zee type chromo-EDM (11.30) and the P, CP-odd 4-quark interaction (11.31). These two processes contribute to the nucleon EDM and the P, CP-odd pion-nucleon interactions, and their relations are given by Eqs. (6.81), (6.82), (6.83) and (6.84). The tightest limits on type 4 RPV bilinears are given by the neutron [12] and the ^{199}Hg EDM [8] experimental data. The relation between the above P, CP-odd hadron level interactions and the ^{199}Hg EDM is given by Eqs. (7.56) [13] and (8.32) [9]. For the dependence of the ^{199}Hg Schiff moment, we have used the average of Table 7.1. We have shown in Table 13.4 the upper limits to the hadronic RPV bilinears from the neutron and ^{199}Hg EDM. We can see the tightness of the constraints by comparing the

Table 13.4 Upper limits on the absolute value of combinations of hadronic RPV bilinears from neutron EDM and ^{199}Hg atomic EDM ($i = 1, 2, 3$)

RPV couplings	$\lvert\mathrm{Im}(\lambda'_{i11}\lambda'^{*}_{i22})\rvert$	$\lvert\mathrm{Im}(\lambda'_{i11}\lambda'^{*}_{i33})\rvert$	$\lvert\mathrm{Im}(\lambda'_{i22}\lambda'^{*}_{i33})\rvert$
Upper limits ($m_{\tilde{\nu}} = 1\,\mathrm{TeV}$)	3.1×10^{-4}	1.1×10^{-5}	1.2×10^{-5}
Upper limits ($m_{\tilde{\nu}} = 5\,\mathrm{TeV}$)	6.5×10^{-3}	2.1×10^{-4}	2.2×10^{-4}
Upper limits ($m_{\tilde{\nu}} = 100\,\mathrm{GeV}$)	4.3×10^{-6}	2.3×10^{-7}	2.4×10^{-7}
Formulae of Ref. [2] ($m_{\tilde{\nu}} = 100\,\mathrm{GeV}$)	2.5×10^{-7}	2.6×10^{-8}	2.7×10^{-8}

limits to the same RPV bilinears given via electromagnetic Barr-Zee (see Table 13.3), which are of 3 to 4 orders larger in magnitudes. This large ratio can be explained by the suppression of the electromagnetic coupling in the Barr-Zee contribution. The ratio between Eqs. (11.30) and (11.26) is suppressed both by the electromagnetic coupling α_{em} and by the square of the d-type quark charge, which renders the chromo-EDM contribution to be important. The neutron and ^{199}Hg EDMs gave close upper limits to the RPV bilinears, except for $\mathrm{Im}(\lambda'_{i22}\lambda'^*_{i33})$. This is because this bilinear are from the chromo-EDM of the strange quark. In our analysis, the strange contribution enters only via the nucleon EDM. As the nucleon EDM contribution is suppressed by Schiff's screening for the ^{199}Hg atom, the strange contribution is relatively suppressed. Our result shows the importance of the strangeness, but this result, together with the u and d contributions, suffers from large theoretical uncertainties, including the uncertainty of the quark mass, value of the quark condensates and also the calculation of the Schiff moment for the ^{199}Hg EDM. Taking all these topics into account, the result may change even by orders of magnitude. The result shows nevertheless the large sensitivity of the neutron and ^{199}Hg EDMs against the purely hadronic RPV bilinears. We should also add some discussions concerning the relative size between the chromo-EDM and the P, CP-odd 4-quark interaction. In this case also, the correct size of the chromo-EDM is smaller than that used in previous analyses [11]. In the present case however, the quantitative comparison of the two contributions is meaningless because of the large theoretical uncertainty of the hadronic and nuclear level calculations.

Type 5

The type 5 RPV bilinears are those constrained by the muon EDM and include $\mathrm{Im}(\lambda_{122}\lambda^*_{133})$, $\mathrm{Im}(\lambda_{i22}\lambda'^*_{i22})$, $\mathrm{Im}(\lambda_{i22}\lambda'^*_{i33})$. These bilinears contribute only via the Barr-Zee type diagram of the muon EDM [see Eq. (11.26)]. The muon EDM is particular in the sense that at the leading order, its dependence on RPV bilinear has no overlap with other available observables. This means that the type 5 RPV bilinears can be constrained only by muon EDM. Unfortunately, constraints from the muon EDM experimental data cannot be discussed, since the present experimental sensitivity is too weak to give any limit on RPV interactions.

Type 6

The last type of RPV bilinears, $\lambda_{i33}\lambda'^*_{i22}$ and $\lambda_{i33}\lambda'^*_{i33}$ cannot be probed by the currently available EDM experimental data, and nor by any projected experiments. One possibility to probe them is the tau EDM, which may be technically difficult.

Table 13.5 Upper limits on the absolute value of the combinations of RPV bilinears from presently available experimental data. Results were given for $m_{SUSY} = 1\text{TeV}$

RPV couplings		Limits from other exp.	Limits from EDMs	EDM observables used
Type 1	$\lvert\text{Im}(\lambda_{311}\lambda^*_{322})\rvert$	0.15	2.1×10^{-3}	YbF molecule [1]
	$\lvert\text{Im}(\lambda_{211}\lambda^*_{233})\rvert$	0.25	1.9×10^{-4}	YbF molecule [1]
Type 2	$\lvert\text{Im}(\lambda_{211}\lambda'^*_{211})\rvert$	2.9×10^{-2}	7.9×10^{-8}	^{199}Hg EDM [8]
	$\lvert\text{Im}(\lambda_{311}\lambda'^*_{311})\rvert$	3.6×10^{-3}	7.9×10^{-8}	^{199}Hg EDM [8]
	$\lvert\text{Im}(\lambda_{211}\lambda'^*_{222})\rvert$	2.9×10^{-2}	3.6×10^{-6}	^{199}Hg EDM [8]
	$\lvert\text{Im}(\lambda_{311}\lambda'^*_{322})\rvert$	1.7×10^{-2}	3.6×10^{-6}	^{199}Hg EDM [8]
	$\lvert\text{Im}(\lambda_{211}\lambda'^*_{233})\rvert$	0.70	1.8×10^{-4}	^{199}Hg EDM [8]
	$\lvert\text{Im}(\lambda_{311}\lambda'^*_{333})\rvert$	0.36	1.8×10^{-4}	^{199}Hg EDM [8]
Type 3	$\lvert\text{Im}(\lambda_{122}\lambda'^*_{111})\rvert$	4.4×10^{-2}	1.2×10^{-1}	neutron EDM [12]
	$\lvert\text{Im}(\lambda_{322}\lambda'^*_{311})\rvert$	6.0×10^{-3}	1.2×10^{-1}	neutron EDM [12]
	$\lvert\text{Im}(\lambda_{133}\lambda'^*_{111})\rvert$	2.6×10^{-2}	1.1×10^{-2}	neutron EDM [12]
	$\lvert\text{Im}(\lambda_{233}\lambda'^*_{211})\rvert$	2.9×10^{-2}	1.1×10^{-2}	neutron EDM [12]
Type 4	$\lvert\text{Im}(\lambda'_{111}\lambda'^*_{122})\rvert$	5.0×10^{-3}	3.1×10^{-4}	neutron EDM [12] or ^{199}Hg EDM [8]
	$\lvert\text{Im}(\lambda'_{211}\lambda'^*_{222})\rvert$	3.2×10^{-3}	3.1×10^{-4}	neutron EDM [12] or ^{199}Hg EDM [8]
	$\lvert\text{Im}(\lambda'_{311}\lambda'^*_{322})\rvert$	6.8×10^{-4}	3.1×10^{-4}	neutron EDM [12] or ^{199}Hg EDM [8]
	$\lvert\text{Im}(\lambda'_{111}\lambda'^*_{133})\rvert$	3.1×10^{-4}	1.1×10^{-5}	neutron EDM [12] or ^{199}Hg EDM [8]
	$\lvert\text{Im}(\lambda'_{211}\lambda'^*_{233})\rvert$	8.0×10^{-2}	1.1×10^{-5}	neutron EDM [12] or ^{199}Hg EDM [8]
	$\lvert\text{Im}(\lambda'_{311}\lambda'^*_{333})\rvert$	1.4×10^{-2}	1.1×10^{-5}	neutron EDM [12] or ^{199}Hg EDM [8]
	$\lvert\text{Im}(\lambda'_{122}\lambda'^*_{133})\rvert$	2.0×10^{-4}	1.2×10^{-5}	neutron EDM [12]
	$\lvert\text{Im}(\lambda'_{222}\lambda'^*_{233})\rvert$	8.0×10^{-2}	1.2×10^{-5}	neutron EDM [12]
	$\lvert\text{Im}(\lambda'_{322}\lambda'^*_{333})\rvert$	6.8×10^{-2}	1.2×10^{-5}	neutron EDM [12]
Type 5	$\lvert\text{Im}(\lambda_{122}\lambda^*_{133})\rvert$	0.15	–	muon EDM
	$\lvert\text{Im}(\lambda_{122}\lambda'^*_{122})\rvert$	2.9×10^{-2}	–	muon EDM
	$\lvert\text{Im}(\lambda_{322}\lambda'^*_{322})\rvert$	2.9×10^{-2}	–	muon EDM
	$\lvert\text{Im}(\lambda_{122}\lambda'^*_{133})\rvert$	1.8×10^{-3}	–	muon EDM
	$\lvert\text{Im}(\lambda_{322}\lambda'^*_{333})\rvert$	0.60	–	muon EDM
Type 6	$\lvert\text{Im}(\lambda_{133}\lambda'^*_{122})\rvert$	1.7×10^{-2}	–	–
	$\lvert\text{Im}(\lambda_{233}\lambda'^*_{222})\rvert$	2.9×10^{-2}	–	–
	$\lvert\text{Im}(\lambda_{133}\lambda'^*_{133})\rvert$	1.0×10^{-3}	–	–
	$\lvert\text{Im}(\lambda_{233}\lambda'^*_{233})\rvert$	0.70	–	–

13.2 Summary

In this chapter, we have discussed the contributions of the R-parity violating (RPV) interactions to the EDM of the neutron, ^{205}Tl, ^{199}Hg atoms, and YbF molecule, and have given limits to the RPV interactions by assuming the dominance of single bilinear of RPV couplings. In doing this, we have calculated the correct contribution of the two-loop level Barr-Zee type EDM diagram, and the result has given one order of magnitude looser limits to products of RPV couplings. This point merits some attention, because the relative size between the contribution from the P, CP-odd 4-fermion interaction is changed, and the phenomenological analysis was altered for the P, CP-odd electron-nucleon interactions. We have also classified the bilinears of RPV couplings into 6 types, and presented their characteristics in detail. This has made clear the interrelations between the RPV couplings and the EDM observables. We give in Table 13.5 the summary of the currently available experimental upper limits to all RPV bilinears considered so far. We show also limits on RPV couplings provided by other experiments.

We note once again that these limits were derived by assuming that only one bilinear of RPV couplings is dominant in each analysis. The phenomenological analysis with the consideration of the whole RPV parameter space should also be done, as was done in the analysis of the EDM constraints within R-parity conserving supersymmetry by Ellis et al. [14]. This is the subject of the Chap. 14.

References

1. J.J. Hudson et al., Nature **473**, 493 (2011)
2. D. Chang, W.-F. Chang, M. Frank, W.-Y. Keung, Phys. Rev. D **62**, 095002 (2000)
3. B.C. Regan et al., Phys. Rev. Lett. **88**, 071805 (2002)
4. P. Herczeg, Phys. Rev. D **61**, 095010 (2000)
5. J.F. Donoghue, C.R. Nappi, Phys. Lett. B **168**, 105 (1986)
6. K. Takeda, S. Aoki, S. Hashimoto, T. Kaneko, T. Onogi, N. Yamada, JLQCD Collaboration, in *PoS LATTICE2010*, 160 (2010) [arXiv:1012.1907 [hep-lat]]
7. T.P. Cheng, L.F. Li, Phys. Rev. Lett. **62**, 1441 (1989)
8. W.C. Griffith et al., Phys. Rev. Lett. **102**, 101601 (2009)
9. V.A. Dzuba, V.V. Flambaum, S.G. Porsev, Phys. Rev. A **80**, 032120 (2009)
10. V.V. Flambaum, I.B. Khriplovich, Zh. Eksp. Teor. Fiz. **89**, 1505 (1985)
11. N. Yamanaka, T. Sato, T. Kubota, Phys. Rev. D **85**, 117701 (2012)
12. C.A. Baker et al., Phys. Rev. Lett. **97**, 131801 (2006)
13. S. Ban, J. Dobaczewski, J. Engel, A. Shukla, Phys. Rev. C **82**, 015501 (2010)
14. J. Ellis, J.S. Lee, A. Pilaftsis, JHEP **1010**, 049 (2010)
15. A.R. Zhitnitsky, Phys. Rev. D **55**, 3006 (1997)
16. H. Ohki et al., JLQCD Collaboration. Phys. Rev. D **78**, 054502 (2008)
17. K. Takeda et al., Phys. Rev. D **83**, 114506 (2011)
18. V.V. Flambaum, I.B. Khriplovich, Sov. Phys. JETP **62**, 872 (1985)
19. J. Ellis, J.S. Lee, A. Pilaftsis, JHEP **1102**, 045 (2011)
20. J. Ellis, J.S. Lee, A. Pilaftsis, arXiv:1009.1151 [math.OC]

Chapter 14
Analysis of the Maximal CP Violation of RPV Interactions within ^{205}Tl, ^{199}Hg, ^{129}Xe and Neutron EDM-Constraints Using Linear Programming Method

The parameter space of the R-parity violation (RPV) is quite large, and the analysis of the full parameter space including the usual R-parity conserving parameters is difficult. In such situations, we often allow only few parameters in the parameter space to vary for tractable phenomenological analyses, as was done in many previous works (assumptions of single coupling dominance) from which many tight constraints of the RPV couplings were derived. This approach assuming the single coupling dominance cannot however exclude regions of the RPV parameter space in which interferences occur, and some couplings may be sufficiently larger than upper limits derived with this assumption. In the R-parity conserving sector, a systematic analysis of the CP phases of the supersymmetry (SUSY) breaking terms was done by Ellis et al., and they obtained a large prediction for many prepared experiments such as the electric dipole moments (EDMs) of the deuteron or ^{225}Ra atom [1–3] (see Chap. 13). If the supersymmetric theory is extended with RPV, an equally full analysis for RPV interactions seems to be needed.

In this chapter, we do a systematic analysis of the full space of the CP violating RPV interactions by using the constraints given by existing experimental data to derive upper limits on RPV couplings without the assumption of the dominance of single RPV bilinears. To do an efficient analysis, we base on the linear programming method to scan the RPV parameter space. The experimental inputs used are the EDMs of the neutron, the ^{205}Tl, ^{199}Hg, and ^{129}Xe atoms, in addition to the other CP conserving experimental data of fundamental precision tests. We first explain the linear programming method. We then give the setup of the supersymmetric parameters used in the calculation, and indicate the set of input inequalities to be used in the linear programming analysis. From this, we derive the limits on the bilinears of RPV couplings, predict the maximal prospects for the P, CP-odd experimental observables in preparation (EDMs of the proton, deuteron, ^{3}He nucleus, ^{211}Rn, ^{225}Ra atoms, muon, and the R-correlation of the neutron beta decay), and analyze the result.

N. Yamanaka, *Analysis of the Electric Dipole Moment in the R-parity Violating Supersymmetric Standard Model*, Springer Theses,
DOI: 10.1007/978-4-431-54544-6_14,

14.1 Linear Programming Method

In a phenomenological perturbative analysis, we often have linearly constrained relations. To derive the maximum of some value (relation) constrained by these relations, we have to solve a set of inequalities in order to find the maximum in the allowed region. To do this, we can do a naïve scan of the full parameter space by dotting and checking each point of the discretized space which contains regions constrained by input inequalities. However, this naïve method looses efficiency when the parameter space becomes large, in particular when the number of dimension increases. An efficient way to derive the maximum is the *linear programming method*. This method is based on the observation that the maximum is located at one of the corners of the multidimensional polygon made of inequality constraints (if the solution exists). This can be understood as follows. The linear relation we want to maximize constitutes a constant gradient. After being located somewhere in the allowed region, we follow the direction of the gradient to increase the objective linear relation. When we reach one of the "wall" (hyperplane) of constraint inequality, we follow then the direction of the projection of the gradient of the objective relation onto the wall. The dimension of the hyperplane we hit in going along the projected gradient diminishes in turn, and we arrive finally at some of the corners of the multidimensional polygonal allowed region. This is the point where the linear relation in question was maximized (of course if the gradient is found to be orthogonal with the final "hyperwall" we striked in going along the projected gradient, the whole hyperwall will be a degenerate solution of the problem). This is exactly the algorithm of the linear programming. The schematic picture of the 2-dimensional example is shown in Fig. 14.1. In our case, we want to predict the maximal value of the prepared

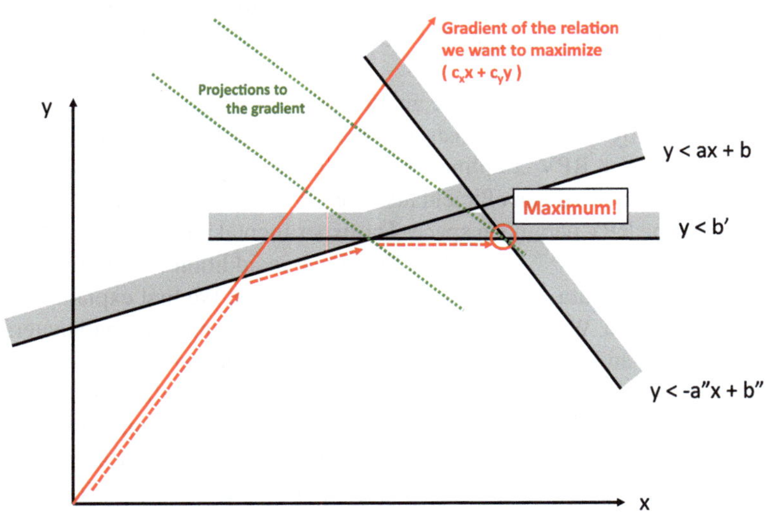

Fig. 14.1 Schematic illustration of linear programming problem in 2-dimension. The maximum of some linear relation is provided by the point whose projection onto the gradient becomes the largest

experimental observables within the EDM-constraints. The linear inequalities and observables must be expressed by variables, which are the combinations of the RPV bilinears. All these inputs will be defined in the next section.

14.2 Setup of Calculation

Before going to the calculation, we must specify parameters we are going to use. This procedure includes the choice of the variables and the limits on the absolute value of them.

SUSY Parameters

In this analysis, the SUSY mass are taken to be 1 TeV, as the excluded region of the LHC experiment (December 2011) [4–8]. The relative size between sparticle masses can be constrained in the CMSSM, from the result of the renormalization group analysis aiming at grand unified theories [9]. In this discussion, we do not consider such constraints. Our present parametrization is just the first step, and in future work, the result of renormalization group analysis will be taken into account. We also assume that the flavor off-diagonal terms of the soft SUSY breaking terms are very suppressed. We assume that the CP violation in the R-parity conserving sector is minimal. The θ-term is minimized with Peccei-Quinn symmetry. The analysis of CP violation including the R-parity conserving sector will be the subject of future works. In addition, we assume that the slepton (sneutrinos and charged sleptons) masses are degenerate. For the RPV sector, we consider only trilinear interactions with no soft breaking terms.

Constraints to RPV Couplings from Other Experiments

In analyzing the full RPV parameter space, we need **to refer to the** constraints provided by other experiments. The bounds on RPV couplings relevant **to** our analysis were essentially derived from the fundamental precision test experiments and are given in Table 14.1.

These limits are quite useful because the multidimensional area of the RPV parameter space is now bounded by them. From these limits, we obtain the 20 linear inequalities which corresponds to a finite “box” in 10 dimensional RPV parameter space as follows:

Table 14.1 Limits to absolute values of RPV couplings from other experimental results than EDMs

RPV coupling	Upper limit	Source
$\lvert\lambda_{211}\rvert$, $\lvert\lambda_{122}\rvert$	$0.05[m_{\tilde{e}_R}]$	Universality ($\Gamma(\tau^- \to \mu^- \bar{\nu}_\mu \nu_\tau)/\Gamma(\mu^- \to e^- \bar{\nu}_e \nu_\mu)$) [10]
$\lvert\lambda_{311}\rvert$, $\lvert\lambda_{133}\rvert$	$0.03[m_{\tilde{e}_R}]$	Universality ($\Gamma(\tau^- \to e^- \bar{\nu}_e \nu_\tau)/\Gamma(\tau^- \to \mu^- \bar{\nu}_\mu \nu_\tau)$) [10]
$\lvert\lambda_{322}\rvert$, $\lvert\lambda_{233}\rvert$	$0.05[m_{\tilde{e}_R}]$	Universality ($\Gamma(\tau^- \to e^- \bar{\nu}_e \nu_\tau)/\Gamma(\tau^- \to \mu^- \bar{\nu}_\mu \nu_\tau)$) [10]
$\lvert\lambda'_{i11}\rvert$, $\lvert\lambda'_{i22}\rvert$ ($i = 1, 2, 3$)	$5.7 \times 10^{-3}[m_{\tilde{d}_R}]$	$K \to \pi \nu \bar{\nu}$ [11–18]
$\lvert\lambda'_{233}\rvert$	$0.14[m_{\tilde{d}_R}]$	$K \to \pi \nu \bar{\nu}$ [11–18]
$\lvert\lambda'_{311}\rvert$	$10^{-3}[m_{\tilde{\nu}_\tau}]$	Sneutrino resonances at collider [19–21]
$\lvert\lambda'_{111}\rvert$	$2.8 \times 10^{-3}[m_{\mathrm{SUSY}}]^{\frac{3}{2}}$	$0\nu\beta\beta$ [22–26]
$\lvert\lambda'_{133}\rvert$	3.5×10^{-3}	ν Majorana mass [27, 28]
$\lvert\lambda'_{333}\rvert$	$0.12[m_{\tilde{d}_R}]$	$B \to X_s \nu \bar{\nu}$ [29, 30]

$[\cdots]$ denotes the mass in unit of 100 GeV

$$
\begin{aligned}
-0.15 &< x_1 < 0.15,\\
-0.25 &< x_2 < 0.25,\\
-3.2\times 10^{-2} &< x_3 < 3.2\times 10^{-2},\\
-4.6\times 10^{-2} &< x_4 < 4.6\times 10^{-2},\\
-1.1 &< x_5 < 1.1,\\
-5.0\times 10^{-2} &< x_6 < 5.0\times 10^{-2},\\
-5.5\times 10^{-2} &< x_7 < 5.5\times 10^{-2},\\
-8.9\times 10^{-3} &< x_8 < 8.9\times 10^{-3},\\
-9.5\times 10^{-2} &< x_9 < 9.5\times 10^{-2},\\
-0.15 &< x_{10} < 0.15,
\end{aligned}
\tag{14.1}
$$

where we have used the convention of Eq. (12.1). We assume the SUSY mass to be equal 1 TeV. In particular, sneutrino masses are degenerate.

EDM-Constraints

The EDM-constraints (^{205}Tl, ^{199}Hg, ^{129}Xe and neutron EDMs) to the leading order in RPV bilinears, can be expressed with eight linear relations of the form

$$
-d_a^{\rm exp} < d_a < d_a^{\rm exp} \Leftrightarrow -d_a^{\rm exp} < \sum_i c_{ai} x_i < d_a^{\rm exp}, \tag{14.2}
$$

where c_{ai}'s are given in Table 12.3. The experimental input is as follows:

$$
\begin{aligned}
d_{\rm Tl}^{\rm exp} &= 9\times 10^{-25} e\,{\rm cm},\\
d_{\rm Hg}^{\rm exp} &= 3.1\times 10^{-29} e\,{\rm cm},\\
d_n^{\rm exp} &= 2.9\times 10^{-26} e\,{\rm cm},\\
d_{\rm Xe}^{\rm exp} &= 4.0\times 10^{-27} e\,{\rm cm}.
\end{aligned}
\tag{14.3}
$$

14.3 Constraints on RPV Couplings from Linear Programming Method

After performing an analysis using the linear programming method, we have found the upper limits on bilinears of RPV couplings listed in Table 14.2. We see that the RPV bilinears $x_2[= {\rm Im}(\lambda_{211}\lambda^*_{233})]$, $x_3[= \sum_{i=2,3} {\rm Im}(\lambda_{i11}\lambda'^*_{i11})]$, $x_4[= \sum_{i=2,3} {\rm Im}(\lambda_{i11}\lambda'^*_{i22})]$, $x_9[= \sum_{i=1,2,3} {\rm Im}(\lambda'_{i11}\lambda'^*_{i33})]$ and $x_{10}[= \sum_{i=1,2,3} {\rm Im}(\lambda'_{i22}\lambda'^*_{i33})]$ can be constrained. The limits obtained are of course looser than those

Table 14.2 Upper limits on the absolute value of bilinears of RPV couplings found by linear programming analysis

RPV bilinears	Limits ($m_{\tilde{\nu},\tilde{e},\chi_0} = 1\,\text{TeV}$)	Limits ($m_{\tilde{\nu},\tilde{e},\chi_0} = 100\,\text{GeV}$)	Limits ($m_{\text{SUSY}} = 5\,\text{TeV}$)
$\lvert x_1\rvert(= \lvert\text{Im}(\lambda_{311}\lambda^*_{322})\rvert)$	0.15	1.5×10^{-3}	3.8
$\lvert x_2\rvert(= \lvert\text{Im}(\lambda_{211}\lambda^*_{233})\rvert)$	0.25	2.5×10^{-3}	6.3
$\lvert x_3\rvert(= \lvert\sum_{i=2,3}\text{Im}(\lambda_{i11}\lambda'^*_{i11})\rvert)$	4.7×10^{-4}	5.1×10^{-5}	9.9×10^{-3}
$\lvert x_4\rvert(= \lvert\sum_{i=2,3}\text{Im}(\lambda_{i11}\lambda'^*_{i22})\rvert)$	4.0×10^{-2}	4.4×10^{-3}	0.81
$\lvert x_5\rvert(= \lvert\sum_{i=2,3}\text{Im}(\lambda_{i11}\lambda'^*_{i33})\rvert)$	1.1	0.11	27
$\lvert x_6\rvert(= \lvert\sum_{i=1,3}\text{Im}(\lambda_{i22}\lambda'^*_{i11})\rvert)$	5.0×10^{-2}	5.0×10^{-3}	2.6
$\lvert x_7\rvert(= \lvert\sum_{i=1,2}\text{Im}(\lambda_{i33}\lambda'^*_{i11})\rvert)$	5.5×10^{-2}	5.5×10^{-3}	2.2
$\lvert x_8\rvert(= \lvert\sum_{i=1,2,3}\text{Im}(\lambda'_{i11}\lambda'^*_{i22})\rvert)$	8.9×10^{-3}	8.9×10^{-3}	0.38
$\lvert x_9\rvert(= \lvert\sum_{i=1,2,3}\text{Im}(\lambda'_{i11}\lambda'^*_{i33})\rvert)$	2.3×10^{-2}	5.3×10^{-3}	0.70
$\lvert x_{10}\rvert(= \lvert\sum_{i=1,2,3}\text{Im}(\lambda'_{i22}\lambda'^*_{i33})\rvert)$	2.0×10^{-2}	4.5×10^{-3}	0.62

obtained in the previous analysis assuming the dominance of single RPV bilinear (see Table 13.5). This illustrates the large degree of freedom of the RPV supersymmetric sector in which rearrangement of the couplings can occur keeping consistency with tight EDM-constraints. Although being looser, this analysis can give tighter upper bounds than the initial input limits of Eq. (14.1) given by other experiments. We can say that limits given in this analysis are the "true" upper bounds on RPV couplings.

Let us do an estimation of the constraints on RPV bilinears when the prospective upper limits of the EDMs of the proton, deuteron ^{3}He nucleus, ^{211}Rn and ^{225}Ra atoms are set. We will set the following limits, regarding the prospects of the experimental sensitivity [31–35]:

$$\begin{aligned} |d_p| &< 10^{-29} e\,\mathrm{cm}, \\ |d_d| &< 10^{-29} e\,\mathrm{cm}, \\ |d_{\mathrm{He}}| &< 10^{-29} e\,\mathrm{cm}, \\ |d_{\mathrm{Rn}}| &< 10^{-29} e\,\mathrm{cm}, \\ |d_{\mathrm{Ra}}| &< 3 \times 10^{-28} e\,\mathrm{cm}. \end{aligned} \tag{14.4}$$

Let us first see the limits on RPV bilinear using the linear programming method within the four EDM-constraints (^{205}Tl, ^{199}Hg, ^{129}Xe and neutron) seen previously with limit from one prospected EDM experiment added. The result is shown in Table 14.3.

By comparing Tables 14.2 and 14.3, we see that many RPV bilinears (x_2, x_3, x_4, x_9 and x_{10}) can be additionally constrained by adding one experimental data (d_p, d_d, $d_{\mathrm{He}}, d_{\mathrm{Rn}}$ or d_{Ra}). This result shows the importance of the prospective experiments. The following observations can be done:

Table 14.3 Upper limits on the absolute value of bilinears of RPV couplings found by linear programming analysis within the EDM-constraints of neutron, ^{205}Tl, ^{199}Hg, ^{129}Xe atoms + one additional limit from prospective EDM experiment (proton, deuteron, ^{3}He nucleus, ^{211}Rn or ^{225}Ra atoms)

RPV bilinears	d_p	d_d	d_{He}	d_{Rn}	d_{Ra}	d_p& d_d& d_{Ra}
x_1	0.15	0.15	0.15	0.15	0.15	0.15
x_2	0.14	0.14	0.14	0.16	0.14	8.1×10^{-2}
x_3	3.6×10^{-4}	3.6×10^{-4}	3.6×10^{-4}	3.9×10^{-4}	3.6×10^{-4}	3.3×10^{-4}
x_4	3.1×10^{-2}	3.0×10^{-2}	3.0×10^{-2}	3.3×10^{-2}	3.0×10^{-2}	2.6×10^{-2}
x_5	1.1	1.1	1.1	1.1	1.1	1.1
x_6	5.0×10^{-2}	5.0×10^{-2}	5.0×10^{-2}	5.0×10^{-2}	5.0×10^{-2}	5.0×10^{-2}
x_7	5.5×10^{-2}	5.5×10^{-2}	5.5×10^{-2}	5.5×10^{-2}	5.5×10^{-2}	5.5×10^{-2}
x_8	8.9×10^{-3}	8.9×10^{-3}	8.9×10^{-3}	8.9×10^{-3}	8.9×10^{-3}	8.6×10^{-3}
x_9	4.8×10^{-4}	3.0×10^{-4}	3.0×10^{-4}	3.7×10^{-3}	3.3×10^{-4}	2.9×10^{-4}
x_{10}	1.5×10^{-4}	1.5×10^{-4}	1.5×10^{-4}	3.1×10^{-3}	1.7×10^{-4}	1.5×10^{-4}

In the final row, we have given the upper limits for RPV bilinears when the constraints from d_p, d_d and d_{Rn} were applied simultaneously. Sparticle mass was set to 1 TeV

- By adding EDM-constraints of proton, deuteron, ^{3}He nucleus and ^{225}Ra, the hadronic RPV bilinears x_9 and x_{10} are constrained by two orders of magnitude tighter. This shows the strong sensitivity of hadronic EDMs (proton, deuteron and ^{3}He) against x_9, x_{10}. The EDM of ^{225}Ra atom is also sensitive to the hadronic R-parity violation, due to the strong enhancement of the nuclear Schiff moment.
- The leptonic and semi-leptonic RPV bilinears x_2, x_3 and x_4, although moderate, can be constrained with any additional prospective EDM constraints. Naïvely, this fact is not obvious, since the purely hadronic EDMs (d_p, d_d and d_{He}) are not sensitive to x_2, x_3 and x_4. This can be understood by the interplay between the additional future EDM-constraints and the existing constraints of diamagnetic atoms (^{199}Hg and ^{129}Xe). This fact shows the importance of giving experimentally the EDM-constraints on many systems.
- It is not possible to constrain x_1, x_5, x_6, x_7 and x_8. This is because the relevant EDMs have too weak sensitivity on them.
- The deuteron and ^{3}He EDMs can give tighter constraint on x_9 than the proton EDM. This shows the non-alignment of coefficients c_{di} and $c_{\text{He}i}$ with those of the neutron c_{ni}. (Recall that the neutron and the proton EDMs have no dependence on P, CP-odd isovector pion-nucleon interaction).
- The experimental limit of the ^{211}Rn EDM can also give tighter constraints on x_2, x_3, x_4, x_9 and x_{10}. This is realized thanks to the strong (prospective) EDM-constraints.

We have also given upper limits when the EDM-constraints of the proton, deuteron and ^{225}Ra atom were applied simultaneously. The result gives even stronger limits. We can conclude from this result that the combination of many EDM-constraints are useful in setting upper bounds to RPV interactions.

14.4 Maximal Prediction of the EDMs of the Proton, Deuteron, ^{3}He Nucleus, ^{211}Rn, ^{225}Ra Atoms and the R-Correlation

Next, we have made a prediction of the maximal value of the EDMs of the proton, deuteron, ^{3}He nucleus, ^{211}Rn, ^{225}Ra atoms and the R-correlation within the linear programming method. This can be done by maximizing the linear relations made by the coefficients of Table 12.4 within EDM-constraints of ^{205}Tl, ^{199}Hg, ^{129}Xe and neutron. The result obtained is summarized in Table 14.4. We show also the result for $m_{\text{SUSY}} = 5$ TeV in Table 14.5. We obtain predictions with the same order of magnitude since, the constraints on biliears of RPV couplings given by EDMs and by other experiments (see Table 14.1) have similar scaling in sparticle masses.

Let us compare our analysis with the previous analysis based on the assumption of the dominance of one single RPV bilinear. In Chap. 3, we have derived the upper limits of the CP violation of the RPV couplings with this assumption. Using these limits, we obtain the upper limits for P, CP-odd observables prepared in next generation experiments listed in Table 14.6 (see also Table 14.7 for $m_{\text{SUSY}} = 5\,\text{TeV}$).

Table 14.4 Maximal predictions of the EDMs of the proton, deuteron, ^{3}He nucleus, ^{211}Rn, ^{225}Ra atoms, and the R-correlation of the neutron beta decay, within the constraints of the ^{205}Tl, ^{199}Hg, ^{129}Xe and neutron EDM experiments

	d_p	d_d	d_{He}	d_{Rn}	d_{Ra}	R
Max.	1.9×10^{-23}	3.9×10^{-22}	6.0×10^{-22}	6.6×10^{-26}	1.5×10^{-22}	5.1×10^{-6}
x_1	-0.15	0.15	0.15	-0.15	-0.15	-0.15
x_2	-0.25	0.25	0.25	-0.25	-0.25	-0.25
x_3	-4.7×10^{-4}	4.7×10^{-4}	4.7×10^{-4}	-4.7×10^{-4}	-4.7×10^{-4}	-4.7×10^{-4}
x_4	-3.8×10^{-2}	3.8×10^{-2}	3.8×10^{-2}	-3.8×10^{-2}	-3.8×10^{-2}	-3.8×10^{-2}
x_5	1.1	-1.1	-1.1	1.1	1.1	1.1
x_6	5.0×10^{-2}	5.0×10^{-2}	-5.0×10^{-2}	5.0×10^{-2}	5.0×10^{-2}	5.0×10^{-2}
x_7	5.5×10^{-2}	5.5×10^{-2}	-5.5×10^{-2}	5.5×10^{-2}	5.5×10^{-2}	5.5×10^{-2}
x_8	-8.9×10^{-3}	8.9×10^{-3}	8.9×10^{-3}	-8.9×10^{-3}	-8.9×10^{-3}	-8.9×10^{-3}
x_9	-2.2×10^{-2}	2.2×10^{-2}	2.2×10^{-2}	-2.2×10^{-2}	-2.2×10^{-2}	-2.2×10^{-2}
x_{10}	2.0×10^{-2}	-2.0×10^{-2}	-2.0×10^{-2}	2.0×10^{-2}	2.0×10^{-2}	2.0×10^{-2}

Coordinates x_i maximizing the observables are also shown. The EDMs are expressed in unit of e cm. The mass of the slepton was taken to be 1 TeV

Table 14.5 Maximal predictions of the EDMs of the proton, deuteron, ^{3}He nucleus, ^{211}Rn, ^{225}Ra atoms, and the R-correlation of the neutron beta decay, within the constraints of the ^{205}Tl, ^{199}Hg, ^{129}Xe and neutron EDM experiments for $m_{\mathrm{SUSY}} = 5$ TeV

	d_p	d_d	d_{He}	d_{Rn}	d_{Ra}	R
Max.	3.1×10^{-23}	6.5×10^{-22}	1.0×10^{-21}	1.1×10^{-25}	2.4×10^{-22}	4.3×10^{-6}

The EDMs are expressed in unit of e cm

We see that all predictions with the assumption of the dominance of one single RPV bilinear are well below our predictions using the linear programming method, by 2 to 4 orders in magnitudes. This huge difference shows the importance of the degree of freedom of the RPV parameters and suggests the possibility of strong cancellation between the RPV contributions within the EDM-constraints. These configurations of RPV couplings were neglected in the previous analyses by the assumption of the

Table 14.6 Upper limits of the prepared experimental observables with the assumption of the single coupling dominance

d_{Xe}	d_p	d_d	d_{He}	d_{Rn}	d_{Ra}	R
1.0×10^{-29}	2.7×10^{-26}	2.7×10^{-26}	9.6×10^{-26}	5.3×10^{-29}	9.9×10^{-27}	5.5×10^{-10}

The unit is e cm for EDM observables. Sparticle masses were taken to 1 TeV

Table 14.7 Upper limits of the prepared experimental observables with the assumption of the single coupling dominance for $m_{\mathrm{SUSY}} = 5$ TeV

d_{Xe}	d_p	d_d	d_{He}	d_{Rn}	d_{Ra}	R
1.0×10^{-29}	2.9×10^{-26}	2.7×10^{-26}	5.3×10^{-26}	5.6×10^{-29}	9.9×10^{-27}	5.5×10^{-10}

The unit is e cm for EDM observables

single coupling dominance. The linear programming method shows efficiency in finding such configurations for RPV parameters. Below, we will try to explain the reasons of this large discrepancy.

Rearrangement of RPV Couplings Due to the EDM-Constraints of ^{205}Tl, ^{199}Hg and ^{129}Xe Atoms

As we have seen in Chap. 13, the atomic EDMs have a strong sensitivity to the type 2 (x_3, x_4 and x_5) semi-leptonic RPV bilinears. Let us see how the RPV bilinears have to arrange themselves to become large and are still consistent with the EDM-constraints. The EDM of the paramagnetic ^{205}Tl atom is sensitive to the type 2 RPV bilinears, and also to the type 1 leptonic RPV bilinears (x_1 and x_2). Thus, to be consistent with the constraint provided by the ^{205}Tl EDM, it suffices to cancel the type 1 and type 2 contributions mutually. As we can see in Table 14.4, this is exactly what happened (for the coefficients, see Tables 12.3 and 12.4).

The constraints from the EDM of diamagnetic atoms (^{199}Hg and ^{129}Xe) have also strong sensitivity to type 2 RPV bilinears, since it is generated by P, CP-odd electron-nucleon interactions. Diamagnetic atoms have however a moderate sensitivity on type 1 bilinears. They receive also contribution from type 4 (x_8, x_9 and x_{10}). To be consistent with experimental upper bounds, we have to cancel the type 2 and type 4 contributions. The largest cancellation occurs between x_3, x_4 and x_5. The remaining small part is cancelled with the type 4 RPV components originating in the nuclear Schiff moment. Within the above constraints, it is possible to enlarge the x_3 ($=\sum \text{Im}(\lambda_{i11}\lambda'^{*}_{i11})$) component up to $\sim 10^{-4}$. The coefficients c_{ia} of the EDM of diamagnetic atoms for type 2 RPV bilinears are aligned. This explains the relatively small maximal prediction for the EDM of ^{211}Rn atom. The same analysis does not hold for the ^{225}Ra atom, since the EDM of ^{225}Ra has a strong sensitivity to the hadronic sector (type 4). The above inspection means that even by introducing the future experimental constraints from ^{211}Rn and ^{225}Ra atoms, it is not possible to give tight upper bounds on RPV bilinears of type 2. To rule out the type 2 bilinears, we need thus another observable with coefficients c_{3a}, c_{4a} and c_{5a} not aligned with those of the diamagnetic atoms. This is possible when we use the R-correlation, which will be seen later.

Limits to Type 3

In this analysis, the type 3 RPV bilinears keep the same value ($|x_6| = 5.0 \times 10^{-2}$, $|x_7| = 5.5 \times 10^{-2}$). This means that the type 3 cannot be constrained from this analysis using linear programming method. This is due to the suppression of the quark EDM by the electromagnetic coupling constant in the Barr-Zee type contribution. The limits coming from other experiments are therefore dominant in this case.

Hadronic Observables

The purely hadronic observables (EDMs of the neutron, proton, deuteron and ^{3}He nucleus) have a large sensitivity against hadronic P, CP violating RPV interactions (type 4). This is due to the absence of the screening electrons and also to the large sensitivity of the prepared experiments with novel techniques using the storage ring [31–34]. The absence of the electrons also suppresses the semi-leptonic contribution, so that it serves to probe a fixed area of the RPV parameter space. One of the important characteristics of the purely hadronic EDMs is that they depend approximately only on type 4 RPV bilinears (type 3 contribution is relatively small). As they have restricted sensitivity against RPV bilinears, they can be used as a good probe to rule out a specific area of the RPV parameter space. This means also that collecting the EDM experimental data of pure hadronic EDMs is an efficient way to constrain the RPV parameter space, since they receive no cancellation from other sector than type 4 RPV bilinears. We must note that the prediction of the hadronic EDMs suffers from large theoretical uncertainty due to the use of model calculations at the hadronic level, and it can change even at the level of the order of magnitude. To do a quantitative analysis, we must improve the accuracy of the QCD level calculation.

R-Correlation

With the assumption of the single coupling dominance, the R-correlation is rather weak, due to the strong constraint of the atomic EDMs against CP violation. For example, within the dominance of one single bilinear of RPV couplings, the EDM of ^{199}Hg atom can constrain the same combination $\mathrm{Im}(\lambda_{i11}\lambda'^*_{i11})$ up to 10^{-8} with the current experimental data, whereas the R-correlation can constrain only up to 10^{-2} in the current experimental prospect of the ^{8}Li [36], well beyond the EDM experimental data. The result of our analysis, however, shows the potential importance of this observable. As seen above, the R-correlation contribution can also be enhanced by using the cancellation mechanism of the atomic EDM-constraints. As the R-correlation depends only on one combination (at least at the tree level), it can avoid the alignment between other EDM observables, and be "safe". The R-correlation is an important probe of the absolute size of $\mathrm{Im}(\lambda_{i11}\lambda'^*_{i11})$. We have seen that $\mathrm{Im}(\lambda_{i11}\lambda'^*_{i11})$ is the most sensitive RPV bilinear for the atomic EDMs, and combined with the EDM-constraints of paramagnetic and diamagnetic atoms, we can fully constrain the type 2 RPV bilinears. If the R-correlation can be measured with sufficient accuracy, it is possible to rule out the large prediction of atomic EDMs from $\mathrm{Im}(\lambda_{i11}\lambda'^*_{i11})$, thus reducing a large portion of the contribution to them. The experimental development for searching R-correlation is therefore strongly recommended.

Muon EDM

In this analysis, the muon EDM was not relevant, but this observable sits in a particular position, so we should add some comment. The muon EDM depends on $\mathrm{Im}(\lambda_{122}\lambda^*_{133})$, $\mathrm{Im}(\lambda_{i22}\lambda'^*_{i22})$, and $\mathrm{Im}(\lambda_{i22}\lambda'^*_{i33})$ via the two-loop level Barr-Zee type diagram, but these combinations, as seen in Chap. 13, do not contribute to the other available P, CP-odd observables. In RPV, the muon EDM is actually a completely independent system which can constrain its own RPV parameter space. It is then important to improve the sensitivity of the muon EDM experiment. As we do not have any EDM experimental data which can constrain the RPV couplings in question, the maximal prediction is just the sum of the upper bounds of RPV bilinears which can be probed with the muon EDM, and is of order 10^{-24}. The present experimental sensitivity ($\sim 10^{-19} e$ cm [37]) is well below the existing limits to the RPV couplings from other experiments. The future muon EDM experiments is prepared to aim at the order of $10^{-24} e$ cm [31–34], but the maximal value predicted in this analysis is also of the same order. Thus it will be difficult to either probe or constrain these RPV bilinears.

Theoretical Uncertainties

Let us mention briefly about the theoretical uncertainties. The first source of large theoretical uncertainty is the nuclear level calculations. The actual results of Schiff moment calculations are not consistent with each other (see Table 7.1). We have tested the dependence of the linear programming analysis on the different results presented in Ref. [38]. The results may change by one or two orders of magnitude. This illustrates the large uncertainty due to the nuclear level calculation of the Schiff moments of nuclei with deformation. The reduction of the theoretical uncertainty due to the difficulty of treating odd numbered nuclei with deformation is one of the outstanding and challenging problem.

The second large theoretical uncertainty is the hadron level calculation. In this analysis, we have used many approximations, which have all a large amount of error. First, to derive the hadron matrix element of the chromo-EDM operator, we have used the relation Eq. 6.30 based on QCD sum rule, which is not accurate better than 100 %. The above relation converted the nucleon matrix element of the chromo-EDM to the quark contents of nucleon, but these quark contents cannot be estimated with high precision, since they have dependence on light quark masses with large uncertainty. We have finally used in deriving the nucleon EDM the approximation taking only the chiral logarithm. This step involves also a considerable uncertainty, since the renormalization of the constant terms which are not necessarily small against terms logarithmic in Nambu-Goldstone boson mass was neglected. We conclude that in the hadron sector, the final result can change by order of magnitude. To reduce these theoretical uncertainties, the lattice QCD study is needed.

By considering these sources of theoretical error, we can say that coefficients c_6, c_7, c_8, c_9 and c_{10} can deviate by orders of magnitude.

Improvement of Constraints on RPV Couplings from Other Experiments

We must note that the limits on RPV couplings can be tightened by improving the constraints provided by other experiments in the linear programming analysis. This is because the reductions of the allowed region of the (absolute values of) RPV couplings can constrain the degrees of freedom left for the rearrangement of parameters within (EDM-)constraints. Here we mention the potential of other experiments.

The first possibility is to improve the experimental accuracy of the measurements of lepton decays (universality test) and hadron decays ($K \to \pi \nu \bar{\nu}$, $B \to X_s \nu \bar{\nu}$) (see Table 14.1).

The second possibility is the constraint from the absolute mass of neutrinos. All lepton number violating RPV interactions contribute to the Majorana mass of the neutrino (see Fig. 4.4), so the improvement of experimental constraints on its mass has a large potential to limit RPV bilinears relevant in our analysis.

The third interesting possibility is the limit from collider experiments. As we have seen in Sect. 4.3, resonances of sneutrino arise in the presence of the RPV interactions at collider experiments (see Fig. 4.8). The parton distribution for strange and bottom quarks allows also to probe the RPV interactions λ'_{i22} and λ'_{i33}. The lepton collider is sensitive to the resonances of sneutrino generated by leptonic RPV interactions λ_{i11} (electron collision), λ_{i22} (muon collision) and λ_{i33} (τ lepton collision).

14.5 Summary

In summary, we have done a wide analysis by taking into account the full CP violating RPV parameter space. For that, we have developed a new calculational technique based on the linear programming method, have given limits on the imaginary parts of RPV bilinears, and predicted the maximal values for observables of prepared or on-going experiments (proton, deuteron, ^{3}He nucleus, ^{211}Rn, ^{225}Ra atoms, muon, and the R-correlation of the neutron beta decay), within the currently available experimental constraints (^{205}Tl, ^{199}Hg, ^{129}Xe and neutron EDMs + other CP conserving experimental data of fundamental precision tests). We have found through this analysis that the RPV bilinears $x_3[=\sum_{i=2,3}\mathrm{Im}(\lambda_{i11}\lambda'^*_{i11})]$, $x_4[=\sum_{i=2,3}\mathrm{Im}(\lambda_{i11}\lambda'^*_{i22})]$, $x_9[=\sum_{i=1,2,3}\mathrm{Im}(\lambda'_{i11}\lambda'^*_{i33})]$ and $x_{10}[=\sum_{i=1,2,3}\mathrm{Im}(\lambda'_{i22}\lambda'^*_{i33})]$ can be constrained within the currently available experimental EDM-constraints. The upper limits on the above RPV bilinears and $x_2[=\mathrm{Im}(\lambda_{211}\lambda^*_{233})]$ can be tightened with additional prospective EDM-constraints of the proton, deuteron, ^{3}He nucleus, ^{211}Rn and ^{225}Ra atoms. In particular, x_9 and x_{10} can be strongly constrained due to the high sensitivity of the planned EDM experiments. RPV bilinears $x_1[=\mathrm{Im}(\lambda_{311}\lambda^*_{322})]$, $x_5[=\sum_{i=2,3}\mathrm{Im}(\lambda_{i11}\lambda'^*_{i33})]$, $x_6[=\sum_{i=1,3}\mathrm{Im}(\lambda_{i22}\lambda'^*_{i11})]$, $x_7[=\sum_{i=1,2}\mathrm{Im}(\lambda_{i33}\lambda'^*_{i11})]$ and $x_8[=\sum_{i=1,2,3}\mathrm{Im}(\lambda'_{i11}\lambda'^*_{i22})]$ could not be constrained due to the weak sensitivity of the EDMs of the relevant systems.

For the prediction of prospective experiments, we have found that very large values are still allowed, although many severe constraints already exist. This result is encouraging for experimentalists, since there is still a possibility to observe large EDM for prospective experiments. We have done a comparison with the "classic" analysis assuming the dominance of only one or some coupling, and have demonstrated the potential importance of the so far omitted area. The interesting observables are the R-correlation, the purely hadronic EDMs and the muon EDM which have sensitivity to the restricted area of the RPV parameter space. We have also obtained the useful information that the R-correlation is an important probe to rule out the type 2 RPV bilinears. The method based on linear programming was a success, but we have also encountered some problems due to the theoretical uncertainties. The reduction of them, in particular at the hadronic and nuclear level, are needed.

We have to mention our future subjects. In this analysis, we have only analyzed the EDM-constraints as linear relations in the linear programming, and the absolute limits on the RPV couplings taken from other experimental data were assumed to hold for single RPV coupling. As a future subject, we have to treat also these absolute limits from other experiments as linear relation inputs in the analysis of linear programming.

As an another future subject, we have to include the effects of the subleading RPV contribution to the EDM observables which contain combinations of RPV couplings and have not been considered so far. In the next chapter, we will consider the one-loop correction analysis of the P, CP-odd 4-fermion interaction in the assumption of the dominance of single bilinear of RPV couplings as a first step.

References

1. J. Ellis, J.S. Lee, A. Pilaftsis, JHEP **1010**, 049 (2010)
2. J. Ellis, J.S. Lee, A. Pilaftsis, JHEP **1102**, 045 (2011)
3. J. Ellis, J.S. Lee, A. Pilaftsis, arXiv:1009.1151 [math.OC]
4. G. Aad et al. (ATLAS Collaboration), Phys. Rev. Lett. **106**, 131802 (2011)
5. G. Aad et al. (ATLAS Collaboration), Phys. Lett. B **701**, 186 (2011)
6. G. Aad et al. (ATLAS Collaboration), arXiv:1110.6189 [hep-ex]
7. V. Khachatryan et al. (CMS Collaboration), Phys. Lett. B **698**, 196 (2011)
8. V. Khachatryan et al. (CMS Collaboration), arXiv:1111.2733.
9. U. Amaldi, W. de Boer, H. Fürstenau, Phys. Lett. B **260**, 447 (1991)
10. Y. Kao, T. Takeuchi, arXiv:0910.4980 [hep-ph]
11. K. Agashe, M. Graesser, Phys. Rev. D **54**, 4445 (1996)
12. D. Choudhury, P. Roy, Phys. Lett. B **378**, 153 (1996)
13. N.G. Deshpande, D.K. Ghosh, X.-G. He, Phys. Rev. D **70**, 093003 (2004)
14. A. Deandrea, J. Welzel, M. Oertel, JHEP **0410**, 38 (2004)
15. A.V. Artamonov et al. (BNL-E949 Collaboration), Phys. Rev. D **79**, 092004 (2009)
16. S. Adler et al., Phys. Rev. D **77**, 052003 (2008)
17. S. Adler et al., Phys. Rev. D **70**, 037102 (2004)
18. S. Adler et al., Phys. Lett. B **537**, 211 (2002)
19. J.R. Ellis, G. Gelmini, C. Jarlskog, G.G. Ross, J.W.F. Valle, Phys. Lett. B **150**, 142 (1985)
20. H.K. Dreiner, G.G. Ross, Nucl. Phys. B **365**, 597 (1991)
21. P.D. Jackson (ATLAS Collaboration), arXiv:1112.0369 [hep-ex]
22. K.S. Babu, R.N. Mohapatra, Phys. Rev. Lett. **75**, 2276 (1995)

23. M. Hirsch, H.V. Klapdor-Kleingrothaus, S.G. Kovalenko, Phys. Rev. Lett. **75**, 17 (1995)
24. A. Faessler, S. Kovalenko, F. Simkovic, J. Schwieger, Phys. Rev. Lett. **78**, 183 (1997)
25. M. Hirsch, H.V. Klapdor-Kleingrothaus, S.G. Kovalenko, Phys. Rev. D **57**, 1947 (1998)
26. A. Faessler, T. Gutsche, S. Kovalenko, F. Simkovic, Phys. Rev. D **77**, 113012 (2008)
27. S. Dimopoulos, L.J. Hall, Phys. Lett. B **207**, 210 (1988)
28. R.M. Godbole, P. Roy, X. Tata, Nucl. Phys. B **401**, 67 (1993)
29. Y. Grossman, Z. Ligeti, E. Nardi, Nucl. Phys. B **465**, 369 (1996)
30. Y. Grossman, Z. Ligeti, E. Nardi, Nucl. Phys. B **480**, 753 (1996) [Erratum]
31. I.B. Khriplovich, Phys. Lett. B **444**, 98 (1998)
32. F.J.M. Farley et al., Phys. Rev. Lett. **93**, 052001 (2004)
33. Y.K. Semertzidis et al., AIP Conf. Proc. **698**, 200 (2004)
34. Y.F. Orlov, W.M. Morse, Y.K. Semertzidis, Phys. Rev. Lett. **96**, 214802 (2006)
35. 5th International conference on "Fundamental Physics Using Atoms", Okayama University, October 2011 (URL: http://xqw.hep.okayama-u.ac.jp/kakenhi/index.php/fpua2011_home/fpua2011_top_e/)
36. J. Murata, private communication
37. G.W. Bennett et al. (Muon $(g-2)$ Collaboration), Phys. Rev. D **80**, 052008 (2009)
38. S. Ban, J. Dobaczewski, J. Engel, A. Shukla, Phys. Rev. C **82**, 015501 (2010)

Chapter 15
Analysis of the RPV Contribution to the P, CP-Odd 4-Fermion Interaction at the One-Loop Level

15.1 Analysis of the RPV Contribution to the P, CP-Odd e-N Interaction at the One-Loop Level

The possible types of one-loop correction are shown in Fig. 15.1. Among the listed diagrams, the vertex corrections are the renormalization of the tree level RPV coupling, so we do not need to consider them. This reduces our analysis only to the box diagrams.

In the evaluation of the box diagrams, we assume that the Yukawa couplings of the 1st and 2nd generations are neglected. The masses of light fermions are neglected. We have also assumed that the soft breaking squark and slepton mass matrices have no off-diagonal components, and diagonal components do not have any CP violating phases. For the RPV interactions, the dominance of single bilinear of RPV interactions is assumed. With these assumptions, there are only two contributing diagrams (with their complex conjugates), which are shown in Fig. 15.2.

The amplitude due to Fig. 15.2 a with its complex conjugate added is:

$$\begin{aligned} i\mathcal{M}_{(a)} = & -8i\,\mathrm{Im}(\lambda_{1i1}^{*}\lambda'_{iam})V_{am}\frac{G_F}{\sqrt{2}}m_W^2 I(m_W^2, m_{u_a}^2, m_{\tilde{e}_{Li}}^2) \\ & \times \left[\bar{e}i\gamma_5 e\cdot\bar{d}_m d_m - \bar{e}e\cdot\bar{d}_m i\gamma_5 d_m + (\text{P-even terms})\right], \end{aligned} \tag{15.1}$$

where we have neglected the external and exchanged momenta. We see that this amplitude is sensitive to the CP phase difference of the RPV coupling λ_{i11} and λ'_{iam}. a, i and m are the flavor indices. The Fermi constant is denoted by G_F, and V_{am} is the CKM matrix element with a and m the flavor indices. Here, m_W, m_{u_a} and $m_{\tilde{e}_{Li}}$

This chapter is based on the article of N. Yamanaka, Phys. Rev. D **85**, 115012 (2012), Copyright

N. Yamanaka, *Analysis of the Electric Dipole Moment in the R-parity Violating Supersymmetric Standard Model*, Springer Theses,
DOI: 10.1007/978-4-431-54544-6_15, © Springer Japan 2014

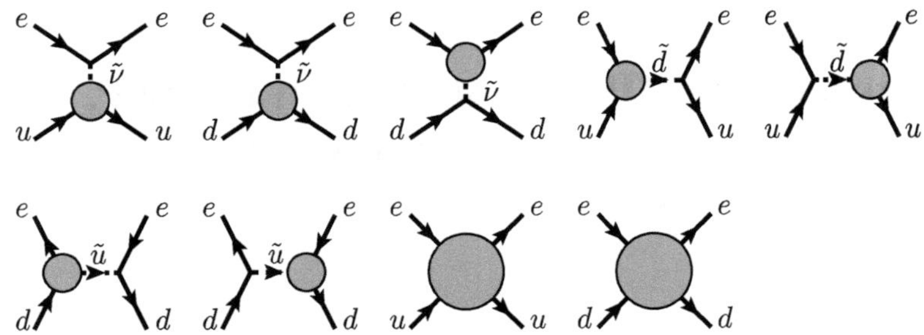

Fig. 15.1 Classification of one-loop correction contributing to P, CP-odd e-N interactions in the R-parity violation

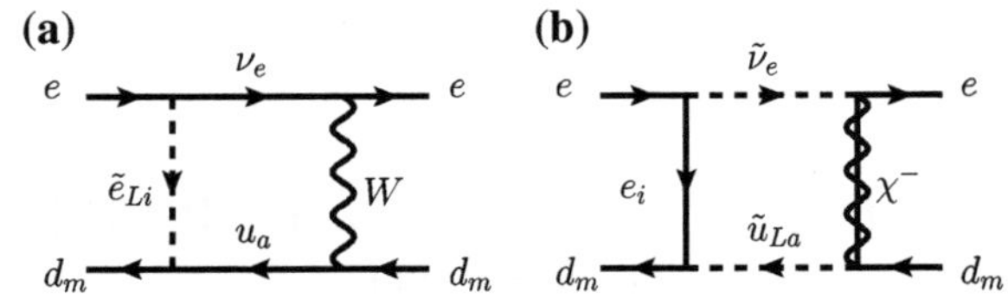

Fig. 15.2 Two diagrams (+ complex conjugates) contribute to the P, CP-odd e-N interactions at the one-loop level in the R-parity violation. The field χ corresponds to the chargino. Indices a, i and m denote the flavor

are the masses of W boson, up type quark (with flavor index a) and charged slepton (with flavor i), respectively. The loop integral I is expressed as follows:

$$I(a, b, c) = \frac{1}{4(4\pi)^2} \frac{1}{a-b} \left[\frac{a}{c-a} \log \frac{c}{a} - \frac{b}{c-b} \log \frac{c}{b} \right]. \tag{15.2}$$

For example, we have for $m_{\tilde{e}_{Li}} = 100$ MeV

$$\begin{aligned} m_W^2 I[m_W^2, m_u^2, m_{\tilde{e}_{Li}}^2 &= (100\,\text{GeV})^2] \approx 1.26 \times 10^{-3} , \\ m_W^2 I[m_W^2, m_c^2, m_{\tilde{e}_{Li}}^2 &= (100\,\text{GeV})^2] \approx 1.26 \times 10^{-3} , \\ m_W^2 I[m_W^2, m_t^2, m_{\tilde{e}_{Li}}^2 &= (100\,\text{GeV})^2] \approx 3.7 \times 10^{-4} , \end{aligned} \tag{15.3}$$

and for $m_{\tilde{e}_{Li}} = 1$ TeV,

$$
\begin{aligned}
m_W^2 I[m_W^2, m_u^2, m_{\tilde{e}_{Li}}^2 &= (1\,\text{TeV})^2] \approx 5.19 \times 10^{-5}\,, \\
m_W^2 I[m_W^2, m_c^2, m_{\tilde{e}_{Li}}^2 &= (1\,\text{TeV})^2] \approx 5.19 \times 10^{-5}\,, \\
m_W^2 I[m_W^2, m_t^2, m_{\tilde{e}_{Li}}^2 &= (1\,\text{TeV})^2] \approx 3.30 \times 10^{-5}\,.
\end{aligned}
\tag{15.4}
$$

We see that this integral has a sharp dependence on the slepton mass.

The amplitude of Fig. 15.2 b (with its complex conjugate added) is:

$$
\begin{aligned}
i\mathcal{M}_{(b)} = &- 8i\text{Im}(\lambda_{1i1}^* \lambda'_{iam}) V_{am} \frac{G_F}{\sqrt{2}} m_W^2 \sum_{j=1,2} |Z_+^{1j}|^2 I(m_{\chi_j}^2, m_{\tilde{\nu}_e}^2, m_{\tilde{u}_{La}}^2) \\
&\times \left[\bar{e} i\gamma_5 e \cdot \bar{d}_m d_m - \bar{e}e \cdot \bar{d}_m i\gamma_5 d_m\right].
\end{aligned}
\tag{15.5}
$$

Here m_{χ_j}, $m_{\tilde{\nu}_e}$ and $m_{\tilde{u}_{La}}$ are respectively the masses of chargino, sneutrino (1st generation) and up type squark (with flavor index a). The mixing matrix elements of the chargino Z_+^{1j} (j=1,2) follow the notation of Rosiek [3]. The important point is that $\mathcal{M}_{(a)}$ and $\mathcal{M}_{(b)}$ have exactly the same combinations of RPV couplings $\text{Im}(\lambda_{i11}^* \lambda'_{iam})$ with the same sign. The diagram (b) of Fig. 15.2 involves three sparticles in the loop, and the integral $I(m_{\chi_j}^2, m_{\tilde{\nu}_e}^2, m_{\tilde{u}_{La}}^2)$ has three unknown variables. The experimental result of the LHC has excluded the squark masses less than 1 TeV [4–6]. We set $m_{\tilde{u}_{La}} = 1$ TeV in this discussion. We show some values of the integral $m_W^2 I(m_{\chi_j}^2, m_{\tilde{\nu}_e}^2, m_{\tilde{u}_{La}}^2)$ for some tentative masses of sneutrino and chargino:

$$
\begin{aligned}
m_W^2 I(m_{\chi_j}^2, m_{\tilde{\nu}_e}^2, m_{\tilde{u}_{La}}^2) &= 3.8 \times 10^{-5} (m_{\chi_j} = m_{\tilde{\nu}_e} = 100\,\text{GeV}), \\
m_W^2 I(m_{\chi_j}^2, m_{\tilde{\nu}_e}^2, m_{\tilde{u}_{La}}^2) &= 9.9 \times 10^{-6} (m_{\chi_j} = 1\,\text{TeV}, m_{\tilde{\nu}_e} = 100\,\text{GeV} \\
&\qquad \text{or } m_{\chi_j} = 100\,\text{GeV}, m_{\tilde{\nu}_e} = 1\,\text{TeV}), \\
m_W^2 I(m_{\chi_j}^2, m_{\tilde{\nu}_e}^2, m_{\tilde{u}_{La}}^2) &= 5.1 \times 10^{-6} (m_{\chi_j} = m_{\tilde{\nu}_e} = 1\,\text{TeV}).
\end{aligned}
\tag{15.6}
$$

The order of magnitude of $m_W^2 I$ stays around $10^{-5} \sim 10^{-6}$ due to the heavy squark mass. The relative size between $\mathcal{M}_{(a)}$ depends then on the slepton mass. We can divide the discussion in two cases. If the slepton mass is around 100 GeV (light slepton), the amplitude $\mathcal{M}_{(a)}$ has a dominant contribution. On the other hand, when we have slepton mass $\approx$ 1 TeV (heavy slepton), the amplitudes $\mathcal{M}_{(a)}$ and $\mathcal{M}_{(b)}$ have same order contribution. We must note that both amplitudes have the same couplings with the same sign, so there is no possibility for cancellation with each other. The constraints on RPV couplings can therefore always be discussed.

To obtain limits on RPV interactions, we must first derive the P, CP-odd e-N interaction from the amplitude previously calculated, and then calculate its contribution to the EDM of ^{199}Hg atom. The derivation of the P, CP-odd e-N interactions [see eq. (8.3)] can be done by multiplying the quark level amplitudes with the quark contents of nucleon. Here we adopt the following results:

$$\langle p|\bar{u}u|p\rangle = 7.7, \tag{15.7}$$
$$\langle p|\bar{d}d|p\rangle = 6.9, \tag{15.8}$$
$$\langle p|\bar{s}s|p\rangle = 0.1, \tag{15.9}$$
$$\langle p|\bar{b}b|p\rangle = 10^{-2}, \tag{15.10}$$

and

$$\langle p|\bar{u}i\gamma_5 u|p\rangle = 248, \tag{15.11}$$
$$\langle p|\bar{d}i\gamma_5 d|p\rangle = -115, \tag{15.12}$$
$$\langle p|\bar{s}i\gamma_5 s|p\rangle = -2.5, \tag{15.13}$$
$$\langle p|\bar{b}i\gamma_5 b|p\rangle = -3\times 10^{-2}. \tag{15.14}$$

Here we have used the scalar and pseudoscalar contents of the nucleon derived in Sect. 6.1 [see Eqs. (6.12), (6.13), (6.17), (6.21) and (6.23)]. The scalar-pseudoscalar coupling C_N^{SP} and pseudoscalar-scalar C_N^{PS} coupling are then

$$C_p^{SP} = \langle p|\bar{d}d|p\rangle\tau_1 + \langle p|\bar{s}s|p\rangle\tau_2 + \langle p|\bar{b}b|p\rangle\tau_3\,, \tag{15.15}$$
$$C_n^{SP} = \langle p|\bar{u}u|p\rangle\tau_1 + \langle p|\bar{s}s|p\rangle\tau_2 + \langle p|\bar{b}b|p\rangle\tau_3\,, \tag{15.16}$$

and

$$C_p^{PS} = -\langle p|\bar{d}i\gamma_5 d|p\rangle\tau_1 - \langle p|\bar{s}i\gamma_5 s|p\rangle\tau_2 - \langle p|\bar{b}i\gamma_5 b|p\rangle\tau_3\,, \tag{15.17}$$
$$C_n^{PS} = -\langle p|\bar{u}i\gamma_5 u|p\rangle\tau_1 - \langle p|\bar{s}i\gamma_5 s|p\rangle\tau_2 - \langle p|\bar{b}i\gamma_5 b|p\rangle\tau_3\,, \tag{15.18}$$

where

$$\tau_m \equiv 8\mathrm{Im}(\lambda^*_{1i1}\lambda'_{iam})V_{am}m_W^2\left[I(m_W^2, m_{u_a}^2, m_{\tilde{e}_{Li}}^2) + \sum_{j=1,2}|Z_+^{1j}|^2 I(m_{\chi_j}^2, m_{\tilde{\nu}_e}^2, m_{\tilde{u}_{La}}^2)\right]. \tag{15.19}$$

Eqs. (15.16) and (15.18) were derived by using isospin symmetry. The dependence of the P, CP-odd e-N couplings on the ^{199}Hg atomic EDM is (see Sect. 8.3)

$$d_{\mathrm{Hg}} = \left[-5.0\times(0.40C_p^{\mathrm{SP}} + 0.60C_n^{\mathrm{SP}}) + 0.6\times(0.09C_p^{\mathrm{PS}} + 0.91C_n^{\mathrm{PS}})\right]\times 10^{-22}e\,\mathrm{cm}\,. \tag{15.20}$$

The experimental upper limit of the ^{199}Hg is $d_{\mathrm{Hg}} < 3.1\times 10^{-29}e$ cm, so we obtain

$$d_{\mathrm{Hg}} = (-166\tau_1 + 1.0\tau_2 - 0.032\tau_3)\times 10^{-22}e\,\mathrm{cm} < 3.1\times 10^{-29}e\,\mathrm{cm} \tag{15.21}$$

Table 15.1 Upper bounds to the RPV couplings given by the ^{199}Hg EDM experimental data via the P, CP-odd e-N interactions for $m_{SUSY} = 1$ TeV

RPV couplings	^{199}Hg EDM	Other experiments
$\lvert Im(\lambda^*_{211}\lambda'_{221})\rvert$	2.0×10^{-5}	2.9×10^{-2}
$\lvert Im(\lambda^*_{311}\lambda'_{321})\rvert$	2.0×10^{-5}	1.7×10^{-2}
$\lvert Im(\lambda^*_{211}\lambda'_{231})\rvert$	8.2×10^{-4}	0.60
$\lvert Im(\lambda^*_{311}\lambda'_{331})\rvert$	8.2×10^{-4}	0.36
$\lvert Im(\lambda^*_{211}\lambda'_{212})\rvert$	3.3×10^{-3}	2.9×10^{-2}
$\lvert Im(\lambda^*_{311}\lambda'_{312})\rvert$	3.3×10^{-3}	1.7×10^{-2}
$\lvert Im(\lambda^*_{211}\lambda'_{232})\rvert$	2.9×10^{-2}	0.60
$\lvert Im(\lambda^*_{311}\lambda'_{332})\rvert$	2.9×10^{-2}	0.36
$\lvert Im(\lambda^*_{211}\lambda'_{213})\rvert$	7	2.9×10^{-2}
$\lvert Im(\lambda^*_{311}\lambda'_{313})\rvert$	7	1.7×10^{-2}
$\lvert Im(\lambda^*_{211}\lambda'_{223})\rvert$	0.6	2.9×10^{-2}
$\lvert Im(\lambda^*_{311}\lambda'_{323})\rvert$	0.6	1.7×10^{-2}

Limits from other experiments [7] are also shown [they can be derived from Eqs. (4.14), (4.20) and (4.26)]

Table 15.2 Upper bounds to the RPV couplings given by the ^{199}Hg EDM experimental data via the P, CP-odd e-N interactions for $m_{\tilde{e}} = 100$ GeV

RPV couplings	^{199}Hg EDM	Other experiments
$\lvert Im(\lambda^*_{211}\lambda'_{221})\rvert$	8.2×10^{-7}	2.9×10^{-3}
$\lvert Im(\lambda^*_{311}\lambda'_{321})\rvert$	8.2×10^{-7}	1.7×10^{-3}
$\lvert Im(\lambda^*_{211}\lambda'_{231})\rvert$	7.3×10^{-5}	6.0×10^{-2}
$\lvert Im(\lambda^*_{311}\lambda'_{331})\rvert$	7.3×10^{-5}	3.6×10^{-2}
$\lvert Im(\lambda^*_{211}\lambda'_{212})\rvert$	1.4×10^{-4}	2.9×10^{-3}
$\lvert Im(\lambda^*_{311}\lambda'_{312})\rvert$	1.4×10^{-4}	1.7×10^{-3}
$\lvert Im(\lambda^*_{211}\lambda'_{232})\rvert$	2.6×10^{-3}	6.0×10^{-2}
$\lvert Im(\lambda^*_{311}\lambda'_{332})\rvert$	2.6×10^{-3}	3.6×10^{-2}
$\lvert Im(\lambda^*_{211}\lambda'_{213})\rvert$	0.3	2.9×10^{-3}
$\lvert Im(\lambda^*_{311}\lambda'_{313})\rvert$	0.3	1.7×10^{-3}
$\lvert Im(\lambda^*_{211}\lambda'_{223})\rvert$	3×10^{-2}	2.9×10^{-3}
$\lvert Im(\lambda^*_{311}\lambda'_{323})\rvert$	3×10^{-2}	1.7×10^{-3}

Limits from other experiments [7] are also shown [they can be derived from Eqs. (4.14), (4.20) and (4.26)]

By solving the above equation for RPV couplings, it is possible to obtain constraints on $Im(\lambda^*_{i11}\lambda'_{iam})$. If there are no accidental cancellations between RPV couplings, we obtain the constraints on bilinears of RPV couplings as shown in Tables. 15.1 (for $m_{SUSY} = 1$ TeV) and 15.2 (for $m_{\tilde{e}} = 100$ GeV).

The limits given for $Im(\lambda^*_{i11}\lambda'_{i21})$, $Im(\lambda^*_{i11}\lambda'_{i31})$, $Im(\lambda^*_{i11}\lambda'_{i12})$ and $Im(\lambda^*_{i11}\lambda'_{i32})$ shows tighter constraints than limits from other experiments [7]. The new constraints were set because of the strong upper limit of the EDM of ^{199}Hg atom and also of the high sensitivity of P, CP-odd e-N interactions to the R-parity violation.

We must pay attention to other processes contributing to the EDM of ^{199}Hg atom induced by the same bilinears of RPV couplings discussed in our analysis. These

RPV couplings contribute to the Barr-Zee type diagram of the electron and quark EDMs with W boson and charged slepton exchange. However these contributions are small for the following reasons. The electron EDM contribution is suppressed since the ^{199}Hg atom is a diamagnetic atom. The quark EDM is also suppressed, since its Barr-Zee type diagram has an electron loop (the Barr-Zee type contribution receives a factor of the mass of the inner loop fermion).

This analysis is expected to be also applicable to the P, CP-odd four-quark interaction within the R-parity violation at the one-loop level. We will see below this discussion.

This result has demonstrated the importance of the subleading order analysis for the EDM, like the well-known analysis of the muon anomalous magnetic moment. It also emphasizes the high accuracy of the ^{199}Hg EDM experimental data and its potential for probing new physics. This analysis has also emphasized the accessibility to a variety of RPV interactions through the subleading loop level contributions, within the assumption of the dominance single bilinear of RPV couplings. We have been able to set new limits to RPV interactions thanks to the combination of the high accuracy of the EDM experimental data and the variety of RPV interactions appearing at the one-loop level.

15.2 Analysis of the RPV Contribution to the P, CP-Odd 4-Quark Interaction at the One-Loop Level

The previous analysis can be extended to the P, CP-odd 4-quark interaction. This discussion is a direct extension of the tree level analysis done by Faessler et al. [8]. Diagrams contributing to the P, CP-odd 4-quark interaction are shown in Fig. 15.3.

The amplitudes are given as follows

$$\begin{aligned} i\mathcal{M}'_{(a)} \approx & 8i\,\mathrm{Im}(\lambda'^{*}_{ib1}\lambda'_{iam}) V_{b1} V_{am} \frac{G_F}{\sqrt{2}} m_W^2 \\ & \times I'_{iab} \cdot \left[\bar{d} i\gamma_5 d \cdot \bar{d}_m d_m - \bar{d} d \cdot \bar{d}_m i\gamma_5 d_m\right] + (\mathrm{P-even\ terms}), \quad (15.22) \end{aligned}$$

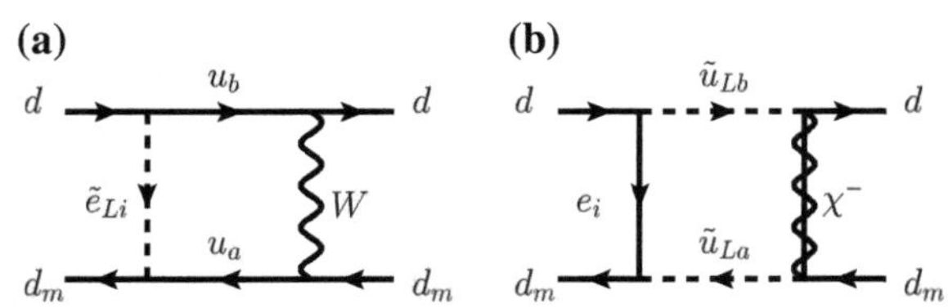

Fig. 15.3 RPV contribution to the P, CP-odd 4-quark interaction at the one-loop level. The structure of the box diagram is exactly the same as the P, CP-odd electron-quark interaction

$$
\begin{aligned}
i\mathcal{M}'_{(\mathrm{b})} \approx & 8i\mathrm{Im}(\lambda'^{*}_{ib1}\lambda'_{iam})V_{b1}V_{am}\frac{G_F}{\sqrt{2}}m_W^2 \sum_{j=1,2} |Z_+^{1j}|^2 I(m_{\chi_j}^2, m_{\tilde{u}_{La}}^2, m_{\tilde{u}_{Lb}}^2) \\
& \times \left[\bar{d}i\gamma_5 d \cdot \bar{d}_m d_m - \bar{d}d \cdot \bar{d}_m i\gamma_5 d_m\right], \qquad (15.23)
\end{aligned}
$$

where $i\mathcal{M}'_{(\mathrm{a})}$ and $i\mathcal{M}'_{(\mathrm{b})}$ are the amplitudes of the diagram with W boson and chargino in the loop, respectively. $a, b = 1, 2, 3$ and $m = 2, 3$ are the flavor indices. I'_{iab} is the loop integral of the amplitude $i\mathcal{M}'_{(\mathrm{a})}$, where

$$
I'_{iab} = \begin{cases} I(m_W^2, 0, m_{\tilde{e}_{Li}}^2) & \text{(a=1,2, b=1,2)} \\ I(m_W^2, m_t^2, m_{\tilde{e}_{Li}}^2) & \text{(a=3, b=1,2)} \\ I(m_W^2, m_t^2, m_{\tilde{e}_{Li}}^2) & \text{(a=1,2,b=3)} \\ J(m_W^2, m_{\tilde{e}_{Li}}^2, m_t^2) & \text{(a=3, b=3)} \end{cases} \qquad (15.24)
$$

with

$$
J(a, b, c) \equiv \frac{1}{4(4\pi)^2}\frac{1}{a-b}\left[\frac{a}{c-a}\left(1-\frac{a}{c-a}\ln\frac{c}{a}\right)-\frac{b}{c-b}\left(1-\frac{b}{c-b}\ln\frac{c}{b}\right)\right].
$$

For example, we have

$$
\begin{aligned}
m_W^2 J[m_W^2, m_{\tilde{e}_{Li}}^2 &= (100\,\mathrm{GeV})^2, m_t^2] \approx 1.9 \times 10^{-4}, \\
m_W^2 J[m_W^2, m_{\tilde{e}_{Li}}^2 &= (1\,\mathrm{TeV})^2, m_t^2] \approx 2.6 \times 10^{-5}, \qquad (15.25)
\end{aligned}
$$

The graph (b) of Fig. 15.3 involves two squarks in the loop, so its contribution is smaller than the amplitude $i\mathcal{M}'_{(\mathrm{a})}$, since $m_W^2 I(m_{\chi_j}^2, m_{\tilde{u}_{La}}^2, m_{\tilde{u}_{Lb}}^2) = 9.9 \times 10^{-6}$ for $m_{\chi_j} = 100$ GeV, and $m_W^2 I(m_{\chi_j}^2, m_{\tilde{u}_{La}}^2, m_{\tilde{u}_{Lb}}^2) = 5.1 \times 10^{-6}$ for $m_{\chi_j} = 1$ TeV (we have assumed $m_{\tilde{u}_L} \approx 1$ TeV). As we have seen for the P, CP-odd e-N interaction, the amplitudes $i\mathcal{M}'_{(\mathrm{a})}$ and $i\mathcal{M}'_{(\mathrm{b})}$ have the same sign and couplings, so there is no possibility of cancellation with each other.

The P, CP-odd 4-quark interaction contributes to the P, CP-odd pion-nucleon interaction. In this discussion, we use the factorization and PCAC reduction to derive the hadron level interaction. This gives the following isovector type P, CP-odd pion-nucleon interaction

$$
\mathcal{L} = \bar{g}_{\pi NN}^{(1)} \bar{N} N \pi^0, \qquad (15.26)
$$

with

$$
\bar{g}_{\pi NN}^{(1)} \approx -\frac{F_\pi m_\pi^2}{2m_d}\frac{G_F}{\sqrt{2}} \sum_{m=2,3} \rho_n \langle p|\bar{d}_m d_m|p\rangle, \qquad (15.27)
$$

where $F_\pi \approx 93$ MeV is the pion decay constant, and $m_\pi = 140$ MeV is the pion mass. The coefficient ρ_m is defined as

$$\rho_m = 8i\mathrm{Im}(\lambda'^*_{ib1}\lambda'_{iam})V_{b1}V_{am}m_W^2\left[I'_{iab} + \sum_{j=1,2}|Z_+^{1j}|^2 I(m^2_{\chi_j}, m^2_{\tilde{u}_{La}}, m^2_{\tilde{u}_{Lb}})\right] . \tag{15.28}$$

We must note that this factorization method has a large uncertainty.

The recent Schiff moment of the ^{199}Hg nucleus was calculated with fully self-consistent many-body treatment taking into account the deformation [9]. The result is [see Eq. (7.56)]

$$S_{\mathrm{Hg}} = 0.007 g_{\pi NN}\bar{g}^{(1)}_{\pi NN} e\,\mathrm{fm}^3 , \tag{15.29}$$

where $g_{\pi NN} \approx 12.9$ is the ordinary pseudoscalar coupling pion-nucleon coupling. We have neglected contributions from the nucleon EDM, isoscalar and isotensor P, CP-odd pion-nucleon interactions. The isovector dependence of ^{199}Hg Schiff moment presented above is the average of five calculations shown in Table. 7.1. Note that this nuclear level calculation has also a large theoretical uncertainty (For some calculational method, the result is of opposite sign).

The dependence of the EDM of the ^{199}Hg atom on the nuclear Schiff moment is given by [10] [see Eq. (8.32)]

$$d_{\mathrm{Hg}} = -2.6 \times 10^{-17}\frac{S_{\mathrm{Hg}}}{e\,\mathrm{fm}^3} e\,\mathrm{cm} . \tag{15.30}$$

The final form of the dependence of ^{199}Hg EDM on the RPV P, CP-odd 4-quark contribution is

$$d_{\mathrm{Hg}} = (3.5\rho_2 + 0.35\rho_3) \times 10^{-25} e\,\mathrm{cm}. \tag{15.31}$$

The constraints on RPV couplings obtained from the experimental data ($d_{\mathrm{Hg}} < 3.1 \times 10^{-29} e$ cm [2]) is shown in Table. 15.3, where $m_{\tilde{e}_{Li}}$ is tentatively taken as 100 GeV and 1 TeV.

The limits obtained are looser than those obtained from other experiments [7], so it is not possible to obtain upper bounds on RPV interactions from the ^{199}Hg EDM experimental data. Nevertheless, this result shows the variety of RPV interactions λ'_{ijk} accessible from the one-loop level P, CP-odd 4-quark interactions.

We should add to this discussion the possibility to constrain RPV interactions from future EDM experiments. The first good candidate is the EDM of ^{225}Ra atom. The ^{225}Ra EDM has a strong sensitivity **to** the P, CP-odd hadronic interactions. The dependence of the P, CP-odd 4-quark interaction on the EDM of ^{225}Ra atom is [10, 11]

$$d_{\mathrm{Ra}} = (-1.0\rho_2 - 0.1\rho_3) \times 10^{-21} e\,\mathrm{cm} . \tag{15.32}$$

We have used the relations (7.59) and (8.35). We see that the sensitivity **to** the P, CP-odd 4-quark interaction is enhanced by a factor of 3000 compared to the EDM

Table 15.3 Upper bounds to the RPV couplings given by the ^{199}Hg EDM experimental data via the P, CP-odd 4-quark interactions

RPV couplings	($m_{\tilde{e}_{Li}}$ =100 GeV)	($m_{\tilde{e}_{Li}}$ =1 TeV)
$\lvert\mathrm{Im}(\lambda'^{*}_{i11}\lambda'_{i12})\rvert$	4.0×10^{-2}	0.97
$\lvert\mathrm{Im}(\lambda'^{*}_{i11}\lambda'_{i32})\rvert$	0.76	8.5
$\lvert\mathrm{Im}(\lambda'^{*}_{i21}\lambda'_{i12})\rvert$	0.17	4.2
$\lvert\mathrm{Im}(\lambda'^{*}_{i21}\lambda'_{i22})\rvert$	4.0×10^{-2}	0.97
$\lvert\mathrm{Im}(\lambda'^{*}_{i21}\lambda'_{i32})\rvert$	3.3	37
$\lvert\mathrm{Im}(\lambda'^{*}_{i31}\lambda'_{i12})\rvert$	15	170
$\lvert\mathrm{Im}(\lambda'^{*}_{i31}\lambda'_{i22})\rvert$	3.5	40
$\lvert\mathrm{Im}(\lambda'^{*}_{i31}\lambda'_{i32})\rvert$	170	1200
$\lvert\mathrm{Im}(\lambda'^{*}_{i11}\lambda'_{i13})\rvert$	25	630
$\lvert\mathrm{Im}(\lambda'^{*}_{i11}\lambda'_{i23})\rvert$	2.2	53
$\lvert\mathrm{Im}(\lambda'^{*}_{i21}\lambda'_{i13})\rvert$	110	2700
$\lvert\mathrm{Im}(\lambda'^{*}_{i21}\lambda'_{i23})\rvert$	9.4	230
$\lvert\mathrm{Im}(\lambda'^{*}_{i21}\lambda'_{i33})\rvert$	1.3	15
$\lvert\mathrm{Im}(\lambda'^{*}_{i31}\lambda'_{i13})\rvert$	9900	1.1×10^5
$\lvert\mathrm{Im}(\lambda'^{*}_{i31}\lambda'_{i23})\rvert$	840	9400
$\lvert\mathrm{Im}(\lambda'^{*}_{i31}\lambda'_{i33})\rvert$	67	490

of ^{199}Hg atom [compare with eq. (15.31)]. This large enhancement is due to the enhancement of the Schiff moment by the octupole deformation of the ^{225}Ra nucleus [11] and the close parity doublet states of the atomic energy level [10]. The ^{225}Ra EDM experiment is being prepared by the group of Argonne National Laboratory aiming at the sensitivity of $O(10^{-28})e$ cm [12]. We can then expect upper bounds of Table. 15.3 to be tightened by several hundred times.

Another experimental candidate is the EDM of the deuteron (nucleus). Recently a new generation of EDM experiment using the storage ring is in preparation , and it offers the possibility to measure the EDM of charged particles with very high sensitivity [13]. With this setup, the high sensitivity of the deuteron EDM is possible, since the suppression of the nuclear level P, CP-odd effect is avoided due to the absence of screening electrons, in addition to the long coherence time during the measurement. The dependence of the P, CP-odd 4-quark interaction on the EDM of the deuteron is [14]

$$d_d = (2.9\rho_2 + 0.29\rho_3) \times 10^{-21} e\,\mathrm{cm}\,. \tag{15.33}$$

We have used the relation (7.54) for the derivation of the dependence given above. We see that the deuteron EDM is 10000 times more sensitive than the EDM of ^{199}Hg atom against P, CP-odd 4-quark interaction. The measurement of deuteron EDM is in preparation at Brookhaven National Laboratory, and the expected experimental sensitivity is $O(10^{-29})e$ cm. We thus expect upper bounds of Table. 15.3 to be tightened by ~10000 times.

We should also point out the potential importance of the nucleon EDMs. In our discussion, we have neglected the contribution of the P, CP-odd 4-quark interaction to the nucleon EDM, since the isovector type P, CP-odd pion-nucleon interaction does not contribute to the nucleon EDM in the approximation which takes only the leading chiral logarithm. This approximation however neglects non-leading terms which, although being model dependent, can involve sizable isovector dependence. This approximation has a large theoretical uncertainty, and accurate evaluation of the dependence of the nucleon EDM on P, CP-odd quark level interactions is needed. We are thus waiting for Lattice QCD calculation. If we assume that the dependences of the nucleon EDM on isovector type and other P, CP-odd pion-nucleon interactions are comparable, the dependence of the P, CP-odd 4-quark interaction on the nucleon EDM will be

$$d_N \sim (0.1\rho_2 + 0.01\rho_3) \times 10^{-20} e\,\mathrm{cm}\,. \quad (15.34)$$

The next generation of the neutron EDM experiments using UCN sources plan to reach the sensitivity of $O(10^{-28})e$ cm [12, 15], which can constrain the RPV interactions listed in Table. 15.3 1000 times tighter than the current ^{199}Hg EDM experimental data. We must also note that the proton EDM can be a strong candidate in limiting these RPV interactions, since its performance should provide the same sensitivity as the deuteron EDM [13].

15.3 Summary

In this chapter we have analyzed the contribution of the R-parity violation to the P, CP-odd e-N interaction at the one-loop level and derived from the recent ^{199}Hg EDM experimental data limits to the imaginary parts of the following products of RPV couplings: $\lambda^*_{i11}\lambda'_{i21}$, $\lambda^*_{i11}\lambda'_{i31}$, $\lambda^*_{i11}\lambda'_{i12}$, $\lambda^*_{i11}\lambda'_{i32}$, $\lambda^*_{i11}\lambda'_{i13}$ and $\lambda^*_{i11}\lambda'_{i23}$ $(i = 2, 3)$. For $\lambda^*_{i11}\lambda'_{i21}$, $\lambda^*_{i11}\lambda'_{i31}$, $\lambda^*_{i11}\lambda'_{i12}$ and $\lambda^*_{i11}\lambda'_{i32}$ $(i = 2, 3)$, we have found that these limits give tighter constraints than those given by other experiments. For $\lambda^*_{i11}\lambda'_{i13}$ and $\lambda^*_{i11}\lambda'_{i23}$ $(i = 2, 3)$, we could not set new limits. The new constraints were set because of the strong upper limit of the ^{199}Hg atom EDM and also of the high sensitivity of the P, CP-odd e-N interaction to the R-parity violation. We have also analyzed the P, CP-odd 4-quark interaction within the R-parity violation at the one-loop level. The current experimental limit of the ^{199}Hg EDM could not set new limits, but new generation of EDM experiments with ^{225}Ra atom, deuteron, neutron and proton has the possibility to constrain the RPV interactions significantly via P, CP-odd 4-quark interaction. This result has demonstrated the importance of the subleading order analysis for the EDM, like the well-known analysis of the muon anomalous magnetic moment. This analysis has also emphasized the accessibility to a variety of RPV interactions through the subleading loop level contributions, within the assumption of the dominance of a single bilinear of RPV couplings. We have been able to set new limits to RPV interactions thanks to the combination of the high accuracy of the EDM experimental data and the variety of RPV interactions appearing at the one-loop level.

We should also note that the limits to RPV interactions given by the analysis of the P, CP-odd 4-quark interactions has a large theoretical uncertainty and model dependence in QCD calculation. To give more definite constraints, accurate calculations are indispensable. The study of the dependence of P, CP-odd hadron level interactions on the P, CP-odd 4-quark interaction within Lattice QCD is therefore required.

References

1. N. Yamanaka, Phys. Rev. D **85**, 115012 (2012)
2. W.C. Griffith et al., Phys. Rev. Lett. **102**, 101601 (2009)
3. J. Rosiek, Phys. Rev. D **41**, 3464 (1990)
4. ATLAS Collaboration (Georges Aad et al.), Phys. Rev. Lett. **106** 131802 (2011)
5. ATLAS Collaboration (Georges Asd et al.), Phys. Lett. B **701** 186 (2011) arXiv:1110.6189 [hep-ex]
6. CMS Collaboration (Vardan Khachatryan et al.), Phys. Lett. B **698**, 196 (2011)
7. G. Bhattacharyya, arXiv:hep-ph/9709395; H. K. Dreiner, arXiv:hep-ph/9707435; R. Barbier et al., Phys. Rept. 420, 1 (2005); M. Chemtob. Prog. Part. Nucl. Phys. **54**, 71 (2005)
8. A. Faessler, T. Gutsche, S. Kovalenko and V. E. Lyubovitskij, Phys. Rev. D 73, 114023 (2006); Phys. Rev. D 74, 074013 (2006)
9. S. Ban, J. Dobaczewski, J. Engel, A. Shukla, Phys. Rev. C **82**, 015501 (2010)
10. V.A. Dzuba, V.V. Flambaum, S.G. Porsev, Phys. Rev. A **80**, 032120 (2009)
11. J. Dobaczewski, J. Engel, Phys. Rev. Lett. **94**, 232502 (2005)
12. 5th International conference on "Fundamental Physics Using Atoms", Okayama University, October 2011 (URL: http://xqw.hep.okayama-u.ac.jp/kakenhi/index.php/fpua2011_home/fpua2011_top_e/)
13. I.B. Khriplovich, Phys. Lett. B 444, 98 (1998); F. J. M. Farley et al., Phys. Rev. Lett. 93, 052001 (2004); Y. K. Semertzidis et al., AIP Conf. Proc. 698, 200 (2004); Y. F. Orlov, W. M. Morse, Y. K. Semertzidis. Phys. Rev. Lett. **96**, 214802 (2006)
14. C.-P. Liu, R.G.E. Timmermans, Phys. Rev. C **70**, 055501 (2004)
15. M. Burghoff et al., arXiv:1110.1505 [nucl-ex]

Chapter 16
Summary and Future Prospects

Here, we would like to summarize our investigations. We have first reviewed the supersymmetry (SUSY) as a candidate of new physics beyond the standard model (SM). The supersymmetry, although not discovered so far, is one of the leading candidates, because it can provide the stability of the electroweak scale physics by the cancellation of the power divergence. The supersymmetry has been discussed and studied extensively, and the phenomenological works on it are still on-going. We have seen that the R-parity violation, which is one possible scenario of SUSY, is also interesting since it can be linked to the Grand unified theories. In this work, we have studied the R-parity violating (RPV) interactions phenomenologically. The SUSY (and other candidates of new physics) can be probed with a very powerful experimental probe, the electric dipole moments (EDMs), which are the second main subject of this thesis. The EDM is a very sensitive probe of the P and CP violations, with a very small SM background, and it can be measured in a variety of systems, which makes it to be a very good tool to detect the new physics. The EDM played an active role in phenomenology, and many parameters of the SUSY and R-parity violation were constrained by it so far.

In this thesis, we have analyzed the R-parity violation within EDM-constraints, and have reported the following five important results:

- We have revised the two-loop level Barr-Zee type RPV contribution to the fermion EDM. Our result gave a smaller value by about one order of magnitude than previous works. This result alters the relative size between the contribution from P, CP-odd electron–nucleon interactions for the EDM of ^{199}Hg atom.
- We have classified the RPV contribution to the EDM observables. RPV bilinears can be classified into six types. The update of EDM-constraints to the RPV interactions were also given (see Table 13.5). These limits were derived by assuming that only one RPV bilinear dominates.
- We have analyzed the EDM-constraints (^{205}Tl, ^{199}Hg, ^{129}Xe and neutron) on the R-parity violation also for the case where all RPV couplings contributing to the leading order are considered simultaneously. For that we have developed a calculational method based on linear programming. Although being softened (due to

N. Yamanaka, *Analysis of the Electric Dipole Moment in the R-parity Violating Supersymmetric Standard Model*, Springer Theses,
DOI: 10.1007/978-4-431-54544-6_16, © Springer Japan 2014

the interference between RPV bilinears), it was possible to set limits on many RPV bilinears. We have also analyzed the limits on R-parity violation when prospective experimental EDM-constraints of proton, deuteron, ^{3}He nucleus, ^{211}Rn, ^{225}Ra atoms were included in evaluations. These planned EDM experiments have high sensitivity on hadronic RPV bilinears, and within the prospective data, it is possible to constrain them by two orders tighter than the currently available EDM-constraints. We must also note that the interplay of several EDM-constraints can significantly limit the RPV parameter space, and this shows the importance of measuring EDMs on many systems.

- We have also predicted the maximal values for the prospective experiments probing P, CP violation (EDMs of the proton, deuteron, ^{3}He nucleus, ^{211}Rn, ^{225}Ra atoms, and the R-correlation of the neutron beta decay). With the linear programming method, all these observables were predicted with large values. The order of magnitude of their maximal prediction is three or four orders larger than the maximal prediction within the assumption of the dominance of single RPV bilinear. This shows the efficiency of the linear programming method in scanning the RPV parameter space. This result is encouraging for experimentalists because EDMs have more chances to be observed than the suggestions made by analyses within the dominance of single RPV bilinear.
- We have also analyzed the P, CP-odd 4-fermion interactions at the one-loop level, and obtained that some RPV interactions can be constrained by the experimental data of the ^{199}Hg atom. We have obtained a new limit to the CP violating RPV couplings, and have shown the importance of the analysis of the subleading order contribution.

We should add some comments on the advantages of the linear programming method. The first advantage is that the full parameter space can be analyzed taking also the area of the parameter space in which interference and cancellation occur. The second advantage is that the linear programming method can be applied to many other new physics. In particular, if the equation is linear in unknown parameters, the systematic phenomenological study is possible.

With the help of the classification of RPV bilinears, some important consequences were obtained. We have learned that the R-correlation is a useful probe of the RPV bilinear $\mathrm{Im}(\lambda_{i11}\lambda'^{*}_{i11})$, and if R can be constrained sufficiently by experiments, the type two semi-leptonic RPV bilinears contributing to the P, CP-odd electron–nucleon interactions will be ruled out. We have also learned that the purely hadronic EDMs and muon EDM have restricted sensitivity against RPV bilinears, so that they can be used as a good probe to rule out a specific area of the RPV parameter space. In particular, the muon EDM is the only observable which can probe its RPV contribution, so the muon EDM experiments will play a very important role in the phenomenology of the R-parity violation. Considering all of these results, we have then analyzed the CP violation of the R-parity violation from a very broad point of view.

We have also found many challenging problems in the course of our study. Let us list them:

- A new way to the RPV scenario: In our analysis, we have given the possibility of scenario with large RPV bilinears with some constrained configuration (due to EDM-constraints), which was not possible within the assumption of single coupling dominance. This is a good challenge for theorists to build scenarios of Grand unification with RPV interactions. Theorists however have also to explain naturally the RPV configuration due to the EDM-constraints, which is a challenge.
- Linear programming analysis including the subleading RPV contribution to the EDM: In our analysis all leading RPV contributions are considered simultaneously within EDM-constraints. But we did not add the contribution from the one-loop level P, CP-odd electron–nucleon interaction. As we have concluded that these subleading contributions can also be important, we should add them into the full analysis using the linear programming method. Actually, there are also another subleading contribution in the fermion EDM. This is the Barr-Zee type two-loop level diagram with W-boson exchange, which involves also RPV interactions not relevant at the leading order. To complete this analysis, we should also evaluate this subleading Barr-Zee type diagram, and do the full RPV analysis together with the one-loop P, CP-odd electron–nucleon interaction and the leading contributions.
- Full analysis of the RPV couplings: In our analysis, we have only analyzed the EDM-constraints as linear relations in the linear programming, and the absolute limits on the RPV couplings taken from other experimental data were assumed to hold for single RPV coupling. We have to treat also these absolute limits from other experiments as linear relation inputs in the analysis of linear programming.
- Full analysis of the supersymmetric parameter space: Ideally, this analysis should be extended to the whole supersymmetric parameter space, including the R-parity conserving sector.
- The uncertainty due to QCD: In deriving the EDM of the nucleon, atoms and nuclei, we have encountered uncertainty due to the model calculations of the QCD. The theoretical uncertainty is difficult to estimate, and the result may change even by orders of magnitude. The development of the Lattice QCD is one of the promising framework to do accurate evaluations of the QCD hadron matrix elements. The calculation of the matrix elements needed in the calculation of EDMs is one of the important subjects.
- The uncertainty of the Schiff moment calculation: As can be seen in Table 7.1, the nuclear Schiff moment calculation of the heavy deformed nuclei is rather unstable. This is mainly due to the fact that the ground state wave functions of deformed odd-nuclei are difficult to construct by using the mean-field method. This subject is also a difficult challenge not only for the EDM calculation but also for the nuclear physics.
- Calculations of the EDMs of few body nuclei: We have seen in our thesis that the purely hadronic EDMs have a strong sensitivity against hadronic RPV bilinears. By looking to the sensitivity of the EDMs of deuteron and ^{3}He nucleus to the P, CP-odd pion-nucleon couplings, we see that the sensitivity increases in the number of nuclei. We can therefore expect to have a more sensitive nuclear EDMs for A = 6, 7, 8, etc. The recent development of nuclear calculation using *ab initio* approaches

for few body systems is remarkable. The author of this thesis plans to do such calculations as future project.

- Study of the molecular P, CP violation: Recently, the new world record of upper limit to electron EDM was made using YbF molecule. The molecular P, CP violation is a promising study in development since the potential sensitivity of them to new physics is huge, and many researches and developments are on-going. In this thesis, we did not include the constraints from the YbF molecule due to the lack of knowledge about relations between the P, CP violation of the YbF molecule and the underlying P, CP-odd mechanisms. The development of theoretical calculation for molecules is also an important subject.
- Final state interaction contribution of the R-correlation: From our analysis of RPV with linear programming, we have concluded that the improvement of sensitivity of the R-correlation is promising in excluding the degree of freedom of the RPV parameter space allowing large atomic EDMs. The R-correlation receives however contribution from the final state interaction (FSI) which is of order $R_{\rm fsi} \sim 10^{-4}$. To improve the sensitivity of the R-correlation, we must carry out an accurate calculation of the FSI contribution. The R-correlation can also be measured with nuclear beta decay, and for some nuclei such as the ^{8}Li, the FSI contribution is known to be smaller than the neutron. The accurate calculation of the FSI for them are an interesting subject for the R-parity violation.

Appendix A
Particle Masses in the MSSM

The MSSM has many additional mass terms and mixings due to the soft breaking of supersymmetry. The spectrum of the particles can be obtained by diagonalizing their mass matrices. Here we present the formulae for the mass eigenvalues and eigenstates of the Higgs bosons, charginos, neutralinos and sfermions.

Higgs Bosons

There are many types of scalar (like squarks, sleptons, Higgs bosons) which can potentially break the symmetry. However, the vacuum is charge, color baryon and lepton number conserving, so that only Higgs bosons can break spontaneously the electroweak symmetry. We then define the vacuum so that the neutral Higgs develops a vacuum expectation value (VEV). It can be shown from the total Higgs lagrangian, that the neutral components of the up-type and down-type Higgs must have finite VEV if the electroweak symmetry is broken. The scalar potential which must be minimized is then

$$\begin{aligned} V = &(m_{H_u}^2 + \mu^2)|h_u^0|^2 + (m_{H_d}^2 + \mu^2)|h_d^0|^2 \\ &- B\mu(h_u^0 h_d^0 + \text{ h.c. }) + \frac{1}{8}(g^2 + g'^2)(|h_u^0|^2 - |h_d^0|^2)^2, \end{aligned} \tag{A.1}$$

with the stabilizing conditions:

$$\begin{cases} (B\mu)^2 > (m_{H_u}^2 + \mu^2)(m_{H_d}^2 + \mu^2) & \text{(Stability of the vacuum)} \\ m_{H_u}^2 + m_{H_d}^2 + 2\mu^2 > 2|B\mu| & \text{(No flat directions)} \end{cases} . \tag{A.2}$$

Here $m_{H_u}^2$ and $m_{H_d}^2$ are the soft SUSY breaking mass shift of the Higgs boson, and $B\mu$ the soft SUSY breaking mixing between up and down type Higgs. The gauge

N. Yamanaka, *Analysis of the Electric Dipole Moment in the R-parity Violating Supersymmetric Standard Model*, Springer Theses,
DOI: 10.1007/978-4-431-54544-6, © Springer Japan 2014

couplings g and g' are the respective coupling constants of the $SU(2)_L$ and $U(1)_Y$ gauge groups, so that we have $g = \frac{e}{\sin\theta_W}$ and $g' = \frac{e}{\cos\theta_W}$.

The minimization condition is

$$\begin{cases} B\mu = \frac{1}{2}(m_{H_u}^2 + m_{H_d}^2 + 2\mu^2)\sin 2\beta \\ \mu^2 = \frac{m_{H_d}^2 - m_{H_u}^2 \tan^2\beta}{(\tan^2\beta - 1)} - \frac{M_Z^2}{2} \end{cases}, \tag{A.3}$$

where β is given by $\tan\beta \equiv \frac{v_u}{v_d}$, and $M_Z^2 = \frac{g^2+g'^2}{2}(v_u^2 + v_d^2)$.
It is important to note that in the supersymmetric gauge theory, the Higgs scalar 4-point interaction is given by gauge couplings, which insures that Higgs fields can be treated perturbatively. The mass of the gauge bosons can be given as in the standard model by replacing the VEV of the single Higgs v^2 by $v_u^2 + v_d^2$.

We now try to give the physical aspect of the Higgs scalar. As discussed before, there are 8 parameters in the Higgs sector, each corresponding to massive and Nambu-Goldstone particles. Masses of these particles are given by the mass matrix of the Higgs bilinear coefficient. Due to the charge conservation and CP invariance of the Higgs sector, Charged Higgs sector, neutral CP-odd and CP-even blocks can be separated.

The charged block uses 4 parameters. If we express it in the SU(2) eigenstate $(h_u^+, h_u^{+*}, h_d^-, h_d^{-*})$, the mass matrix is

$$\mathscr{M}_{h^\pm}^2 = \begin{pmatrix} B\mu\cot\beta + \frac{g^2}{2}v_d^2 & -B\mu - \frac{g^2}{2}v_u v_d \\ -B\mu - \frac{g^2}{2}v_u v_d & B\mu\tan\beta + \frac{g^2}{2}v_u^2 \end{pmatrix}. \tag{A.4}$$

The mass eigenvalue is

$$m_G^\pm = 0\ ,\ m_{H^\pm}^2 = B\mu(\cot\beta + \tan\beta) + M_W^2. \tag{A.5}$$

Here, the zero mass eigenstates are the Nambu-Goldstone modes, which will give masses to W bosons.

The rotation we have used is then

$$\begin{pmatrix} G^+ \\ H^+ \end{pmatrix} = \begin{pmatrix} \cos\beta & \sin\beta \\ -\sin\beta & \cos\beta \end{pmatrix} \begin{pmatrix} h_d^{-*} \\ h_u^+ \end{pmatrix}. \tag{A.6}$$

Next, we treat the neutral CP-odd (pseudo-scalar) Higgs field (h_{uI}^0, h_{dI}^0).

$$\mathscr{M}_{h_{u,dI}^0}^2 = \begin{pmatrix} B\mu\cot\beta & B\mu \\ B\mu & B\mu\tan\beta \end{pmatrix}. \tag{A.7}$$

The mass eigenvalue is

$$m_{G^0} = 0\,, \quad m_A^2 = B\mu(\cot\beta + \tan\beta). \tag{A.8}$$

The zero mass eigenstate is the Nambu-Goldstone mode, which will give masses to Z bosons. The rotation is

$$\begin{pmatrix} G^0 \\ A \end{pmatrix} = \begin{pmatrix} \sin\beta & -\cos\beta \\ \cos\beta & \sin\beta \end{pmatrix} \begin{pmatrix} h_{uI}^0 \\ h_{dI}^0 \end{pmatrix}. \tag{A.9}$$

The remaining sector is the neutral CP-even Higgs (h_{uR}^0, h_{dR}^0):

$$\mathcal{M}_{h_{u,dI}^0}^2 = \begin{pmatrix} m_A^2\cos^2\beta + M_Z^2\sin^2\beta & -(m_A^2 + M_Z^2)\sin\beta\cos\beta \\ -(m_A^2 + M_Z^2)\sin\beta\cos\beta & m_A^2\sin^2\beta + M_Z^2\cos^2\beta \end{pmatrix}. \tag{A.10}$$

The mass eigenvalue is

$$m_{h,H}^2 = \frac{1}{2}\left[(m_A^2 + M_Z^2) \mp \sqrt{(m_A^2 + M_Z^2)^2 - 4m_A^2 M_Z^2\cos^2 2\beta}\,\right]. \tag{A.11}$$

There are 2 massive scalars, with one heavy and one light. The rotation is

$$\begin{pmatrix} h \\ H \end{pmatrix} = \begin{pmatrix} \cos\alpha & \sin\alpha \\ -\sin\alpha & \cos\alpha \end{pmatrix} \begin{pmatrix} h_{uR}^0 \\ h_{dR}^0 \end{pmatrix}, \tag{A.12}$$

with

$$\tan\alpha = \frac{(m_A^2 - M_Z^2)\cos 2\beta + \sqrt{(m_A^2 - M_Z^2)^2 - 4m_A^2 M_Z^2\cos^2 2\beta}}{(m_A^2 + M_Z^2)\sin\beta}. \tag{A.13}$$

Technically, these rotation matrices are incorporated in vertices which create or annihilate these Higgs for using mass eigenstate propagators.

Charginos and Neutralinos

Supersymmetric partners of Higgs scalars and gauge bosons have the same quantum number, so they can mix with each other. Within the charge conserving theory, 2 types of particles are possible. The first one is the mixing between supersymmetric partners of charged Higgs and W bosons: the chargino. The other one is the mixing between supersymmetric partners of the neutral Higgs, the neutral $U(1)_Y$ gauge boson, and the t_3 component of the $SU(2)_L$ gauge boson: the neutralino.

There are 3 distinct sources in the mass matrix:

- Mass terms from the superpotential (Higgs bilinear):

$$\mathscr{L}_1 = -\frac{\mu}{2}\left(\overline{\psi}_{h_u^0}\psi_{h_d^0} + \overline{\psi}_{h_d^0}\psi_{h_u^0} + \overline{\psi}_{h_+^0}\psi_{h_d^-} + \overline{\psi}_{h_d^-}\psi_{h_u^+}\right). \tag{A.14}$$

- Gaugino-Higgsino mixing due to the breakdown of electroweak symmetry:

$$\begin{aligned}\mathscr{L}_2 = &- g v_u \overline{\lambda}\frac{1+\gamma_5}{2}\psi_h - g v_d \overline{\lambda}\frac{1-\gamma_5}{2}\psi_h + \text{h.c.}\\ &- \frac{g v_u}{\sqrt{2}}\overline{\lambda}_3\psi_{h_u^0} + \frac{g' v_u}{\sqrt{2}}\overline{\lambda}_0\psi_{h_u^0} + \frac{g v_d}{\sqrt{2}}\overline{\lambda}_3\psi_{h_d^0} - \frac{g' v_d}{\sqrt{2}}\overline{\lambda}_0\psi_{h_d^0},\end{aligned} \tag{A.15}$$

where $\lambda = \frac{\lambda_1 + i\lambda_2}{\sqrt{2}}$ and $\psi_h = P_L\psi_{h_d^-} - P_R\psi_{h_u^+}$. ($P_L \equiv \frac{1-\gamma_5}{2}$ and $P_R \equiv \frac{1+\gamma_5}{2}$)

- Supersymmetry breaking gaugino mass:

$$\mathscr{L} = -\frac{1}{2}M_1\overline{\lambda}_0\lambda_0 - \frac{1}{2}M_2\overline{\lambda}_3\lambda_3 - M_2\overline{\lambda}\lambda. \tag{A.16}$$

Chargino Mass Matrix

The total chargino mass term can be written as

$$\mathscr{L}_{\text{chargino}} = -\left(\overline{\lambda}, \overline{\psi}_h\right)\left(\mathscr{M}_c P_L + \mathscr{M}_c^T P_R\right)\begin{pmatrix}\lambda\\ \psi_h\end{pmatrix}, \tag{A.17}$$

where,

$$\mathscr{M}_c = \begin{pmatrix} M_2 & g v_d \\ g v_u & -\mu \end{pmatrix}. \tag{A.18}$$

To find its physical mass eigenstate, we must diagonalize this mass matrix. We see that the chargino mass matrix has CP-odd mass terms, so chiral rotation is needed to diagonalize it. This can be achieved by applying different unitary matrices Z_+ and Z_- to the left and right. The physical chargino mass matrix can then be written as

$$\mathscr{M}_D \equiv Z_-^T \mathscr{M}_c Z_+ \equiv \begin{pmatrix} m_{\chi_2} & 0 \\ 0 & m_{\chi_1}\end{pmatrix}, \quad \mathscr{M}_D^\dagger \equiv Z_+^\dagger \mathscr{M}_c^T Z_-^* \equiv \begin{pmatrix} m_{\chi_2} & 0 \\ 0 & m_{\chi_1}\end{pmatrix}, \tag{A.19}$$

where the physical chargino (mass) eigenstate χ satisfies

$$P_L\begin{pmatrix}\lambda\\ \psi_h\end{pmatrix} \equiv Z_+ P_L \chi \equiv Z_+ P_L \begin{pmatrix}\chi_2\\ \chi_1\end{pmatrix}, \quad P_R\begin{pmatrix}\lambda\\ \psi_h\end{pmatrix} \equiv Z_-^* P_R \chi \equiv Z_-^* P_R\begin{pmatrix}\chi_2\\ \chi_1\end{pmatrix}. \tag{A.20}$$

Here Z_+ and Z_-^* are unitary matrix which diagonalize $\mathscr{M}_c^T \mathscr{M}_c$ and $\mathscr{M}_c \mathscr{M}_c^T$, respectively. By solving the eigenvalue problem, we find

$$Z_+ = \begin{pmatrix} \frac{1}{\sqrt{1+x_2^2}} & \frac{1}{\sqrt{1+x_1^2}} \\ \frac{x_2}{\sqrt{1+x_2^2}} & \frac{x_1}{\sqrt{1+x_1^2}} \end{pmatrix}, \quad Z_-^* = \begin{pmatrix} \frac{1}{\sqrt{1+y_2^2}} & \frac{1}{\sqrt{1+y_1^2}} \\ \frac{y_2}{\sqrt{1+y_2^2}} & \frac{y_1}{\sqrt{1+y_1^2}} \end{pmatrix}, \tag{A.21}$$

with

$$\begin{aligned} x_1 &= \frac{\mu^2 - M_2^2 + 2m_W^2 \cos 2\beta - \sqrt{(\mu^2 - M_2^2)^2 + 4m_W^2(m_W^2 \cos^2 2\beta + \mu^2 + M_2^2 - 2\mu M_2 \sin 2\beta)}}{2\sqrt{2}m_W(-M_2 \cos\beta + \mu \sin\beta)}, \\ x_2 &= \frac{\mu^2 - M_2^2 + 2m_W^2 \cos 2\beta + \sqrt{(\mu^2 - M_2^2)^2 + 4m_W^2(m_W^2 \cos^2 2\beta + \mu^2 + M_2^2 - 2\mu M_2 \sin 2\beta)}}{2\sqrt{2}m_W(-M_2 \cos\beta + \mu \sin\beta)}, \\ y_1 &= \frac{\mu^2 - M_2^2 + 2m_W^2 \cos 2\beta - \sqrt{(\mu^2 - M_2^2)^2 + 4m_W^2(m_W^2 \cos^2 2\beta + \mu^2 + M_2^2 - 2\mu M_2 \sin 2\beta)}}{2\sqrt{2}m_W(-M_2 \sin\beta + \mu \cos\beta)}, \\ y_2 &= \frac{\mu^2 - M_2^2 + 2m_W^2 \cos 2\beta + \sqrt{(\mu^2 - M_2^2)^2 + 4m_W^2(m_W^2 \cos^2 2\beta + \mu^2 + M_2^2 - 2\mu M_2 \sin 2\beta)}}{2\sqrt{2}m_W(-M_2 \sin\beta + \mu \cos\beta)}, \end{aligned} \tag{A.22}$$

where m_W is the mass of the W boson.

Neutralino Mass Matrix

The total neutralino mass term can be written as

$$\mathscr{L}_{\text{neutralino}} = -\frac{1}{2}\left(\overline{\psi}_{h_u^0}, \overline{\psi}_{h_d^0}, \overline{\lambda}_3, \overline{\lambda}_0\right) \begin{pmatrix} 0 & \mu & \frac{gv_u}{\sqrt{2}} & -\frac{g'v_u}{\sqrt{2}} \\ \mu & 0 & -\frac{gv_d}{\sqrt{2}} & \frac{g'v_d}{\sqrt{2}} \\ \frac{gv_u}{\sqrt{2}} & -\frac{gv_d}{\sqrt{2}} & M_2 & 0 \\ -\frac{g'v_u}{\sqrt{2}} & \frac{g'v_d}{\sqrt{2}} & 0 & M_1 \end{pmatrix} \begin{pmatrix} \psi_{h_u^0} \\ \psi_{h_d^0} \\ \lambda_3 \\ \lambda_0 \end{pmatrix}. \tag{A.23}$$

This matrix must also be diagonalized to find the physical mass eigenstates. The neutrinalino mass matrix is real and hermitian, so it can be diagonalized by orthogonal matrix. If some mass eigenvalue are negative, we can always redefine the eigenstate like $\xi' = (-i\gamma_5)\xi$ to make the mass positive. The diagonalization of 4×4 matrix is tedious, and we generally diagonalize it numerically.

Squarks and Sleptons

Squarks and sleptons have inter-generation (flavor) and intra-generation ($L \leftrightarrow R$) mixings. Thus, there are 6 eigenstates for each squark and slepton. Their mass matrices have 4 distinct sources:

- Superpotential terms:
  Sfermions get the same contribution as their fermion partners from the superpotential, from the auxiliary field bilinear:

$$\mathscr{F}_i^\dagger \mathscr{F}_i = -\sum_i \left|\frac{\partial \hat{f}}{\partial \hat{\mathscr{S}}_i}\right|^2_{\hat{\mathscr{S}}=\mathscr{S}} \quad \text{with } \hat{f} = \mu \hat{h}_u^0 \hat{h}_d^0 + f_i \hat{f}_i \hat{h}_{u,d}^0 \hat{f}_i^c. \tag{A.24}$$

  Here, $\hat{f}_i$ is the fermion field, f_i its Yukawa coupling. Derivation of each type of scalar (Higgs, $SU(2)_L$ singlet and doublet fermions) gives the diagonal mass contributions

$$\mathscr{L}_1 = \sum_{\text{flavor, R/L}} m_i^2 - \tilde{f}_i^\dagger \tilde{f}_i, \tag{A.25}$$

  and also the intra-generational ($L \leftrightarrow R$) mixing

$$\mathscr{L}_2 = \sum_{\text{flavor}} -(\mu f_i h_{u/d}^0)(\tilde{f}_{iL}^\dagger \tilde{f}_{iR} + \tilde{f}_{iR}^\dagger \tilde{f}_{iL}), \tag{A.26}$$

  where $h_{u/d}^0 = h_u^0$ for up type sfermions, and $h_{u/d}^0 = h_d^0$ for down type sfermions. The intra-generation mixing is realized when the Higgs fields get vacuum expectation value, since $f_i v_{u/d}^0 = m_i$.
- Soft supersymmetry breaking scalar masses:
  The first and second line of Eq. (3.4). There are flavor mixing contributions, but no intra-generation mixings.
- Soft supersymmetry breaking trilinear terms:
  The fifth line of Eq. (3.4). There are only intra-generation mixings, with flavor changing contributions. The electroweak symmetry must be broken.
- D-term contributions:
  The scalar 4 point interaction present in the D-term ($\propto \mathscr{S}_i^\dagger g t_A \mathscr{S}_i$) can generate mass contributions, since it contains also Higgs scalars.

$$\begin{aligned}\mathscr{L} &= -\frac{1}{2}\sum_A \left|\sum_i \mathscr{S}_i^\dagger g t_A \mathscr{S}_i\right|^2 \\ &= -\frac{1}{2}g^2 \left|\tilde{Q}^\dagger T_{3Q}\tilde{Q} + H_u^\dagger \frac{\tau_3}{2} H_u - H_d^\dagger \frac{\tau_3}{2} H_d\right|^2 \\ &\quad -\left(\frac{g'}{2}\right)^2 \left|H_u^\dagger Y_{H_u} H_u + H_d^\dagger Y_{H_d} H_d + \tilde{Q}^\dagger Y_Q \tilde{Q} + \tilde{u}_{Ri}^\dagger Y_{U^c} \tilde{u}_{Ri} + \tilde{d}_{Ri}^\dagger Y_{D^c} \tilde{d}_{Ri}\right|^2.\end{aligned} \tag{A.27}$$

Y are the hypercharges of each scalars. Cross terms between spontaneously broken Higgs and sfermions generate mass contributions. Note that hypercharges of right-handed sfermions are in fact those of left-handed anti-fermions. We can rewrite the corresponding mass by using the electric charge (instead of hypercharge)

$$m^2_{\text{D-term}} = M_Z^2 \cos 2\beta (T_3 - Q \sin^2 \theta_W). \tag{A.28}$$

It is important to note that there are no inter/intra-generational mixing from the D-term.

Generally, the flavor mixing contributions are strongly suppressed due to the phenomenology of the flavor changing neutral current. If the flavor mixing is neglected, we can construct the sfermion mass matrix by treating each flavor independently. The sfermion mass matrix (only for one generation) can be given as follows:

$$\mathscr{M}_{\text{sfermion}} = \begin{pmatrix} m^2_{\tilde{f}_{Li}} + m^2_{f_i} + \frac{e^2(v_u^2 - v_d^2)(T_{3f_i} - Q_{f_i} \sin^2\theta_W)}{4\sin^2\theta_W \cos^2\theta_W} & -m_f(R_f\mu + A_{fi}/Y_{f_i}) \\ -m_f(R_f\mu + A_{fi}/Y_{f_i}) & m^2_{\tilde{f}_{Ri}} + m^2_{f_i} + Q_{fi}\frac{e^2(v_u^2 - v_d^2)}{4\cos^2\theta_W} \end{pmatrix}, \tag{A.29}$$

where $R_f = \cot\beta$ for up type sfermions and $R_f = \tan\beta$ for down type sfermions. Parameters μ and A_{f_i} may have CP violating phases.

Appendix B
Fierz Transformation

The main purpose of the Fierz transformation is to convert a product of two fermion bilinear into another with one member of the bilinear interchanged.

$$(\overline{u}_1 O_i u_2)(\overline{u}_3 O'_i u_4) = \sum_k C_{ik}(\overline{u}_1 O_k u_4)(\overline{u}_3 O'_k u_2), \tag{B.1}$$

where $O_i \times O'_i$ are matrices which obey the same Lorentz transformation. In the case of 4 dimensional field theory, there are 16 independent combinations of these matrices and some of them are summed each other to give the same Lorentz transformation.

To obtain the Fierz transformation coefficients, we use the completeness of the matrices O_i. We take the trace of the matrices in equation (B.1) with every combinations $O_j \times O'_j$:

$$(O_i)_{\alpha\beta}(O'_i)_{\gamma\delta} = \sum_k C_{ik}(O_k)_{\alpha\delta}(O'_k)_{\gamma\beta}. \tag{B.2}$$

If we introduce $(O_j)_{\delta\alpha}(O'_j)_{\beta\gamma}$, this equation becomes

$$Tr(O_j O_i O'_j O'_i) = \sum_k C_{ik} Tr(O_k O_j) Tr(O'_k O'_j). \tag{B.3}$$

We must be careful in the order of the matrices.

Defining $A_{kj} \equiv Tr(O_k O_j)Tr(O'_k O'_j)$, $B_{ij} \equiv Tr(O_j O_i O'_j O'_i)$, and using the fact that $A_{kj} = A_j$ is diagonal, we have finally

$$C_{ij} = B_{ij}/A_j. \tag{B.4}$$

For Lorentz scalar combinations, we have

$$O_i \times O'_i \;=\; \mathbf{1} \times \mathbf{1},\ \gamma_5 \times \gamma_5,\ \gamma^\mu \times \gamma_\mu,\ \gamma^\mu\gamma_5 \times \gamma_\mu\gamma_5,\ \sigma^{\mu\nu} \times \sigma_{\mu\nu}. \tag{B.5}$$

N. Yamanaka, *Analysis of the Electric Dipole Moment in the R-parity Violating Supersymmetric Standard Model*, Springer Theses,
DOI: 10.1007/978-4-431-54544-6, © Springer Japan 2014

The Fierz transformation coefficients are given by

$$C_{ij} = \begin{pmatrix} 1/4 & 1/4 & 1/4 & -1/4 & 1/8 \\ 1/4 & 1/4 & -1/4 & 1/4 & 1/8 \\ 1 & -1 & -1/2 & -1/2 & 0 \\ -1 & 1 & -1/2 & -1/2 & 0 \\ 3 & 3 & 0 & 0 & -1/2 \end{pmatrix}. \quad \text{(B.6)}$$

Appendix C
Condensates and QCD Correlators

Mixed Condensate in Nucleon ($\langle N | g_s \overline{q} \sigma_{\mu\nu} G_a^{\mu\nu} t_a q | N \rangle$)

By reducing the matrix element involving the chromo-EDM with PCAC reduction, we obtain the mixed condensate in nucleon $\langle N | g_s \overline{q} \sigma_{\mu\nu} G_a^{\mu\nu} t_a q | N \rangle$.

By taking the lightest 0^{++} state as leading, we have

$$\frac{\langle N | \hat{O}_s | N \rangle}{\langle N | \theta_\mu^\mu | N \rangle} \approx \frac{\langle 0 | \hat{O}_s | \sigma \rangle \langle \sigma | N \overline{N} \rangle}{\langle 0 | \theta_\mu^\mu | \sigma \rangle \langle \sigma | N \overline{N} \rangle}, \tag{C.1}$$

where θ_μ^μ is the trace of the energy-momentum tensor, $\hat{O}_s$ is the scalar operator for which the condensate will be taken, and σ the lowest 0^{++} flavor $SU(3)$ singlet state. Here the detail of the state σ is not important. By applying the above relation for $\hat{O}_s = g_s \overline{q} \sigma^{\mu\nu} G_{\mu\nu}^a t_a q$, $\hat{O}_s = g_s \overline{q} q$, and taking their ratio, we obtain

$$\frac{\langle N | g_s \overline{q} \sigma^{\mu\nu} G_{\mu\nu}^a t_a q | N \rangle}{\langle N | \overline{q} q | N \rangle} \approx \frac{\langle 0 | g_s \overline{q} \sigma^{\mu\nu} G_{\mu\nu}^a t_a q | \sigma \rangle}{\langle 0 | \overline{q} q | \sigma \rangle}. \tag{C.2}$$

The matrix element $\langle 0 | \hat{O}_s | \sigma \rangle$ can be evaluated by using the low energy theorem.

$$\lim_{q \to 0} i \int dx\, e^{iq(x-y)} \left\langle 0 \left| T \left\{ \hat{O}_s(x) \frac{\beta(\alpha_s)}{\alpha_s} G_{\mu\nu}^a G_a^{\mu\nu}(y) \right\} \right| 0 \right\rangle \approx -d \langle 0 | \hat{O}_s | 0 \rangle + \cdots, \tag{C.3}$$

where $\beta(\alpha_s) = -(\frac{11}{3} N_c - \frac{2}{3} N_f)(\alpha_s^2 / 2\pi) + O(\alpha_s^3)$ and d is the canonical dimension of the operator $\hat{O}_s$. The ellipses denote terms linear in quark masses. We have therefore

$$\frac{\langle N | g_s \overline{q} \sigma^{\mu\nu} G_{\mu\nu}^a t_a q | N \rangle}{\langle N | \overline{q} q | N \rangle} \approx \frac{5}{3} m_0^2. \tag{C.4}$$

N. Yamanaka, *Analysis of the Electric Dipole Moment in the R-parity Violating Supersymmetric Standard Model*, Springer Theses,
DOI: 10.1007/978-4-431-54544-6, © Springer Japan 2014

where $m_0^2 = \langle 0| g_s\overline{q}\sigma^{\mu\nu}G^a_{\mu\nu}t_a q |0\rangle / \langle 0| \overline{q}q |0\rangle$. We must note that if the vacuum saturation approximation is taken,

$$\frac{\langle N| g_s\overline{q}\sigma^{\mu\nu}G^a_{\mu\nu}t_a q |N\rangle}{\langle N| \overline{q}q |N\rangle} \approx m_0^2. \tag{C.5}$$

Topological Susceptibility

The correlator we want to evaluate is

$$K = -i \lim_{k\to 0} \int d^4x\, e^{ik(x-y)} \left\langle 0 \left| T \left\{ \frac{\alpha_s}{8\pi} G^a_{\mu\nu}\tilde{G}^{\mu\nu,a}(x)\, \frac{\alpha_s}{8\pi} G^b_{\rho\sigma}\tilde{G}^{\rho\sigma,b}(y) \right\} \right| 0 \right\rangle . \tag{C.6}$$

Consider first the polarization operator

$$P_{\mu\nu}(q) = i \int d^4x\, e^{iq(x-y)} \langle 0 | T \{ j_{5\mu}(x) j_{5\nu}(y) \} | 0 \rangle , \tag{C.7}$$

where $j_{5\mu}(x) \equiv \sum_q^{N_f} \overline{q}(x)\gamma_\mu\gamma_5 q(x)$. By putting $q^\mu q^\nu$, we obtain

$$\begin{aligned}
\lim_{q\to 0} q^\mu q^\nu P_{\mu\nu}(q) &= \lim_{q\to 0} i \int d^4x\, q^\mu q^\nu e^{iq(x-y)} \langle 0 | T \{ j_{5\mu}(x) j_{5\nu}(y) \} | 0 \rangle \\
&= -\lim_{q\to 0} i \int d^4x \left(\partial_x^\mu \partial_x^\nu e^{iq(x-y)} \right) \langle 0 | T \{ j_{5\mu}(x) j_{5\nu}(y) \} | 0 \rangle \\
&= -i \int d^4x\, \partial_x^\mu \partial_x^\nu \langle 0 | T \{ j_{5\mu}(x) j_{5\nu}(y) \} | 0 \rangle \\
&= i \int d^4x\, \partial_x^\mu \partial_y^\nu \langle 0 | T \{ j_{5\mu}(x) j_{5\nu}(y) \} | 0 \rangle \\
&= i \int d^4x \langle 0 | T \{ \partial_x^\mu j_{5\mu}(x) \partial_y^\nu j_{5\nu}(y) \} \\
&\qquad + [j_{50}(x), \partial_y^\nu j_{5\nu}(y)] \delta(x^0 - y^0) \\
&\qquad - [\partial_x^\mu j_{5\mu}(x), j_{50}(y)] \delta(x^0 - y^0) \\
&\qquad - [j_{50}(x), j_{50}(y)] \partial^0 \delta(x^0 - y^0) |0\rangle \\
&= i \int d^4x \langle 0 | T \{ \partial_x^\mu j_{5\mu}(x) \partial_y^\nu j_{5\nu}(y) \} \\
&\qquad + [j_{50}(x), \partial_y^\nu j_{5\nu}(y)] \delta(x^0 - y^0) |0\rangle \\
&= i \int d^4x \left\langle 0 \right| T \left\{ N_f \frac{\alpha_s}{4\pi} G^a_{\mu\nu}\tilde{G}^{\mu\nu,a}(x)\, N_f \frac{\alpha_s}{4\pi} G^b_{\rho\sigma}\tilde{G}^{\rho\sigma,b}(y) \right.
\end{aligned}$$

$$
\begin{aligned}
&\quad + \sum_q^{N_f} 2m_q \overline{q} i\gamma_5 q(x)\, N_f \frac{\alpha_s}{4\pi} G^a_{\mu\nu}\tilde{G}^{\mu\nu,a}(y) \\
&\quad + N_f \frac{\alpha_s}{4\pi} G^a_{\mu\nu}\tilde{G}^{\mu\nu,a}(x) \sum_q^{N_f} 2m_q \overline{q} i\gamma_5 q(y) \\
&\quad + \sum_q^{N_f} 2m_q \overline{q} i\gamma_5 q(x) \sum_{q'}^{N_f} 2m_{q'} \overline{q}' i\gamma_5 q'(y)\}|0\rangle \\
&+ \sum_q^{N_f} 4m_q \langle 0|\overline{q}q|0\rangle \\
&+ i \int d^4x\, \langle 0|\, [j_{50}(x), N_f \frac{\alpha_s}{4\pi} G^a_{\mu\nu}\tilde{G}^{\mu\nu,a}(y)]\delta(x^0 - y^0)\, |0\rangle \\
&= 0, \qquad \text{(C.8)}
\end{aligned}
$$

where $\partial^\mu_x \equiv \frac{\partial}{\partial x_\mu}$ and $\partial^\nu_y \equiv \frac{\partial}{\partial y_\nu}$. In the fourth equality, we have used the translational invariance $\partial^\nu_x f(x-y) = -\partial^\nu_y f(x-y)$. The third term of the fifth equality can be integrated by part, thus leading to the form $[j_{50}(x), j_{50}(y)]\partial^0\delta(x^0 - y^0)$ which cancels with the last term of the fifth equality. Note that in the fifth equality, the time ordering T applies only to the first term. In the seventh equality, we have used the anomaly relation $\partial_\mu j^\mu_5 = \sum_q^{N_f} 2m_q\overline{q}i\gamma_5 q + N_f \frac{\alpha_s}{4\pi} G^a_{\mu\nu}\tilde{G}^{\mu\nu,a}$. To derive the term with $\langle 0|\overline{q}q|0\rangle$ in the seventh equality, we have used the standard anti-commutation rule between quark field operators to reduce the commutator $[j_{50}(x), \sum_q^{N_f} 2m_q\overline{q}i\gamma_5 q(y)]$. The polarization $q^\mu q^\nu P_{\mu\nu}(q)$ vanishes in the limit $q \to 0$ since there are no singularity in $q^2 = 0$ (there are no massless particles in QCD). The last term of the seventh equality can be omitted since $\langle 0|\frac{\alpha_s}{4\pi} G^a_{\mu\nu}\tilde{G}^{\mu\nu,a}|n\rangle$ (with $|n\rangle$ an arbitrary hadronic state) can be rotated away with the chiral rotation, just like when we remove the θ-term.

Consider next the following correlator

$$
P_\mu(q) = i \int d^4x\, e^{iq(x-y)} \left\langle 0 \left| T \left\{ j_{5\mu}(x)\, N_f \frac{\alpha_s}{4\pi} G^a_{\mu\nu}\tilde{G}^{\mu\nu,a}(y) \right\} \right| 0 \right\rangle. \qquad \text{(C.9)}
$$

As in Eq. (C.8), we put q^μ into the integral:

$$
\begin{aligned}
\lim_{q\to 0} q^\mu P_\mu(q) &= \lim_{q\to 0} i \int d^4x\, q^\mu e^{iq(x-y)} \left\langle 0 \left| T \left\{ j_{5\mu}(x)\, N_f \frac{\alpha_s}{4\pi} G^a_{\rho\sigma}\tilde{G}^{\rho\sigma,a}(y) \right\} \right| 0 \right\rangle \\
&= - \lim_{q\to 0} \int d^4x\, e^{iq(x-y)} \partial^\mu \left\langle 0 \left| T \left\{ j_{5\mu}(x)\, N_f \frac{\alpha_s}{4\pi} G^a_{\rho\sigma}\tilde{G}^{\rho\sigma,a}(y) \right\} \right| 0 \right\rangle \\
&= - \int d^4x \left\langle 0 \right| T \left\{ \sum_q^{N_f} 2m_q \overline{q} i\gamma_5 q(x)\, N_f \frac{\alpha_s}{4\pi} G^a_{\mu\nu}\tilde{G}^{\mu\nu,a}(y) \right.
\end{aligned}
$$

$$
\begin{aligned}
& \qquad\qquad +N_f \frac{\alpha_s}{4\pi} G^a_{\mu\nu} \tilde{G}^{\mu\nu,a}(x)\, N_f \frac{\alpha_s}{4\pi} G^b_{\rho\sigma} \tilde{G}^{\rho\sigma,b}(y) \Big\} \Big| 0 \Big\rangle \\
& - \int d^4x \, \Big\langle 0 \Big| \Big[j_{50}(x)\, ,\, N_f \frac{\alpha_s}{4\pi} G^a_{\mu\nu} \tilde{G}^{\mu\nu,a}(y) \Big] \delta(x^0 - y^0) \Big| 0 \Big\rangle \\
= & - \int d^4x \, \Big\langle 0 \Big| T \Big\{ \sum_q^{N_f} 2m_q \overline{q} i\gamma_5 q(x)\, N_f \frac{\alpha_s}{4\pi} G^a_{\mu\nu} \tilde{G}^{\mu\nu,a}(y) \\
& \qquad\qquad +N_f \frac{\alpha_s}{4\pi} G^a_{\mu\nu} \tilde{G}^{\mu\nu,a}(x)\, N_f \frac{\alpha_s}{4\pi} G^b_{\rho\sigma} \tilde{G}^{\rho\sigma,b}(y) \Big\} \Big| 0 \Big\rangle \\
= & \; 0. \qquad\qquad\qquad\qquad\qquad\qquad\qquad\qquad\qquad\qquad\qquad\qquad \text{(C.10)}
\end{aligned}
$$

This time again the above relation vanishes, since there are no massless hadrons. We have also omitted the term with $[j_{50}(x)\, ,\, N_f \frac{\alpha_s}{4\pi} G^a_{\mu\nu} \tilde{G}^{\mu\nu,a}(0)]$ using the chiral rotation as in Eq. (C.8). By taking $x - y \to y - x$ and using the hermiticity of the above relation, we have also

$$
\begin{aligned}
& \int d^4x \, \Big\langle 0 \Big| T \Big\{ N_f \frac{\alpha_s}{4\pi} G^a_{\mu\nu} \tilde{G}^{\mu\nu,a}(x) \sum_q^{N_f} 2m_q \overline{q} i\gamma_5 q(y) \\
& \qquad\qquad +N_f \frac{\alpha_s}{4\pi} G^a_{\mu\nu} \tilde{G}^{\mu\nu,a}(x)\, N_f \frac{\alpha_s}{4\pi} G^b_{\rho\sigma} \tilde{G}^{\rho\sigma,b}(y) \Big\} \Big| 0 \Big\rangle = 0\,. \qquad \text{(C.11)}
\end{aligned}
$$

By substituting Eqs. (C.10) and (C.11) to Eq. (C.8), we obtain the following sum rule:

$$
\begin{aligned}
& i \int d^4x \, \Big\langle 0 \Big| T \Big\{ - N_f \frac{\alpha_s}{4\pi} G^a_{\mu\nu} \tilde{G}^{\mu\nu,a}(x)\, N_f \frac{\alpha_s}{4\pi} G^b_{\rho\sigma} \tilde{G}^{\rho\sigma,b}(y) \\
& \quad + \sum_q^{N_f} 2m_q \overline{q} i\gamma_5 q(x) \sum_{q'}^{N_f} 2m_{q'} \overline{q}' i\gamma_5 q'(y) \Big\} \Big| 0 \Big\rangle + \sum_q^{N_f} 4m_q \langle 0 | \overline{q} q | 0 \rangle = 0\,.
\end{aligned}
\tag{C.12}
$$

The final step is the evaluation of $T \equiv i \int d^4x \, \langle 0 | T\{ \sum_q^{N_f} 2m_q \overline{q} i\gamma_5 q(x) \sum_{q'}^{N_f} 2m_{q'} \overline{q}' i\gamma_5 q'(y) \} | 0 \rangle$. From now we consider only u and d quarks, which give the leading contribution.

$$
\begin{aligned}
T = & \lim_{q \to 0} i \int d^4x \, e^{iq(x-y)} \Big\langle 0 | \, T \Big\{ \sum_{q=u,d} 2m_q \overline{q} i\gamma_5 q(x) \sum_{q'=u,d} 2m_{q'} \overline{q}' i\gamma_5 q'(y) \Big\} \Big| 0 \Big\rangle \\
\approx & \lim_{q \to 0} \langle 0 | \, 2(m_u \overline{u} i\gamma_5 u + m_d \overline{d} i\gamma_5 d) \, | \pi \rangle \frac{-1}{q^2 - m_\pi^2} \langle \pi | \, 2(m_u \overline{u} i\gamma_5 u + m_d \overline{d} i\gamma_5 d) \, | 0 \rangle \\
= & \frac{4}{m_\pi^2} | \langle 0 | \, m_u \overline{u} i\gamma_5 u + m_d \overline{d} i\gamma_5 d \, | \pi \rangle |^2.
\end{aligned}
\tag{C.13}
$$

In the second equality, we have taken the pion propagation as the leading contribution, since it is the lightest mode. The matrix element $\langle 0|\, m_u \overline{u} i\gamma_5 u + m_d \overline{d} i\gamma_5 d\, |\pi\rangle$ is calculated as follows:

$$\begin{aligned}\langle 0|\, m_u \overline{u} i\gamma_5 u + m_d \overline{d} i\gamma_5 d\, |\pi\rangle &= \frac{1}{2}\langle 0|\, (m_u + m_d)(\overline{u} i\gamma_5 u + \overline{d} i\gamma_5 d) \\ &\quad +(m_u - m_d)(\overline{u} i\gamma_5 u - \overline{d} i\gamma_5 d)\, |\pi\rangle \\ &\approx \frac{1}{2}(m_u - m_d)\langle 0|\, \overline{u} i\gamma_5 u - \overline{d} i\gamma_5 d\, |\pi\rangle. \end{aligned} \tag{C.14}$$

We have neglected the isovector matrix element $\langle 0|\, \overline{u} i\gamma_5 u + \overline{d} i\gamma_5 d\, |\pi\rangle$. The isoscalar matrix element $\langle 0|\, \overline{u} i\gamma_5 u - \overline{d} i\gamma_5 d\, |\pi\rangle$ can be obtained via the relation for the divergence of the isovector axial current as follows:

$$\begin{aligned}\sqrt{2} f_\pi m_\pi^2 &= q_\mu \frac{1}{\sqrt{2}}\langle 0|\, \overline{u}\gamma^\mu\gamma_5 u - \overline{d}\gamma^\mu\gamma_5 d\, |\pi\rangle \\ &= \sqrt{2}\langle 0|\, m_u \overline{u} i\gamma_5 u - m_d \overline{d} i\gamma_5 d\, |\pi\rangle \\ &= \frac{1}{\sqrt{2}}\langle 0|\, (m_u + m_d)(\overline{u} i\gamma_5 u - \overline{d} i\gamma_5 d) + (m_u - m_d)(\overline{u} i\gamma_5 u + \overline{d} i\gamma_5 d)\, |\pi\rangle \\ &\approx \frac{1}{\sqrt{2}}(m_u + m_d)\langle 0|\, \overline{u} i\gamma_5 u - \overline{d} i\gamma_5 d\, |\pi\rangle, \end{aligned} \tag{C.15}$$

where we have again neglected the isovector matrix element $\langle 0|\, \overline{u} i\gamma_5 u + \overline{d} i\gamma_5 d\, |\pi\rangle$. We then obtain

$$T = -4\frac{(m_u - m_d)^2}{m_u + m_d}\langle 0|\overline{q}q|0\rangle, \tag{C.16}$$

where the Gell-Mann-Oakes-Renner relation $\langle 0|\overline{q}q|0\rangle = -\frac{f_\pi^2 m_\pi^2}{m_u + m_d}$ was used. The vacuum condensates for light quarks (u and d) are approximately the same ($\langle 0|\overline{u}u|0\rangle \approx \langle 0|\overline{d}d|0\rangle \approx \langle 0|\overline{q}q|0\rangle$).

Combining Eqs. (C.12) and (C.16), we obtain

$$i\int d^4x \left\langle 0 \left| T\left\{\frac{\alpha_s}{2\pi} G^a_{\mu\nu}\tilde{G}^{\mu\nu,a}(x)\, \frac{\alpha_s}{2\pi} G^b_{\rho\sigma}\tilde{G}^{\rho\sigma,b}(y)\right\}\right| 0\right\rangle = 16\frac{m_u m_d}{m_u + m_d}\langle 0|\overline{q}q|0\rangle, \tag{C.17}$$

for $N_f = 2$. The topological susceptibility is then

$$K \approx -\frac{m_u m_d}{m_u + m_d}\langle 0|\overline{q}q|0\rangle. \tag{C.18}$$

Correlation Between the Anomaly and the Chromo-EDM

The factor K_1 is expressed in terms of the correlator between the anomaly and the P, CP-odd operator as follows

$$
\begin{aligned}
K_1(O_{CP}) &= -i \lim_{k\to 0} \int d^4x e^{ik(x-y)} \left\langle 0 \left| T\left\{ \frac{\alpha_s}{8\pi} G^a_{\mu\nu}\tilde{G}^{\mu\nu,a}(x) O_{CP}(y) \right\} \right| 0 \right\rangle \\
&= -i \int d^4x \left\langle 0 \left| T\left\{ \left(\frac{1}{2N_f}\partial_\mu j_5^\mu(x) - \frac{1}{N_f}\sum_q^{N_f} m_q \bar{q} i\gamma_5 q(x) \right) O_{CP}(y) \right\} \right| 0 \right\rangle \\
&= -i \int d^4x \left\{ \frac{1}{2N_f}\partial_\mu \langle 0|T\{ j_5^\mu(x) O_{CP}(y) \}|0\rangle \right. \\
&\qquad - \frac{1}{2N_f}\delta(x_0 - y_0)\langle 0|[j_5^0(x), O_{CP}(y)]|0\rangle \\
&\qquad \left. - \frac{1}{N_f}\sum_q^{N_f} \langle 0| T\{ m_q \bar{q} i\gamma_5 q(x) O_{CP}(y) \}|0\rangle \right\} \\
&= i \int d^4x \left\{ \frac{1}{2N_f}\delta(x_0 - y_0)\langle 0|[j_5^0(x), O_{CP}(y)]|0\rangle \right. \\
&\qquad \left. + \frac{1}{N_f}\sum_q^{N_f} \langle 0| T\{ m_q \bar{q} i\gamma_5 q(x) O_{CP}(y) \}|0\rangle \right\}. \qquad \text{(C.19)}
\end{aligned}
$$

The third equality was obtained by integration by part. In the last equality, the total derivative (the first term of the third equality) was omitted.

For the chromo-EDM ($O_{CP} = -\frac{i}{2} d_q^c \bar{q} g_s \sigma^{\mu\nu} G^a_{\mu\nu} t_a \gamma_5 q$), the first term is

$$
\begin{aligned}
\langle 0|[j_5^0(x), O_{CP}(y)]|0\rangle &= -\frac{i}{2}\sum_{q'}^{N_f} d_q^c \langle 0|[q'^\dagger \gamma_5 q'(x)\, ,\, \bar{q} g_s \sigma^{\mu\nu} G^a_{\mu\nu} t_a \gamma_5 q(y)]|0\rangle \\
&= i \sum_q^{N_f} d_q^c \langle 0|\bar{q} g_s \sigma^{\mu\nu} G^a_{\mu\nu} t_a q|0\rangle \delta(\mathbf{x} - \mathbf{y})\, . \qquad \text{(C.20)}
\end{aligned}
$$

Here we have used the commutation relation of the quark bilinear

$$
\begin{aligned}
[(q^\dagger_\alpha q_\beta)(x), (q^\dagger_\gamma q_\delta)(y)] &\equiv q^\dagger_\alpha(x) q_\beta(x) q^\dagger_\gamma(y) q_\delta(y) - q^\dagger_\gamma(y) q_\delta(y) q^\dagger_\alpha(x) q_\beta(x) \\
&= -q^\dagger_\alpha(x) q^\dagger_\gamma(y) q_\beta(x) q_\delta(y) + q^\dagger_\gamma(y) q^\dagger_\alpha(x) q_\delta(y) q_\beta(x) \\
&\quad + \delta(\mathbf{x} - \mathbf{y})[q^\dagger_\alpha(x) q_\delta(y)\delta_{\beta\gamma} - q^\dagger_\gamma(y) q_\beta(x)\delta_{\delta\alpha}] \\
&= \delta(\mathbf{x} - \mathbf{y})[q^\dagger_\alpha(x) q_\delta(y)\delta_{\beta\gamma} - q^\dagger_\gamma(y) q_\beta(x)\delta_{\delta\alpha}], \qquad \text{(C.21)}
\end{aligned}
$$

where α, β, γ and δ denote the indices of all internal degrees of freedom including the Dirac spinor index.

The time ordered correlation $T_{CP} \equiv \sum_q^{N_f} \langle 0| T\{ m_q \overline{q} i\gamma_5 q(x) O_{CP}(y) \}|0\rangle$ [second term of the last equality of Eq. (C.19)] can be transformed as follows:

$$
\begin{aligned}
T_{CP} &= \lim_{k\to 0} \sum_{q'}^{N_f} i \int d^4x e^{ik(x-y)} \langle 0| T\{m_{q'} \overline{q}' i\gamma_5 q'(x) O_{CP}(y)\}|0\rangle \\
&\approx -\frac{i}{2} \sum_{q,q'}^{N_f} \langle 0| m_{q'} \overline{q}' i\gamma_5 q' |\pi\rangle \frac{1}{m_\pi^2} \langle \pi| d_q^c \overline{q} g_s \sigma^{\mu\nu} G^a_{\mu\nu} t_a \gamma_5 q |0\rangle \\
&\approx -\frac{i}{2m_\pi^2} \sum_q^{N_f} \langle 0| m_u \overline{u} i\gamma_5 u + m_d \overline{d} i\gamma_5 d |\pi\rangle \langle \pi| d_q^c \overline{q} g_s \sigma^{\mu\nu} G^a_{\mu\nu} t_a \gamma_5 q |0\rangle \\
&\approx -\frac{i}{4m_\pi^2} \sum_q^{N_f} (m_u - m_d) \langle 0| \overline{u} i\gamma_5 u - \overline{d} i\gamma_5 d |\pi\rangle \langle \pi| d_q^c \overline{q} g_s \sigma^{\mu\nu} G^a_{\mu\nu} t_a \gamma_5 q |0\rangle \\
&\approx -\frac{i}{8m_\pi^2} (m_u - m_d) \langle 0| \overline{u} i\gamma_5 u - \overline{d} i\gamma_5 d |\pi\rangle \\
&\quad \times \sum_{q=u,d} \langle \pi| (d_u^c + d_d^c) \overline{q} g_s \sigma^{\mu\nu} G^a_{\mu\nu} t_a \gamma_5 q + (d_u^c - d_d^c) \overline{q} \tau_3 g_s \sigma^{\mu\nu} G^a_{\mu\nu} t_a \gamma_5 q) |0\rangle \\
&\approx -i \frac{f_\pi}{4} \frac{m_u - m_d}{m_u + m_d} \\
&\quad \times \sum_{q=u,d} \frac{i}{f_\pi} \langle 0| (d_u^c + d_d^c) \overline{q} \tau_3 g_s \sigma^{\mu\nu} G^a_{\mu\nu} t_a q + (d_u^c - d_d^c) \overline{q} g_s \sigma^{\mu\nu} G^a_{\mu\nu} t_a q) |0\rangle \\
&= \frac{1}{2} \frac{m_u - m_d}{m_u + m_d} \langle 0| d_u^c \overline{u} g_s \sigma^{\mu\nu} G^a_{\mu\nu} t_a u - d_d^c \overline{d} g_s \sigma^{\mu\nu} G^a_{\mu\nu} t_a d |0\rangle
\end{aligned}
\tag{C.22}
$$

where $\tau_3 \equiv \begin{pmatrix} 1 & 0 \\ 0 & -1 \end{pmatrix}$. In the fourth equality, we have used neglected the isovector matrix element as in Eq. (C.14). In the sixth equality, we have used Eq. (C.15) and the PCAC relation

$$
\lim_{k\to 0} \langle \pi^c| O_{CP} |0\rangle \approx -\frac{i}{f_\pi} \int d^3x \langle 0| [q^\dagger \gamma_5 T_c q(\mathbf{x}), O_{CP}] |0\rangle , \tag{C.23}
$$

where T_c is the generator of the $SU(2)$ group for $N_f = 2$ (for example, $T_3 = \frac{\tau_3}{2}$).

By substituting Eqs. (C.20) and (C.22) to Eq. (C.19), we obtain

$$\begin{aligned}
K_1(O_{CP}) &\approx -\frac{1}{4}\langle 0|d_u^c\overline{u}g_s\sigma^{\mu\nu}G^a_{\mu\nu}t_a u + d_d^c\overline{d}g_s\sigma^{\mu\nu}G^a_{\mu\nu}t_a d|0\rangle \\
&+\frac{1}{4}\frac{m_u - m_d}{m_u + m_d}\langle 0|\, d_u^c\overline{u}g_s\sigma^{\mu\nu}G^a_{\mu\nu}t_a u - d_d^c\overline{d}g_s\sigma^{\mu\nu}G^a_{\mu\nu}t_a d\,|0\rangle \\
&= -\frac{1}{2(m_u + m_d)}\langle 0|\, m_d d_u^c\overline{u}g_s\sigma^{\mu\nu}G^a_{\mu\nu}t_a u + m_u d_d^c\overline{d}g_s\sigma^{\mu\nu}G^a_{\mu\nu}t_a d\,|0\rangle \\
&= -\frac{m_*}{2}\sum_{q=u,d}\frac{d_q^c}{m_q}\langle 0|\,\overline{q}g_s\sigma^{\mu\nu}G^a_{\mu\nu}t_a q\,|0\rangle\,, \qquad \text{(C.24)}
\end{aligned}$$

for $N_f = 2$.

Appendix D
EDM One-loop Diagram

The leading contribution to the nucleon EDM is given by the meson one-loop diagrams as shown in Fig. D.1.

The chiral $SU(3)_L \times SU(3)_R$ invariant effective lagrangian is

$$\begin{aligned}\mathscr{L}_0 = & \frac{f_\pi^2}{4}\mathrm{Tr}(\partial_\mu U \partial^\mu U^\dagger) + \mathrm{Tr}[\overline{B}(i\partial\!\!\!/ - m)B] \\ & + \frac{i}{2}\mathrm{Tr}[\overline{B}\gamma_\mu(\xi\partial^\mu\xi^\dagger + \xi^\dagger\partial^\mu\xi)B] + \frac{i}{2}\mathrm{Tr}[\overline{B}\gamma_\mu B(\partial^\mu\xi\xi^\dagger + \partial^\mu\xi^\dagger\xi)] \\ & + \frac{i}{2}(D+F)\mathrm{Tr}[\overline{B}\gamma_\mu\gamma_5(\xi\partial^\mu\xi^\dagger - \xi^\dagger\partial^\mu\xi)B] \\ & - \frac{i}{2}(D-F)\mathrm{Tr}[\overline{B}\gamma_\mu\gamma_5 B(\partial^\mu\xi\xi^\dagger - \partial^\mu\xi^\dagger\xi)],\end{aligned} \tag{D.1}$$

where M is the meson field, B is the baryon field, and

$$\xi = \exp\left(\frac{iM}{\sqrt{2}f_\pi}\right), \tag{D.2}$$

with $U = \xi^2$. The meson field is defined by

$$M = \begin{pmatrix} \frac{\pi^0}{\sqrt{2}} + \frac{\eta^0}{\sqrt{6}} & \pi^+ & K^+ \\ \pi^- & -\frac{\pi^0}{\sqrt{2}} + \frac{\eta^0}{\sqrt{6}} & K^0 \\ K^- & \overline{K}^0 & -2\frac{\eta^0}{\sqrt{6}} \end{pmatrix}, \tag{D.3}$$

N. Yamanaka, *Analysis of the Electric Dipole Moment in the R-parity Violating Supersymmetric Standard Model*, Springer Theses,
DOI: 10.1007/978-4-431-54544-6,

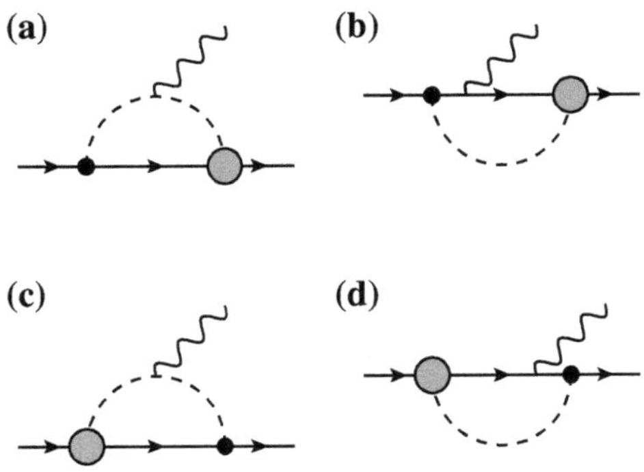

Fig. D.1 Meson loop contribution to the nucleon EDM. The *grey blob* represents the P, CP-odd meson-baryon interaction

and the baryon field by

$$B = \begin{pmatrix} \frac{\Sigma^0}{\sqrt{2}} + \frac{\Lambda^0}{\sqrt{6}} & \Sigma^+ & p \\ \Sigma^- & -\frac{\Sigma^0}{\sqrt{2}} + \frac{\Lambda^0}{\sqrt{6}} & n \\ \Xi^- & \Xi^0 & -2\frac{\Lambda^0}{\sqrt{6}} \end{pmatrix}. \tag{D.4}$$

The second line of Eq. (D.1) does not contribute to the EDM at the one-loop order, so we will neglect it. The third line can be rewritten to the zeroth order in the meson field expansion as

$$\begin{aligned} \mathscr{L}_{(M^1)} &= \frac{D+F}{\sqrt{2}f_\pi}\mathrm{Tr}\,[\overline{B}\gamma_\mu\gamma_5(\partial^\mu M)B] + \frac{D-F}{\sqrt{2}f_\pi}\mathrm{Tr}\,[\overline{B}\gamma_\mu\gamma_5 B(\partial^\mu M)] \\ &\simeq \frac{D+F}{\sqrt{2}f_\pi}\Big[\overline{p}\gamma_\mu\gamma_5 n\partial^\mu\pi^+ + \overline{n}\gamma_\mu\gamma_5 p\partial^\mu\pi^- + \overline{p}\gamma_\mu\gamma_5 p\partial^\mu\left(\frac{\pi^0}{\sqrt{2}} + \frac{\eta^0}{\sqrt{6}}\right) \\ &\quad - \frac{2}{\sqrt{6}}\overline{p}\gamma_\mu\gamma_5\Lambda\partial^\mu K^+ - \frac{2}{\sqrt{6}}\overline{\Lambda}\gamma_\mu\gamma_5 p\partial^\mu K^-\Big] \\ &\quad + \frac{D-F}{\sqrt{2}f_\pi}\Big[\overline{\Sigma}^+\gamma_\mu\gamma_5 p\partial^\mu\overline{K}^0 + \overline{p}\gamma_\mu\gamma_5\Sigma^+\partial^\mu K^0 + \overline{\Sigma}^-\gamma_\mu\gamma_5 n\partial^\mu K^- \\ &\quad + \overline{n}\gamma_\mu\gamma_5\Sigma^-\partial^\mu K^+ + \frac{1}{\sqrt{2}}\overline{\Sigma}^0\gamma_\mu\gamma_5 p\partial^\mu K^- + \frac{1}{\sqrt{2}}\overline{p}\gamma_\mu\gamma_5\Sigma^0\partial^\mu K^+ \\ &\quad + \frac{1}{\sqrt{6}}\overline{\Lambda}\gamma_\mu\gamma_5 p\partial^\mu K^- + \frac{1}{\sqrt{6}}\overline{p}\gamma_\mu\gamma_5\Lambda\partial^\mu K^+ - \frac{2}{\sqrt{6}}\overline{p}\gamma_\mu\gamma_5 p\partial^\mu\eta^0\Big], \end{aligned} \tag{D.5}$$

where $\simeq$ means that we have taken only the relevant terms. Terms without nucleon field operators were omitted since they do not appear in the nucleon EDM diagrams at the one-loop level. Terms with only electrically neutral field operators were also omitted since no photon can be attached to the one-loop graph. If we take the on-shell contribution of the baryon in the loop as dominating, we can approximate the above lagrangian by using the equation of motion $\overline{B}\gamma_\mu\gamma_5 B'\partial^\mu M \approx -i(m_B + m_{B'})\overline{B}\gamma_5 B' M$, the lagrangian can be approximated to

$$\begin{aligned}\mathcal{L}_{mBB} \approx & -g_{\pi NN}\overline{p}i\gamma_5 p\pi^0 - \sqrt{2}g_{\pi NN}(\overline{p}i\gamma_5 n\pi^+ + \text{h.c.}) - g_{\eta NN}\overline{p}i\gamma_5 p\eta^0 \\ & + g_{K\Lambda N}(\overline{p}i\gamma_5 \Lambda K^+ + \text{h.c.}) - g_{K\Sigma N}(\overline{\Sigma}^+ i\gamma_5 p\overline{K}^0 + \overline{\Sigma}^- i\gamma_5 nK^- \\ & + \frac{1}{\sqrt{2}}\overline{\Sigma}^0 i\gamma_5 pK^- + \text{h.c.}),\end{aligned} \tag{D.6}$$

where $g_{\pi NN} = \frac{m_N}{f_\pi}(D+F) \approx 12.6$, $g_{\eta NN} = \frac{m_N}{\sqrt{3}f_\pi}(3F-D) \approx 3.0$, $g_{K\Lambda N} = \frac{m_N+m_\Lambda}{2\sqrt{3}f_\pi}(D+3F) \approx 6.4$ and $g_{K\Sigma N} = \frac{m_N+m_\Sigma}{\sqrt{2}f_\pi}(D-F) \approx 6.0$, with $D = 0.81$ and $F = 0.44$. In deriving these values, we neglected the isospin splitting.

The CP-odd meson-baryon lagrangian is

$$\begin{aligned}\mathcal{L}_{CPV} = & \frac{1}{\sqrt{2}f_\pi}\Big[X\text{Tr}\,(\overline{B}B\{M,A\}) + Y\text{Tr}\,(\overline{B}\{M,A\}B) \\ & + \frac{2}{3}(Z-X-Y)\text{Tr}\,(AM)\text{Tr}\,(\overline{B}B)\Big] \\ = & \frac{X}{\sqrt{2}f_\pi}\Big[(A_d+A_s)\overline{\Sigma}^+ p\overline{K}^0 + (A_u+A_s)\left(\frac{\overline{\Sigma}^0}{\sqrt{2}} + \frac{\overline{\Lambda}}{\sqrt{6}}\right)pK^- \\ & + (A_u+A_s)\overline{\Sigma}^- nK^- + (\text{h.c.}) - \frac{4}{\sqrt{6}}A_s\overline{p}p\eta^0\Big] \\ & + \frac{Y}{\sqrt{2}f_\pi}\Big[(A_u+A_d)\overline{p}n\pi^+ - \frac{2}{\sqrt{6}}(A_u+A_s)\overline{p}\Lambda K^+ + (\text{h.c.}) \\ & + 2A_u\overline{p}p\left(\frac{\pi^0}{\sqrt{2}} + \frac{\eta^0}{\sqrt{6}}\right)\Big] \\ & + \frac{\sqrt{2}}{3f_\pi}(Z-X-Y)\left[\frac{A_u-A_d}{\sqrt{2}}\overline{p}p\pi^0 + \frac{A_u+A_d-2A_s}{\sqrt{6}}\overline{p}p\eta^0\right],\end{aligned} \tag{D.7}$$

where +(h.c.) means that we add the hermitian conjugates of the terms of its left.

One-loop Diagram with Yukawa Coupling (Photon Attached to Scalar)

The one-loop diagrams we want to calculate are shown in Fig. D.2.

Let us assume the following lagrangian with the meson ϕ, baryon ψ and nucleon ψ':

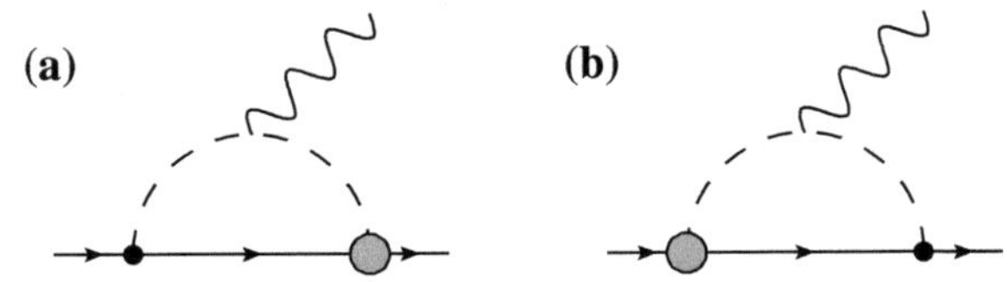

Fig. D.2 One-loop contribution to the fermion EDM, with scalar-fermion vertex. The *grey blob* represents the P, CP-odd scalar-fermion interaction. The black dot denotes either the Yukawa coupling, or the vertex with derivatives in chiral perturbation theory

$$\mathcal{L}_{\text{a,b}} = (D^{(1)}_\mu \phi)(D^{(1)\mu}\phi)^\dagger - m_\phi^2 \phi\phi^\dagger + \overline{\psi}'(i\partial\!\!\!/ - m_{\psi'})\psi' + \overline{\psi}(iD\!\!\!\!/^{(2)} - m_\psi)\psi$$
$$+(-A\phi\overline{\psi}' i\gamma_5 \psi - B\phi\overline{\psi}'\psi + \text{h.c.}), \tag{D.8}$$

with $D^{(1)}_\mu \equiv \partial_\mu - iQ_\phi e A_\mu$, $Q_\phi e$ being the charge of the meson, $D^{(2)}_\mu \equiv \partial_\mu - iQ_\psi e A_\mu$ with $Q_\psi e$ the charge of the baryon. Note that in this convention, the electromagnetic coupling e is negative. The charge (in unit of e) Q_ϕ and Q_ψ would be $Q_{\pi^+} = +1$, $Q_p = +1$, $Q_{K^-} = -1$, etc. A and B are the coupling constants for P, CP-even and P, CP-odd meson-baryon interactions, respectively, and are taken to be purely real.

The amplitudes of the diagrams (a) and (b) are

$$i\mathcal{M}_{(a)} = -ABQ_\phi e\varepsilon^*_\mu(q) \int \frac{d^4k}{(2\pi)^4} \frac{\overline{u}_{\psi'}(p-q)(p\!\!\!/ - k\!\!\!/ + m_\psi) i\gamma_5 u_{\psi'}(p)\cdot(2k^\mu - q^\mu)}{[k^2 - m_\phi^2][(k-q)^2 - m_\phi^2][(p-k)^2 - m_\psi^2]}, \tag{D.9}$$

$$i\mathcal{M}_{(b)} = -ABQ_\phi e\varepsilon^*_\mu(q) \int \frac{d^4k}{(2\pi)^4} \frac{\overline{u}_{\psi'}(p-q) i\gamma_5 (p\!\!\!/ - k\!\!\!/ + m_\psi) u_{\psi'}(p)\cdot(2k^\mu - q^\mu)}{[k^2 - m_\phi^2][(k-q)^2 - m_\phi^2][(p-k)^2 - m_\psi^2]}. \tag{D.10}$$

The addition is then

$$\begin{aligned} i\mathcal{M}_{(a)+(b)} &\equiv i\mathcal{M}_{(a)} + i\mathcal{M}_{(b)} \\ &= -2ABQ_\phi e\varepsilon^*_\mu(q) \int \frac{d^4k}{(2\pi)^4} \frac{m_\psi \overline{u}_{\psi'}(p-q) i\gamma_5 u_{\psi'}(p)\cdot(2k^\mu - q^\mu)}{[k^2 - m_\phi^2][(k-q)^2 - m_\phi^2][(p-k)^2 - m_\psi^2]}. \end{aligned} \tag{D.11}$$

The denominator can be written using the Feynman parameter as

$$\begin{aligned} I &\equiv \frac{1}{[k^2 - m_\phi^2][(k-q)^2 - m_\phi^2][(p-k)^2 - m_\psi^2]} \\ &= \int_0^1 \frac{2\delta(1-x-y-z)dxdydz}{\left[k^2 - 2y(k\cdot q) + yq^2 - 2z(p\cdot k) + zp^2 - (x+y)m_\phi^2 - zm_\psi^2\right]^3} \end{aligned}$$

$$= \int_0^1 \frac{2\delta(1-x-y-z)dxdydz}{\left[(k-yq-zp)^2-(yq+zp)^2+yq^2+zp^2-(x+y)m_\phi^2-zm_\psi^2\right]^3}$$

$$= \int_0^1 dz \int_0^{1-z} dy \frac{2}{\left[(k-yq-zp)^2-(yq+zp)^2+yq^2+zp^2-(1-z)m_\phi^2-zm_\psi^2\right]^3}$$

$$= \int_0^1 dz \int_0^{1-z} dy \times$$

$$\frac{2}{\left[(k-yq-zp)^2-2yz(q\cdot p)+yq^2-y^2q^2-z^2m_{\psi'}^2+z(m_{\psi'}^2-m_\psi^2+m_\phi^2)-m_\phi^2\right]^3}$$

$$= \int_0^1 dz \int_0^{1-z} dy \frac{2}{\left[(k-yq-zp)^2-z^2m_{\psi'}^2+z(m_{\psi'}^2-m_\psi^2+m_\phi^2)-m_\phi^2\right]^3}, \tag{D.12}$$

We have used the fact that $p^2 = m_{\psi'}^2$ and $(p-q)^2 = m_{\psi'}^2$ from the on-shell condition. In the soft photon limit ($q^2 \approx 0$), $(p\cdot q) = 0$ can be set.

Returning to the loop diagram,

$$i\mathcal{M}_{(a)+(b)} = \int_0^1 dz \int_0^{1-z} dy \int \frac{d^4k}{(2\pi)^4}$$

$$\frac{-4ABQ_\phi\, e\varepsilon_\mu^*(q)\, m_\psi\, \overline{u}_{\psi'}(p-q)i\gamma_5 u_{\psi'}(p)\cdot(2k^\mu-q^\mu)}{\left[(k-yq-zp)^2-2yz(q\cdot p)+y(1-y)q^2-z^2m_{\psi'}^2+z(m_{\psi'}^2-m_\psi^2+m_\phi^2)-m_\phi^2\right]^3}$$

$$= -4ABQ_\phi e \int_0^1 dz \int_0^{1-z} dy \int \frac{d^4k}{(2\pi)^4}$$

$$\times \frac{\varepsilon_\mu^*(q)\, m_\psi\, \overline{u}_{\psi'}(p-q)i\gamma_5 u_{\psi'}(p)\cdot(2zp^\mu+(2y-1)q^\mu)}{\left[k^2-2yz(q\cdot p)+y(1-y)q^2-z^2m_{\psi'}^2+z(m_{\psi'}^2-m_\psi^2+m_\phi^2)-m_\phi^2\right]^3}$$

$$= -4ABQ_\phi e\, m_\psi \frac{-i}{2(4\pi)^2}\varepsilon_\mu^*(q) \int_0^1 dz \int_0^{1-z} dy$$

$$\times \frac{\overline{u}_{\psi'}(p-q)i\gamma_5 u_{\psi'}(p)\cdot(2zp^\mu+(2y-1)q^\mu)}{2yz(q\cdot p)-y(1-y)q^2+z^2m_{\psi'}^2-z(m_{\psi'}^2-m_\psi^2+m_\phi^2)+m_\phi^2}$$

$$= \frac{-2i}{(4\pi)^2}ABQ_\phi e\, m_\psi\, \varepsilon_\mu^*(q)\overline{u}_{\psi'}(p-q)\sigma^{\mu\nu}q_\nu\gamma_5 u_{\psi'}(p)$$

$$\times \int_0^1 \frac{z(1-z)dz}{z^2m_{\psi'}^2-z(m_{\psi'}^2-m_\psi^2+m_\phi^2)+m_\phi^2} + (\propto \varepsilon_\mu^* q^\mu \text{ term})\,. \tag{D.13}$$

We have used (the γ_5-version of) the Gordon identity:

$$\overline{u}_{\psi'}(p-q)\sigma^{\mu\nu}q_\nu\gamma_5 u_{\psi'}(p) = -(2p^\mu-q^\mu)\overline{u}_{\psi'}(p-q)i\gamma_5 u_{\psi'}(p). \tag{D.14}$$

The integral is

$$\int_0^1 \frac{z(1-z)dz}{az^2+bz+c} = \frac{a+b}{2a^2}\ln\left|\frac{a+b+c}{c}\right| - \frac{1}{a} - \frac{b^2-2ac+ba}{2a^2}\left(I_B(1)-I_B(0)\right), \tag{D.15}$$

where

$$I_B(x) = \begin{cases} \dfrac{2}{\sqrt{4ac-b^2}}\arctan\dfrac{2ax+b}{\sqrt{4ac-b^2}} & (b^2 < 4ac) \\ \dfrac{-2}{\sqrt{b^2-4ac}}\mathrm{arctanh}\dfrac{\mathrm{2ax+b}}{\sqrt{\mathrm{b^2-4ac}}} & (b^2 > 4ac) \end{cases}. \tag{D.16}$$

(The second case ($b^2 - 4ac > 0$) applies when η-meson propagates in the loop.)

The nucleon EDM with meson loop, approximated to the leading logarithm (the so-called chiral limit), is

$$i\mathcal{M}_{(a)+(b)} = \frac{-2i}{(4\pi)^2} ABQ_\phi e\, \varepsilon^*_\mu(q)\overline{u}_{\psi'}(p-q)\sigma^{\mu\nu}q_\nu\gamma_5 u_{\psi'}(p)\frac{m_\psi^3}{m_{\psi'}^4}\ln\frac{m_\psi}{m_\phi}. \tag{D.17}$$

From the relation of the lagrangian and the amplitude

$$\mathcal{L} = -\frac{i}{2}d_{\psi'}\overline{\psi}'\sigma^{\mu\nu}\gamma_5\psi' F_{\mu\nu} \leftrightarrow \mathcal{M} = -d_{\psi'}\overline{u}_{\psi'}(p-q)\sigma^{\mu\nu}q_\nu\gamma_5 u(p)\varepsilon^*_\mu(q), \tag{D.18}$$

where $d_{\psi'}$ is the EDM of the fermion ψ', we obtain

$$d_{\psi'} = \frac{ABQ_\phi e}{8\pi^2}\frac{m_\psi^3}{m_{\psi'}^4}\ln\frac{m_\psi}{m_\phi}. \tag{D.19}$$

One-loop Diagram with Yukawa Coupling (Photon Attached to Fermion)

The one-loop diagrams we want to calculate are shown in Fig. D.3.

We assume the following lagrangian with the meson ϕ, baryon ψ and nucleon ψ':

$$\begin{aligned}\mathcal{L}_{\mathrm{c,d}} = &(D_\mu^{(1)}\phi)(D^{(1)\mu}\phi)^\dagger - m_\phi^2\phi\phi^\dagger + \overline{\psi}'(i\partial\!\!\!/ - m_{\psi'})\psi' + \overline{\psi}(iD\!\!\!\!/^{(2)} - m_\psi)\psi \\ &+(-A'\phi\overline{\psi}'i\gamma_5\psi - B'\phi\overline{\psi}'\psi + \mathrm{h.c.}),\end{aligned} \tag{D.20}$$

with $D_\mu^{(1)} \equiv \partial_\mu - iQ_\phi eA_\mu$, $Q_\phi e$ being the charge of the meson, $D_\mu^{(2)} \equiv \partial_\mu - iQ_\psi eA_\mu$ with $Q_\psi e$ the charge of the baryon. Coefficients A' and B' are the coupling constants for the P, CP-even and P, CP-odd meson-baryon interactions, respectively, and are

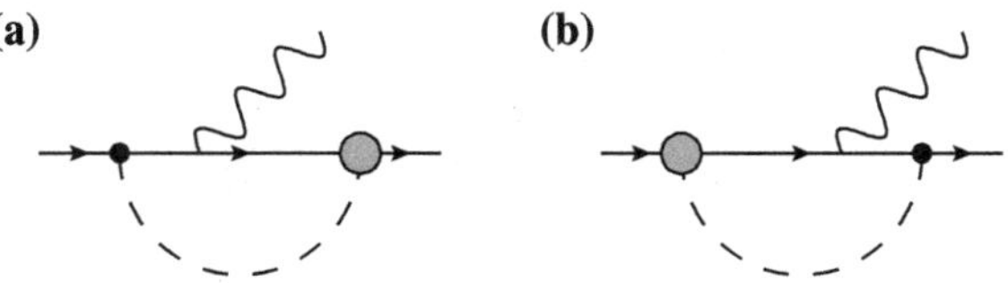

Fig. D.3 One-loop contribution to the fermion EDM, with scalar-fermion vertex. The *grey blob* represents the P, CP-odd scalar-femrion interaction. The *black dot* denotes either the Yukawa coupling, or the vertex with derivatives in chiral perturbation theory

taken to be purely real. Note that the meson ϕ and the baryon ψ are different from the calculation of the one-loop diagram with the photon attached to the meson.

The amplitudes of the diagrams (c) and (d) are

$$i\mathcal{M}_{(c)} = -A'B'Q_\psi e\varepsilon^*_\mu(q)\int \frac{d^4k}{(2\pi)^4}\frac{\bar{u}_{\psi'}(p-q)(\not{k}-\not{q}+m_\psi)\gamma^\mu(\not{k}+m_\psi)i\gamma_5 u_{\psi'}(p)}{[(p-k)^2-m_\phi^2][(k-q)^2-m_\psi^2][k^2-m_\psi^2]}, \tag{D.21}$$

$$i\mathcal{M}_{(d)} = -A'B'Q_\psi e\varepsilon^*_\mu(q)\int \frac{d^4k}{(2\pi)^4}\frac{\bar{u}_{\psi'}(p-q)i\gamma_5(\not{k}-\not{q}+m_\psi)\gamma^\mu(\not{k}+m_\psi)u_{\psi'}(p)}{[(p-k)^2-m_\phi^2][(k-q)^2-m_\psi^2][k^2-m_\psi^2]}, \tag{D.22}$$

The addition is then

$$\begin{aligned}
i\mathcal{M}_{(c)+(d)} &\equiv i\mathcal{M}_{(c)} + i\mathcal{M}_{(d)} \\
&= -2A'B'Q_\psi e\varepsilon^*_\mu(q)\int \frac{d^4k}{(2\pi)^4}\frac{m_\psi\bar{u}_{\psi'}(p-q)\left[(\not{k}-\not{q})\gamma^\mu+\gamma^\mu\not{k}\right]i\gamma_5 u_{\psi'}(p)}{[(p-k)^2-m_\phi^2][(k-q)^2-m_\psi^2][k^2-m_\psi^2]} \\
&= -2A'B'Q_\psi e\varepsilon^*_\mu(q)\int \frac{d^4k}{(2\pi)^4}\frac{m_\psi\bar{u}_{\psi'}(p-q)\left[-\not{q}\gamma^\mu+2k^\mu\right]i\gamma_5 u_{\psi'}(p)}{[(p-k)^2-m_\phi^2][(k-q)^2-m_\psi^2][k^2-m_\psi^2]} \\
&= \int_0^1 dz\int_0^{1-z} dy\int\frac{d^4k}{(2\pi)^4} \\
&\quad \frac{-4A'B'Q_\psi e\varepsilon^*_\mu(q)m_\psi\bar{u}_{\psi'}(p-q)\left[-\not{q}\gamma^\mu+2k^\mu\right]i\gamma_5 u_{\psi'}(p)}{\left[(k-yq-zp)^2-2yz(q\cdot p)+y(1-y)q^2-z^2m_{\psi'}^2+z(m_{\psi'}^2-m_\phi^2)-(1-z)m_\psi^2\right]^3} \\
&= \int_0^1 dz\int_0^{1-z} dy\int\frac{d^4k}{(2\pi)^4} \\
&\quad \times\frac{-4A'B'Q_\psi e\varepsilon^*_\mu(q)m_\psi\bar{u}_{\psi'}(p-q)\left[-\not{q}\gamma^\mu+2zp^\mu+2yq^\mu\right]i\gamma_5 u_{\psi'}(p)}{\left[k^2-2yz(q\cdot p)+y(1-y)q^2-z^2m_{\psi'}^2+z(m_{\psi'}^2-m_\phi^2+m_\psi^2)-m_\psi^2\right]^3} \\
&= -4A'B'Q_\psi e m_\psi\frac{-i}{2(4\pi)^2}\varepsilon^*_\mu(q)\int_0^1 dz\int_0^{1-z} dy \\
&\quad \times\frac{\bar{u}_{\psi'}(p-q)\left[-\not{q}\gamma^\mu+2zp^\mu+2yq^\mu\right]i\gamma_5 u_{\psi'}(p)}{2yz(q\cdot p)-y(1-y)q^2+z^2m_{\psi'}^2-z(m_{\psi'}^2-m_\phi^2+m_\psi^2)+m_\psi^2}
\end{aligned}$$

$$
\begin{aligned}
&= \frac{2i}{(4\pi)^2} A' B' Q_\psi e m_\psi \varepsilon^*_\mu(q) \overline{u}_{\psi'}(p-q) \sigma^{\mu\nu} q_\nu \gamma_5 u_{\psi'}(p) \\
&\times \int_0^1 dz \frac{(1-z)(1-z)}{z^2 m^2_{\psi'} - z(m^2_{\psi'} - m^2_\phi + m^2_\psi) + m^2_\psi} + (\varepsilon^*_\mu q^\mu \text{ term}). \qquad \text{(D.23)}
\end{aligned}
$$

Again, the last line was obtained by taking the soft photon limit.
The integral is

$$
\begin{aligned}
\int_0^1 \frac{(1-z)(1-z)dz}{az^2+bz+c} &= \int_0^1 dz \frac{(1-z)-(1-z)z}{az^2+bz+c} \\
&= -\frac{2a+b}{2a^2} \ln\left|\frac{a+b+c}{c}\right| + \frac{1}{a} \\
&\quad + \frac{b^2+2a(a+b-c)}{2a^2} \left(I_B(1) - I_B(0)\right), \qquad \text{(D.24)}
\end{aligned}
$$

where we have used I_B defined in Eq. (D.16). Note that the leading logarithmic term disappears for the nucleon EDM graph with pion loop in the chiral limit.

Appendix E
Nucleon EDM in the Non-Relativistic Constituent Quark Model

We calculate the electric dipole moment (EDM) of the nucleon (proton and neutron) in the non-relativistic constituent quark model. We assume that the nucleon is made of three constituent quarks with intrinsic EDM. We also assume an s-wave nucleon without spin and isospin dependent interactions.

The general wave function of the nucleon in the non-relativistic quark model is given by

$$| N : j_z, I_z \rangle = \Big[\, | \, \mathrm{spherical} \, \rangle \otimes | \, \mathrm{spin} \, \rangle \Big]_{j=\frac{1}{2}, j_z} \otimes | \, \mathrm{isospin} \, \rangle_{I=\frac{1}{2}, I_z} \otimes | \, \mathrm{color} \, \rangle_{\mathrm{singlet}} . \tag{E.1}$$

The quarks are fermions, so the wave function of the system must be totally antisymmetric in the interchange of the constituents, due to the exclusion principle of Pauli. The color $SU(3)_c$ part of the nucleon wave function $| \, \mathrm{color} \, \rangle_{\mathrm{singlet}}$ is totally antisymmetric, since the nucleon is a singlet state of the color $SU(3)_c$ group due to the confinement. We have also assumed that the nucleon is an s-wave state, so the spherical part of the nucleon wave function $| \, \mathrm{spherical} \, \rangle$ is totally symmetric and can be factored out (in other word, $\Big[\, | \, \mathrm{spherical} \, \rangle \otimes | \, \mathrm{spin} \, \rangle \Big]_{j=\frac{1}{2}, j_z} = | \, \mathrm{spherical} \, \rangle_{l=0} \otimes | \, \mathrm{spin} \, \rangle_{s=\frac{1}{2}, s_z}$, with $s_z = j_z$). The remaining spin and isospin parts must then be combined to be totally symmetric in the interchange of particles.

In the non-relativistic quark model, the nucleon is a three-body system of up and down quarks. In the three-body system, we have three ways to combine the spins (and the isospin). The three ways are

$$\Big[\big[\chi_1 \otimes \chi_2 \big]_{\sigma_1} \otimes \chi_3 \Big]_{s=\frac{1}{2}, s_z}, \; \Big[\big[\chi_2 \otimes \chi_3 \big]_{\sigma_2} \otimes \chi_1 \Big]_{s=\frac{1}{2}, s_z}, \; \text{and} \Big[\big[\chi_3 \otimes \chi_1 \big]_{\sigma_3} \otimes \chi_2 \Big]_{s=\frac{1}{2}, s_z}, \tag{E.2}$$

where χ_i $(i = 1, 2, 3)$ is the spin wave function for the ith constituent quark. The spin of the coupled two-quark system is denoted by $\sigma_i = 0, 1$ $(i = 1, 2, 3)$. Similar combinations can be written for the isospin part. The general form of the spin/isospin part of the wave function is then given by

N. Yamanaka, *Analysis of the Electric Dipole Moment in the R-parity Violating Supersymmetric Standard Model*, Springer Theses,
DOI: 10.1007/978-4-431-54544-6, © Springer Japan 2014

$$| \text{spin} \rangle_{s=\frac{1}{2},s_z} \otimes | \text{isospin} \rangle_{I=\frac{1}{2},I_z} = \sum_{ijk} \sum_{\sigma_i=0,1} \sum_{T_i=0,1} C_i^{(\sigma_i,T_i)} \Big[\big[\chi_i \otimes \chi_j \big]_{\sigma_i} \otimes \chi_k \Big]_{s=\frac{1}{2},s_z} \times \Big[\big[\eta_i \otimes \eta_j \big]_{T_i} \otimes \eta_k \Big]_{I=\frac{1}{2},I_z}, \tag{E.3}$$

where (i, j, k) are even permutations of $(1, 2, 3)$, and η_i the isospin part of the nucleon wave function. The isospin of the coupled two-quark system is given by $T_i = 0, 1$. The symmetry or the antisymmetry in the interchange of quarks are determined by the spin σ_i and the isospin T_i of the coupled two-quark system. To keep the system symmetric in the interchange of quarks, we must have $C_1^{(\sigma_1,T_1)} = C_2^{(\sigma_2,T_2)} = C_3^{(\sigma_3,T_3)} \equiv C^{(\sigma,T)}$, $\sigma_1 = \sigma_2 = \sigma_3 \equiv \sigma$, $T_1 = T_2 = T_3 \equiv T$, and, the spin and the isospin must be combined such that $\sigma = T = 0$ or $\sigma = T = 1$. This can be easily verified by using the symmetric and antisymmetric relations $[\chi_i \otimes \chi_j]_{\sigma=1} = [\chi_j \otimes \chi_i]_{\sigma=1}$ and $[\chi_i \otimes \chi_j]_{\sigma=0} = -[\chi_j \otimes \chi_i]_{\sigma=0}$ (similar relations hold for the isospin wave function η_i). As we are not considering spin and isospin dependent interactions, the coefficient $C^{(\sigma,T)}$ must satisfy $C^{(0,0)} = C^{(1,1)} = C$. The spin and isospin parts of the nucleon wave function can then be rewritten as

$$| \text{spin} \rangle_{s=\frac{1}{2},s_z} \otimes | \text{isospin} \rangle_{I=\frac{1}{2},I_z} = C \sum_{ijk} \sum_{\sigma=0,1} \Big[\big[\chi_i \otimes \chi_j \big]_{\sigma} \otimes \chi_k \Big]_{s=\frac{1}{2},s_z} \times \Big[\big[\eta_i \otimes \eta_j \big]_{T=\sigma} \otimes \eta_k \Big]_{I=\frac{1}{2},I_z}. \tag{E.4}$$

The next step is to recombine the three channels where the spin and the isospin of the quarks are combined in different ways, to one definite channel. The recombination of the spin part of the coupled two-body subsystem in the three-body system is written as

$$\Big[\big[\chi_i \otimes \chi_k \big]_{\sigma'} \otimes \chi_j \Big]_{s=\frac{1}{2},s_z} = \sum_{\sigma=0,1} 2\sqrt{(2\sigma+1)(2\sigma'+1)} \begin{Bmatrix} \frac{1}{2} & \frac{1}{2} & \sigma \\ \frac{1}{2} & 0 & \frac{1}{2} \\ \sigma' & \frac{1}{2} & \frac{1}{2} \end{Bmatrix} \times \Big[\big[\chi_i \otimes \chi_j \big]_{\sigma} \otimes \chi_k \Big]_{s=\frac{1}{2},s_z}, \tag{E.5}$$

where the expression with the curly bracket is Wigner's 9-j symbol, given by the following relation

$$\langle ((j_1, j_2)j_{12}, (j_3, j_4)j_{34})j, j_z \,|\, ((j_1, j_3)j_{13}, (j_2, j_4)j_{24})j, j_z \rangle = \sqrt{(2j_{12}+1)(2j_{13}+1)(2j_{24}+1)(2j_{34}+1)} \begin{Bmatrix} j_1 & j_2 & j_{12} \\ j_3 & j_4 & j_{34} \\ j_{13} & j_{24} & j \end{Bmatrix}. \tag{E.6}$$

Here $(j_n, j_m)j_l$ means that we have coupled the angular momenta j_n and j_m to obtain j_l. This is just the coefficient appearing when the pairs of angular momenta

(j_1, j_2) and (j_3, j_4) are recombined into (j_1, j_3) and (j_2, j_4). The coefficients needed in this section are given explicitly as follows:

$$\begin{Bmatrix} \frac{1}{2} & \frac{1}{2} & 0 \\ \frac{1}{2} & 0 & \frac{1}{2} \\ 0 & \frac{1}{2} & \frac{1}{2} \end{Bmatrix} = \frac{1}{4}, \begin{Bmatrix} \frac{1}{2} & \frac{1}{2} & 0 \\ \frac{1}{2} & 0 & \frac{1}{2} \\ 1 & \frac{1}{2} & \frac{1}{2} \end{Bmatrix} = \begin{Bmatrix} \frac{1}{2} & \frac{1}{2} & 1 \\ \frac{1}{2} & 0 & \frac{1}{2} \\ 0 & \frac{1}{2} & \frac{1}{2} \end{Bmatrix} = \frac{1}{4}, \text{ and } \begin{Bmatrix} \frac{1}{2} & \frac{1}{2} & 1 \\ \frac{1}{2} & 0 & \frac{1}{2} \\ 1 & \frac{1}{2} & \frac{1}{2} \end{Bmatrix} = -\frac{1}{12}. \quad \text{(E.7)}$$

The relation (E.5) can then be rewritten as

$$\begin{pmatrix} \left[[\chi_k \otimes \chi_i]_{\sigma'=0} \otimes \chi_j \right]_{s=\frac{1}{2}, s_z} \\ \left[[\chi_k \otimes \chi_i]_{\sigma'=1} \otimes \chi_j \right]_{s=\frac{1}{2}, s_z} \end{pmatrix} = \begin{pmatrix} -\frac{1}{2} & -\frac{\sqrt{3}}{2} \\ \frac{\sqrt{3}}{2} & -\frac{1}{2} \end{pmatrix} \begin{pmatrix} \left[[\chi_i \otimes \chi_j]_{\sigma=0} \otimes \chi_k \right]_{s=\frac{1}{2}, s_z} \\ \left[[\chi_i \otimes \chi_j]_{\sigma=1} \otimes \chi_k \right]_{s=\frac{1}{2}, s_z} \end{pmatrix}, \quad \text{(E.8)}$$

where i, j and k are even permutation of 1, 2 and 3. Furthermore, we have

$$\begin{aligned} \begin{pmatrix} \left[[\chi_j \otimes \chi_k]_{\sigma'=0} \otimes \chi_i \right]_{s=\frac{1}{2}, s_z} \\ \left[[\chi_j \otimes \chi_k]_{\sigma'=1} \otimes \chi_i \right]_{s=\frac{1}{2}, s_z} \end{pmatrix} &= \begin{pmatrix} -\frac{1}{2} & -\frac{\sqrt{3}}{2} \\ \frac{\sqrt{3}}{2} & -\frac{1}{2} \end{pmatrix}^2 \begin{pmatrix} \left[[\chi_i \otimes \chi_j]_{\sigma=0} \otimes \chi_k \right]_{s=\frac{1}{2}, s_z} \\ \left[[\chi_i \otimes \chi_j]_{\sigma=1} \otimes \chi_k \right]_{s=\frac{1}{2}, s_z} \end{pmatrix} \\ &= \begin{pmatrix} -\frac{1}{2} & \frac{\sqrt{3}}{2} \\ -\frac{\sqrt{3}}{2} & -\frac{1}{2} \end{pmatrix} \begin{pmatrix} \left[[\chi_i \otimes \chi_j]_{\sigma=0} \otimes \chi_k \right]_{s=\frac{1}{2}, s_z} \\ \left[[\chi_i \otimes \chi_j]_{\sigma=1} \otimes \chi_k \right]_{s=\frac{1}{2}, s_z} \end{pmatrix}. \end{aligned} \quad \text{(E.9)}$$

The recombination of the isospin wave function goes exactly in the same way. By combining the spin and isospin parts, we obtain

$$\begin{aligned} \chi_0(kij)\eta_0(kij) &= \left[-\frac{1}{2}\chi_0(ijk) - \frac{\sqrt{3}}{2}\chi_1(ijk) \right] \left[-\frac{1}{2}\eta_0(ijk) - \frac{\sqrt{3}}{2}\eta_1(ijk) \right] \\ &= \frac{1}{4}\chi_0(ijk)\eta_0(ijk) + \frac{\sqrt{3}}{4}\chi_0(ijk)\eta_1(ijk) \\ &\quad + \frac{\sqrt{3}}{4}\chi_1(ijk)\eta_0(ijk) + \frac{3}{4}\chi_1(ijk)\eta_1(ijk), \end{aligned} \quad \text{(E.10)}$$

$$\begin{aligned} \chi_1(kij)\eta_1(kij) &= \left[\frac{\sqrt{3}}{2}\chi_0(ijk) - \frac{1}{2}\chi_1(ijk) \right] \left[\frac{\sqrt{3}}{2}\eta_0(ijk) - \frac{1}{2}\eta_1(ijk) \right] \\ &= \frac{3}{4}\chi_0(ijk)\eta_0(ijk) - \frac{\sqrt{3}}{4}\chi_0(ijk)\eta_1(ijk) \\ &\quad - \frac{\sqrt{3}}{4}\chi_1(ijk)\eta_0(ijk) + \frac{1}{4}\chi_1(ijk)\eta_1(ijk), \end{aligned} \quad \text{(E.11)}$$

and

$$\begin{aligned}\chi_0(jki)\eta_0(jki) &= \left[-\frac{1}{2}\chi_0(ijk)+\frac{\sqrt{3}}{2}\chi_1(ijk)\right]\left[-\frac{1}{2}\eta_0(ijk)+\frac{\sqrt{3}}{2}\eta_1(ijk)\right]\\ &= \frac{1}{4}\chi_0(ijk)\eta_0(ijk)-\frac{\sqrt{3}}{4}\chi_0(ijk)\eta_1(ijk)\\ &\quad -\frac{\sqrt{3}}{4}\chi_1(ijk)\eta_0(ijk)+\frac{3}{4}\chi_1(ijk)\eta_1(ijk)\,, \end{aligned}\tag{E.12}$$

$$\begin{aligned}\chi_1(jki)\eta_1(jki) &= \left[-\frac{\sqrt{3}}{2}\chi_0(ijk)-\frac{1}{2}\chi_1(ijk)\right]\left[-\frac{\sqrt{3}}{2}\eta_0(ijk)-\frac{1}{2}\eta_1(ijk)\right]\\ &= \frac{3}{4}\chi_0(ijk)\eta_0(ijk)+\frac{\sqrt{3}}{4}\chi_0(ijk)\eta_1(ijk)\\ &\quad +\frac{\sqrt{3}}{4}\chi_1(ijk)\eta_0(ijk)+\frac{1}{4}\chi_1(ijk)\eta_1(ijk)\,, \end{aligned}\tag{E.13}$$

where we have defined

$$\chi_\sigma(ijk)\equiv\Big[\big[\chi_i\otimes\chi_j\big]_\sigma\otimes\chi_k\Big]_{s=\frac{1}{2},s_z},\ \text{and}\ \eta_\sigma(ijk)\equiv\Big[\big[\eta_i\otimes\eta_j\big]_\sigma\otimes\eta_k\Big]_{I=\frac{1}{2},I_z}.\tag{E.14}$$

The spin/isospin wave function is then

$$\begin{aligned}|\,\text{spin}\,\rangle_{s=\frac{1}{2},s_z}\otimes|\,\text{isospin}\,\rangle_{I=\frac{1}{2},I_z} &= C\sum_{\sigma=0,1}\sum_{ijk}\Big[\big[\chi_i\otimes\chi_j\big]_\sigma\otimes\chi_k\Big]_{s=\frac{1}{2},s_z}\\ &\quad\times\Big[\big[\eta_i\otimes\eta_j\big]_\sigma\otimes\eta_k\Big]_{I=\frac{1}{2},I_z}\\ &= C\sum_{\sigma=0,1}\frac{3}{2}\Big[\big[\chi_l\otimes\chi_m\big]_\sigma\otimes\chi_n\Big]_{s=\frac{1}{2},s_z}\\ &\quad\times\Big[\big[\eta_l\otimes\eta_m\big]_\sigma\otimes\eta_n\Big]_{I=\frac{1}{2},I_z}\\ &= \frac{1}{\sqrt{2}}\sum_{\sigma=0,1}\Big[\big[\chi_l\otimes\chi_m\big]_\sigma\otimes\chi_n\Big]_{s=\frac{1}{2},s_z}\\ &\quad\times\Big[\big[\eta_l\otimes\eta_m\big]_\sigma\otimes\eta_n\Big]_{I=\frac{1}{2},I_z}, \end{aligned}\tag{E.15}$$

where (l, m, n) are arbitrary even permutation of $(1, 2, 3)$. The coefficient $\frac{1}{\sqrt{2}}=\frac{3}{2}C$ of the final equality was obtained just by noticing that the spin/isospin wave function is composed of $\sigma = T = 0$ and $\sigma = T = 1$ states with equal weight. The nucleon wave function in the non-relativistic constituent quark model without spin and isospin dependent interactions can now be written as

$$|\,N : s_z, I_z\,\rangle = |\,\text{spherical}\,\rangle_{l=0} \otimes |\,\text{color}\,\rangle_{\text{singlet}} \otimes \frac{1}{\sqrt{2}} \sum_{\sigma=0,1} \Big|\,(\sigma, 1/2)1/2; s_z\Big\rangle \otimes \Big|\,(\sigma, 1/2)1/2; I_z\Big\rangle, \quad \text{(E.16)}$$

where we have defined $|\,(\sigma, 1/2)1/2; s_z\,\rangle \equiv l[[\,\chi_l \otimes \chi_m]_\sigma \otimes \chi_n r]_{s=\frac{1}{2}, s_z}$ and $|\,(\sigma, 1/2)1/2; I_z\,\rangle \equiv l[[\,\eta_l \otimes \eta_m]_\sigma \otimes \eta_n r]_{I=\frac{1}{2}, I_z}$. The spherical part $|\,\text{spherical}\,\rangle_{l=0}$ and the color part $|\,\text{color}\,\rangle_{\text{singlet}}$ are each normalized to 1.

The quark EDM contribution to the nucleon EDM is given by

$$\begin{aligned} d_N(d_q) &= \sum_{q=u,d} \langle\, N : s_z = 1/2, I_z\,|\,\hat{d}_q\,|\,N : s_z = 1/2, I_z\,\rangle \\ &= \sum_{q=u,d}\sum_{i=1}^{3} \langle\, N : 1/2, I_z\,|\,\hat{d}_{qi}\,|\,N : 1/2, I_z\,\rangle \\ &= \sum_{q=u,d} 3\langle\, N : 1/2, I_z\,|\,\hat{d}_{qn}\,|\,N : 1/2, I_z\,\rangle, \end{aligned} \quad \text{(E.17)}$$

where $\hat{d}_q$ is the quark EDM operator, defined by $\hat{d}_u \equiv \frac{d_u}{2}(1+\tau_z)\sigma_z$ and $\hat{d}_d \equiv \frac{d_d}{2}(1-\tau_z)\sigma_z$ with σ_z the usual Pauli matrix acting on the spin part, and τ_z the Pauli matrix acting on the isospin space of the up and down quarks. The index $n = 1, 2, 3$ labeling the quarks in the third equality can be specified arbitrarily, since the nucleon wave function is invariant under any even permutations of the quark labels [see Eq. (E.15)].

The calculation of the isoscalar matrix element goes as follows:

$$\begin{aligned} M_0 &\equiv \langle\, N : s_z = 1/2, I_z = \pm 1/2\,|\,\sigma_z\,|\,N : s_z = 1/2, I_z = \pm 1/2\,\rangle \\ &= \frac{1}{2}\Big\langle\,(0, 1/2)1/2; 1/2\,\Big|\,\sigma_z\,\Big|\,(0, 1/2)1/2; 1/2\Big\rangle \\ &\quad + \frac{1}{2}\Big\langle\,(1, 1/2)1/2; 1/2\,\Big|\,\sigma_z\,\Big|\,(1, 1/2)1/2; 1/2\Big\rangle \\ &= \frac{1}{2}\begin{pmatrix} 1/2 & 1 & 1/2 \\ -1/2 & 0 & 1/2 \end{pmatrix}\Big[\Big\langle\,(0, 1/2)1/2\,\Big|\Big|\,\boldsymbol{\sigma}\,\Big|\Big|\,(0, 1/2)1/2\Big\rangle \\ &\qquad\qquad + \Big\langle\,(1, 1/2)1/2\,\Big|\Big|\,\boldsymbol{\sigma}\,\Big|\Big|\,(1, 1/2)1/2\Big\rangle\Big] \\ &= \frac{1}{2} \times \frac{1}{\sqrt{6}}\left[\sqrt{6} - \frac{\sqrt{6}}{3}\right] = \frac{1}{3}, \end{aligned} \quad \text{(E.18)}$$

where the expression with the parenthesis in the third equality is the 3-j symbol. For the isoscalar matrix element M_0, the isospin contribution is trivial, since it has no isospin dependent operator. The result is of course independent on the isospin of the nucleon. Note that the transition elements from $\sigma = 0$ to $\sigma = 1$ vanish. The reduced matrix elements of the third equality were calculated as

$$\big\langle (0,1/2)1/2 \,\big|\big|\, \boldsymbol{\sigma} \,\big|\big|\, (0,1/2)1/2 \big\rangle = \langle\, 1/2 \,||\, \boldsymbol{\sigma} \,||\, 1/2 \,\rangle = \sqrt{6}\,, \tag{E.19}$$

$$\begin{aligned}\big\langle (1,1/2)1/2 \,\big|\big|\, \boldsymbol{\sigma} \,\big|\big|\, (1,1/2)1/2 \big\rangle &= -2 \begin{Bmatrix} 1/2 & 1/2 & 1 \\ 1/2 & 1/2 & 1 \end{Bmatrix} \langle\, 1/2 \,||\, \boldsymbol{\sigma} \,||\, 1/2 \,\rangle \\ &= -2 \times \frac{1}{6} \times \sqrt{6} = -\frac{\sqrt{6}}{3}\,,\end{aligned} \tag{E.20}$$

where the expression with the curly bracket is the 6-j symbol. The calculation of the isovector matrix element goes in a similar way:

$$\begin{aligned}M_{1\pm} &\equiv \langle\, N : s_z = 1/2,\, I_z = \pm 1/2 \,|\, \sigma_z \tau_z \,|\, N : s_z = 1/2,\, I_z = \pm 1/2 \,\rangle \\ &= \frac{1}{2} \big\langle (0,1/2)1/2; 1/2 \,\big|\, \sigma_z \,\big|\, (0,1/2)1/2; 1/2 \big\rangle \\ &\quad \times \big\langle (0,1/2)1/2; \pm 1/2 \,\big|\, \tau_z \,\big|\, (0,1/2)1/2; \pm 1/2 \big\rangle \\ &\quad + \frac{1}{2} \big\langle (1,1/2)1/2; 1/2 \,\big|\, \sigma_z \,\big|\, (1,1/2)1/2; 1/2 \big\rangle \\ &\quad \times \big\langle (1,1/2)1/2; \pm 1/2 \,\big|\, \tau_z \,\big|\, (1,1/2)1/2; \pm 1/2 \big\rangle \\ &= \frac{1}{2} \begin{pmatrix} 1/2 & 1 & 1/2 \\ -1/2 & 0 & 1/2 \end{pmatrix} \begin{pmatrix} 1/2 & 1 & 1/2 \\ \mp 1/2 & 0 & \pm 1/2 \end{pmatrix} \\ &\quad \times \Big[\big\langle (0,1/2)1/2 \,\big|\big|\, \boldsymbol{\sigma} \,\big|\big|\, (0,1/2)1/2 \big\rangle \big\langle (0,1/2)1/2 \,\big|\big|\, \boldsymbol{\tau} \,\big|\big|\, (0,1/2)1/2 \big\rangle \\ &\quad + \big\langle (1,1/2)1/2 \,\big|\big|\, \boldsymbol{\sigma} \,\big|\big|\, (1,1/2)1/2 \big\rangle \big\langle (1,1/2)1/2 \,\big|\big|\, \boldsymbol{\tau} \,\big|\big|\, (1,1/2)1/2 \big\rangle \Big] \\ &= \pm \frac{1}{2} \times \frac{1}{\sqrt{6}} \times \frac{1}{\sqrt{6}} \left[\sqrt{6}^{\,2} + \left(-\frac{\sqrt{6}}{3} \right)^2 \right] = \pm \frac{5}{9}\,.\end{aligned} \tag{E.21}$$

We see that the isovector matrix element M_1 changes its sign when the nucleon isospin flips. In this case, we also find that the transition elements from $\sigma = 0$ to $\sigma = 1$ vanish.

Combining M_0 and $M_{1\pm}$, we obtain the final formulae for the nucleon EDM as

$$\begin{aligned}d_n(d_q) &= \frac{3}{2}(M_0 + M_{1-})d_u + \frac{3}{2}(M_0 - M_{1-})d_d \\ &= \left(\frac{1}{2} - \frac{5}{6}\right) d_u + \left(\frac{1}{2} + \frac{5}{6}\right) d_d = -\frac{1}{3} d_u + \frac{4}{3} d_d\,,\end{aligned} \tag{E.22}$$

$$\begin{aligned}d_p(d_q) &= \frac{3}{2}(M_0 + M_{1+})d_u + \frac{3}{2}(M_0 - M_{1+})d_d \\ &= \frac{4}{3} d_u - \frac{1}{3} d_d\,,\end{aligned} \tag{E.23}$$

which give Eqs. (6.52) and (6.53).

Appendix F
Nuclear EDM and Schiff Moment in the Simple Shell Model

We show in this section the detailed derivation of the nuclear EDM and the nuclear Schiff moment in the simple shell model. In the simple shell model, the valence nucleon becomes the only degree of freedom, and the nucleons of the core are irrelevant. Therefore, the nuclear EDM and the nuclear Schiff moment receive contributions only from the valence nucleon, and this picture makes possible to avoid the impossible ab initio many-body calculation for heavy nuclei. Note that in this appendix the electric charge e is defined as $e = |e| > 0$, the convention of Chap. 7.

Nuclear EDM Generated by the P, CP-odd N-N Interaction in the Simple Shell Model

The EDM of odd-nuclei generated by the nuclear polarization in the simple shell model is given by

$$
\begin{aligned}
d_A(\xi) &\equiv \langle \tilde{\psi} : j, j_z = j \,|\, qez \,|\, \tilde{\psi} : j, j_z = j \rangle \\
&= qe\xi \langle \psi : j, j_z = j \,|\, z\,(\boldsymbol{\sigma} \cdot \nabla) \,|\, \psi : j, j_z = j \rangle + \text{c.c.} \\
&= 2qe\xi \begin{pmatrix} j & 1 & j \\ -j & 0 & j \end{pmatrix} \langle \psi : j \,||\, z\,(\boldsymbol{\sigma} \cdot \nabla) \,||\, \psi : j \rangle \\
&= 2qe\xi \sqrt{\frac{j}{(2j+1)(j+1)}} \langle \psi : j \,||\, z\,(\boldsymbol{\sigma} \cdot \nabla) \,||\, \psi : j \rangle \\
&= 2qe\xi \sqrt{\frac{j}{(2j+1)(j+1)}} \int_0^\infty r^3 R(r) \langle (l, s)j \,||\, \hat{r}\,(\boldsymbol{\sigma} \cdot \nabla) \,||\, (l, s)j \rangle R(r)\, dr,
\end{aligned}
\tag{F.1}
$$

where z (and r) is the coordinate of the external (valence) nucleon with the charge qe ($q = 1$ for the proton and $q = 0$ for the neutron). The unit vector is given by $\hat{r}$. The

N. Yamanaka, *Analysis of the Electric Dipole Moment in the R-parity Violating Supersymmetric Standard Model*, Springer Theses,
DOI: 10.1007/978-4-431-54544-6, © Springer Japan 2014

state $\left|\tilde{\psi}\right\rangle = (1+\xi\boldsymbol{\sigma}\cdot\nabla)\left|\psi\right\rangle$ is the nuclear ground state perturbed with the small P, CP-odd state $\xi\boldsymbol{\sigma}\cdot\nabla\left|\psi\right\rangle$, and has the total angular momentum j. The unperturbed P, CP-even state $\left|\psi\right\rangle$ has the quantum numbers l (orbital angular momentum) and $s=\frac{1}{2}$ (spin). The radial part $R(r)$ satisfies $\int_0^\infty r^2R^2(r)dr = \int|\psi(r)|^2d^3r = 1$. In the last line, the order of the elements in the integral is important, since the angular matrix element involves a gradient. The coupling of angular momenta is expressed by $(l,s)j$ which means that we have coupled l and s to obtain the total angular momentum j.

Let us now calculate the angular reduced matrix element.

$$\left\langle (l,s)j \,\|\, \hat{r}\,(\boldsymbol{\sigma}\cdot\nabla) \,\|\, (l,s)j \right\rangle = \sqrt{3}(-1)^{2j+1}\sum_{j''}\begin{Bmatrix} 1 & 0 & 1 \\ j & j & j'' \end{Bmatrix}\left\langle (l,s)j \,\|\, \hat{r} \,\|\, j'' \right\rangle \times\left\langle j'' \,\|\, (\boldsymbol{\sigma}\cdot\nabla) \,\|\, (l,s)j \right\rangle, \tag{F.2}$$

where the sum over j'' is taken over all possible intermediate states. In this case there is only one possible intermediate state. For the external state with $j = l \mp \frac{1}{2}$, we have $j'' = j = l' \pm \frac{1}{2}$ with $l' = l \mp 1$ (only the orbital angular momentum changes). We thus obtain

$$\begin{aligned}
\left\langle (l,s)j \,\|\, \hat{r}\,(\boldsymbol{\sigma}\cdot\nabla) \,\|\, (l,s)j \right\rangle &= \sqrt{3}\begin{Bmatrix} 1 & 0 & 1 \\ j & j & j \end{Bmatrix}\left\langle (l,s)j \,\|\, \hat{r} \,\|\, (l',s)j \right\rangle \\
&\quad\times\left\langle (l',s)j \,\|\, (\boldsymbol{\sigma}\cdot\nabla) \,\|\, (l,s)j \right\rangle \\
&= \frac{1}{\sqrt{2j+1}}\left\langle (l,s)j \,\|\, \hat{r} \,\|\, (l',s)j \right\rangle\left\langle (l',s)j \,\|\, (\boldsymbol{\sigma}\cdot\nabla) \,\|\, (l,s)j \right\rangle \\
&= \frac{1}{\sqrt{2j+1}}(-1)^{l+s+j+1}(2j+1)\begin{Bmatrix} l & l' & 1 \\ j & j & s \end{Bmatrix}\left\langle l \,\|\, \hat{r} \,\|\, l' \right\rangle \\
&\quad\times(-1)^{s+l+j}\sqrt{2j+1}\begin{Bmatrix} l' & s & j \\ s & l & 1 \end{Bmatrix}\left\langle l' \,\|\, \nabla \,\|\, l \right\rangle\left\langle s \,\|\, \boldsymbol{\sigma} \,\|\, s \right\rangle \\
&= \frac{-1}{\sqrt{j(j+1)(2j+1)}}\left\langle l \,\|\, \hat{r} \,\|\, l' \right\rangle\left\langle l' \,\|\, \nabla \,\|\, l \right\rangle \\
&= \begin{cases} \dfrac{1}{2}\sqrt{\dfrac{2j+1}{j(j+1)}}\left[\dfrac{\partial}{\partial r} - \dfrac{1}{r}\left(j-\dfrac{1}{2}\right)\right] & (\text{for } j = l + \dfrac{1}{2} \text{ nuclei}) \\ \dfrac{1}{2}\sqrt{\dfrac{2j+1}{j(j+1)}}\left[\dfrac{\partial}{\partial r} + \dfrac{1}{r}\left(j+\dfrac{3}{2}\right)\right] & (\text{for } j = l - \dfrac{1}{2} \text{ nuclei}) \end{cases}.
\end{aligned} \tag{F.3}$$

In the fourth equality, we have used the following relations of the 6-j symbol:

$$\begin{Bmatrix} 1 & b & c \\ 1/2 & c+1/2 & b-1/2 \end{Bmatrix} = (-1)^{1+b+c}\times\frac{1}{2}\sqrt{\frac{(1+b-c)(2-b+c)}{b(2b+1)(2c+1)(c+1)}}, \tag{F.4}$$

and

$$\begin{Bmatrix} b+1 & b & 1 \\ c & c & a \end{Bmatrix} = (-1)^{a+b+c+1} \times \frac{1}{2}\sqrt{\frac{(a+b+c+2)(a+b-c+1)(a-b+c)(-a+b+c+1)}{(2b+1)(b+1)(2b+3)c(2c+1)(c+1)}}. \tag{F.5}$$

By substituting Eq. (F.3) into Eq. (F.1), we obtain

$$\begin{aligned} d_A(\xi) &= 2qe\xi\sqrt{\frac{j}{(2j+1)(j+1)}} \int_0^\infty r^3\, R(r) \big\langle (l,s)j \,\big|\big|\, \hat{r}\,(\boldsymbol{\sigma}\cdot\nabla) \,\big|\big|\, (l,s)j \big\rangle R(r)\, dr \\ &= \frac{\xi q e}{j+1}\int_0^\infty r^3 dr\, R(r) \times \begin{cases} \left[\frac{\partial}{\partial r} - \frac{1}{r}\left(j-\frac{1}{2}\right)\right] R(r) \;(\text{for } j = l + \frac{1}{2} \text{ nuclei}) \\ \left[\frac{\partial}{\partial r} + \frac{1}{r}\left(j+\frac{3}{2}\right)\right] R(r) \;(\text{for } j = l - \frac{1}{2} \text{ nuclei}) \end{cases} \\ &= \frac{\xi e q}{j+1} \times \begin{cases} -\frac{3}{2} - j + \frac{1}{2} \;(\text{for } j = l + \frac{1}{2} \text{ nuclei}) \\ -\frac{3}{2} + j + \frac{3}{2} \;(\text{for } j = l - \frac{1}{2} \text{ nuclei}) \end{cases} \\ &= \begin{cases} -\xi e q \quad (\text{for } j = l + \frac{1}{2} \text{ nuclei}) \\ \xi e q \frac{j}{j+1} \;(\text{for } j = l - \frac{1}{2} \text{ nuclei}) \end{cases}, \end{aligned} \tag{F.6}$$

where we have used the following relation in the third equality:

$$\begin{aligned} &\int_0^\infty dr\; R(r) r^n \frac{\partial}{\partial r} R(r) = -n \int_0^\infty dr\; R(r) r^n \frac{\partial}{\partial r} R(r) - \int_0^\infty dr\; R^2(r) r^{n-1} \\ \Rightarrow &\int_0^\infty dr\; R(r) r^n \frac{\partial}{\partial r} R(r) = -\frac{n}{2}\int_0^\infty dr\; R^2(r) r^{n-1}. \end{aligned} \tag{F.7}$$

If we consider the shift of the center of mass, we obtain

$$d_A(\xi) = -e\xi t_j \left(q - \frac{Z}{A}\right), \tag{F.8}$$

with $t_j = 1$ for $j = l + \frac{1}{2}$ nuclei and $t_j = \frac{-j}{j+1}$ for $j = l - \frac{1}{2}$ nuclei, which is the expression of Eq. (7.33).

Nuclear Schiff Moment Generated by the P, CP-odd N-N Interaction in the Simple Shell Model

The Schiff moment is a rank 1 operator, so calculation similar to the nuclear EDM can be performed. In particular, the angular matrix element (F.3) is exactly the same, as the Schiff moment and the dipole operator have both rank 1. Actually, we have just to replace one r in Eq. (F.1) by $\frac{e}{10}\left(r^2 - \frac{5}{3}\langle r^2\rangle_{\rm ch}\right) r$, with $\langle r^2\rangle_{\rm ch} = \int_0^\infty R^2(r) r^4 dr$. We then obtain the following nuclear Schiff moment:

$$
\begin{aligned}
S_A(\xi) &= 2qe\xi\sqrt{\frac{j}{(2j+1)(j+1)}} \int_0^\infty r^3\, R(r)\frac{e}{10}\left(r^2 - \frac{5}{3}\langle r^2\rangle_{\rm ch}\right) \\
&\quad \times \left\langle (l,s)j \,\middle|\middle|\, \hat{r}\,(\boldsymbol{\sigma}\cdot\nabla) \,\middle|\middle|\, (l,s)j \right\rangle R(r)\, dr \\
&= \frac{\xi q e}{j+1}\int_0^\infty r^3 dr\, R(r)\frac{e}{10}\left(r^2 - \frac{5}{3}\langle r^2\rangle_{\rm ch}\right) \\
&\quad \times \begin{cases} \left[\dfrac{\partial}{\partial r} - \dfrac{1}{r}\left(j - \dfrac{1}{2}\right)\right] R(r) \ (\text{for } j = l + \frac{1}{2} \text{ nuclei}) \\ \left[\dfrac{\partial}{\partial r} + \dfrac{1}{r}\left(j + \dfrac{3}{2}\right)\right] R(r) \ (\text{for } j = l - \frac{1}{2} \text{ nuclei}) \end{cases} \\
&= \begin{cases} -\dfrac{1}{2}\xi e q\left[\dfrac{1}{5}\dfrac{j+2}{j+1} r_{ex}^2 - \dfrac{1}{3}\langle r^2\rangle_{\rm ch}\right] & (\text{for } j = l + \dfrac{1}{2} \text{ nuclei}) \\ \dfrac{1}{2}\xi e q\left[\dfrac{1}{5}\dfrac{j-1}{j+1} r_{ex}^2 - \dfrac{1}{3}\dfrac{j}{j+1}\langle r^2\rangle_{\rm ch}\right] & (\text{for } j = l - \dfrac{1}{2} \text{ nuclei}) \end{cases}, \qquad \text{(F.9)}
\end{aligned}
$$

where $r_{ex}^2 \equiv \int_0^\infty R^2(r) r^4 dr$ is the mean square radius of the valence nucleon. This gives the expression of Eq. (7.34).

Nuclear EDM Generated by the Valence Nucleon EDM in the Simple Shell Model

The nuclear EDM generated by the EDM of the valence nucleon N in the simple shell model is given by

$$
\begin{aligned}
d_A(d_N) &\equiv \left\langle \tilde{\psi} : j, j_z = j \,\middle|\, (d_N)_z \,\middle|\, \tilde{\psi} : j, j_z = j \right\rangle \\
&= d_N \left\langle \psi : j, j_z = j \,\middle|\, \sigma_z \,\middle|\, \psi : j, j_z = j \right\rangle \\
&= d_N \begin{pmatrix} j & 1 & j \\ -j & 0 & j \end{pmatrix} \left\langle \psi : j \,\middle|\middle|\, \boldsymbol{\sigma} \,\middle|\middle|\, \psi : j \right\rangle \\
&= d_N \sqrt{\frac{j}{(2j+1)(j+1)}} \left\langle \psi : j \,\middle|\middle|\, \boldsymbol{\sigma} \,\middle|\middle|\, \psi : j \right\rangle
\end{aligned}
$$

$$= d_N \sqrt{\frac{j}{(2j+1)(j+1)}} \left\langle (l,s)j \,||\, \boldsymbol{\sigma} \,||\, (l,s)j \right\rangle \int_0^\infty r^2 R^2(r)\, dr$$
$$= d_N \sqrt{\frac{j}{(2j+1)(j+1)}} \left\langle (l,s)j \,||\, \boldsymbol{\sigma} \,||\, (l,s)j \right\rangle. \tag{F.10}$$

The angular matrix element can be written as

$$\begin{aligned}
\left\langle (l,s)j \,||\, \boldsymbol{\sigma} \,||\, (l,s)j \right\rangle &= (-1)^{l+s+j+1}(2j+1) \begin{Bmatrix} s & s & 1 \\ j & j & l \end{Bmatrix} \left\langle s \,||\, \boldsymbol{\sigma} \,||\, s \right\rangle \\
&= (-1)^{l+s+j+1}(2j+1) \\
&\quad \times (-1)^{s+j+l+1} \frac{s(s+1)+j(j+1)-l(l+1)}{2\sqrt{s(s+1)(2s+1)j(j+1)(2j+1)}} \times \sqrt{6} \\
&= \sqrt{\frac{2j+1}{j(j+1)}} \left[s(s+1) + j(j+1) - l(l+1) \right] \\
&= \sqrt{\frac{2j+1}{j(j+1)}} \\
&\quad \times \begin{cases} \frac{3}{4} + j(j+1) - j(j+1) + j + \frac{1}{4} & (\text{for} j = l + \frac{1}{2} \text{ nuclei}) \\ \frac{3}{4} + j(j+1) - j(j+1) - j - \frac{3}{4} & (\text{for} j = l - \frac{1}{2} \text{ nuclei}) \end{cases} \\
&= \sqrt{\frac{2j+1}{j(j+1)}} \times \begin{cases} j+1 & (\text{for } j = l + \frac{1}{2} \text{ nuclei}) \\ -j & (\text{for } j = l - \frac{1}{2} \text{ nuclei}) \end{cases}.
\end{aligned} \tag{F.11}$$

By substituting Eq. (F.11) into Eq. (F.10), we obtain

$$d_A(d_N) = d_N t_j = \begin{cases} d_N & (\text{for } j = l + \frac{1}{2} \text{ nuclei}) \\ -\frac{j}{j+1} d_N & (\text{for } j = l - \frac{1}{2} \text{ nuclei}) \end{cases}. \tag{F.12}$$

This gives the formula of Eq. (7.37)

Nuclear Schiff Moment Generated by the Valence Nucleon EDM in the Simple Shell Model

We should first derive the nuclear Schiff moment operator with the nucleon EDM contribution. The nuclear Schiff moment operator with the charge distribution of the nucleon is given by

$$\begin{aligned}
\hat{\mathbf{S}} &= \frac{1}{10}\sum_{N=1}^{A}\sum_{i_N} e_{i_N}\left[(\mathbf{r}_N+\boldsymbol{\rho}_{i_N})^2-\frac{5}{3}\langle r^2\rangle_{\rm ch}\right](\mathbf{r}_N+\boldsymbol{\rho}_{i_N}) \\
&= \frac{1}{10}\sum_{N=1}^{A}\sum_{i_N} e_{i_N}\left[r_N^2+2(\mathbf{r}_N\cdot\boldsymbol{\rho}_{i_N})+\rho_{i_N}^2-\frac{5}{3}\langle r^2\rangle_{\rm ch}\right](\mathbf{r}_N+\boldsymbol{\rho}_{i_N}) \\
&\approx \frac{1}{10}\sum_{N=1}^{A}\sum_{i_N} e_{i_N}\left[r_N^2\mathbf{r}_N-\frac{5}{3}\langle r^2\rangle_{\rm ch}\,\mathbf{r}_N+r_N^2\boldsymbol{\rho}_{i_N}+2(\mathbf{r}_N\cdot\boldsymbol{\rho}_{i_N})\mathbf{r}_N-\frac{5}{3}\langle r^2\rangle_{\rm ch}\,\boldsymbol{\rho}_{i_N}\right] \\
&= \hat{\mathbf{S}}^{\rm ch}+\frac{1}{10}\sum_{N=1}^{A}\left[r_N^2\mathbf{d}_N-\frac{5}{3}\langle r^2\rangle_{\rm ch}\,\mathbf{d}_N+2(\mathbf{r}_N\cdot\mathbf{d}_N)\mathbf{r}_N\right], \qquad \text{(F.13)}
\end{aligned}$$

where $\boldsymbol{\rho}_{i_N}$ is the coordinate of the constituents of the nucleon N. In the third line, we have used the relation $d_N \equiv \sum_{i_N} e_{i_N}\rho_{i_N}$, and we have neglected $O(\rho_{i_N}^2)$ terms. The nucleon EDM contribution to the nuclear Schiff moment operator is then

$$\hat{\mathbf{S}}^{\rm nucleon} \equiv \frac{1}{10}\sum_{N=1}^{A}\left[r_N^2\mathbf{d}_N-\frac{5}{3}\langle r^2\rangle_{\rm ch}\,\mathbf{d}_N+2(\mathbf{r}_N\cdot\mathbf{d}_N)\mathbf{r}_N\right]. \qquad \text{(F.14)}$$

This is the formula of Eq. (7.27).

The nuclear Schiff moment generated by the EDM of the valence nucleon N in the simple shell model is given by

$$\begin{aligned}
S_A(d_N) &\equiv \left\langle \tilde{\psi}: j, j_z=j \,\middle|\, (\hat{S}^{\rm nucleon})_z \,\middle|\, \tilde{\psi}: j, j_z=j \right\rangle \\
&= \begin{pmatrix} j & 1 & j \\ -j & 0 & j \end{pmatrix}\left\langle \psi: j \,\|\, \hat{\mathbf{S}}^{\rm nucleon} \,\|\, \psi: j \right\rangle \\
&= \sqrt{\frac{j}{(2j+1)(j+1)}}\left\langle \psi: j \,\|\, \hat{\mathbf{S}}^{\rm nucleon} \,\|\, \psi: j \right\rangle \\
&= \sqrt{\frac{j}{(2j+1)(j+1)}} \\
&\quad \times \int_0^\infty r^2 R^2(r)\left\langle (l,s)j \,\middle\|\, \frac{1}{10}r^2\mathbf{d}_N-\frac{1}{6}\langle r^2\rangle_{\rm ch}\,\mathbf{d}_N+\frac{1}{5}(\mathbf{r}\cdot\mathbf{d}_N)\mathbf{r} \,\middle\|\, (l,s)j \right\rangle dr \\
&= d_N\sqrt{\frac{j}{(2j+1)(j+1)}} \\
&\quad \times \int_0^\infty r^2 R^2(r)\left\langle (l,s)j \,\middle\|\, \frac{1}{10}r^2\boldsymbol{\sigma}-\frac{1}{6}\langle r^2\rangle_{\rm ch}\,\boldsymbol{\sigma}+\frac{1}{5}r^2(\hat{r}\cdot\boldsymbol{\sigma})\hat{r} \,\middle\|\, (l,s)j \right\rangle dr \\
&= d_N\sqrt{\frac{j}{(2j+1)(j+1)}}\left\{ \left(\frac{1}{10}r_{ex}^2-\frac{1}{6}\langle r^2\rangle_{\rm ch}\right)\left\langle (l,s)j \,\|\, \boldsymbol{\sigma} \,\|\, (l,s)j \right\rangle \right. \\
&\quad \left. +\frac{1}{5}r_{ex}^2\left\langle (l,s)j \,\|\, (\hat{r}\cdot\boldsymbol{\sigma})\hat{r} \,\|\, (l,s)j \right\rangle \right\}, \qquad \text{(F.15)}
\end{aligned}$$

where d_N is the EDM of the valence nucleon. Here we need the expression of two reduced matrix elements. The first reduced matrix element $\left\langle (l,s)j \,||\, \boldsymbol{\sigma} \,||\, (l,s)j \right\rangle$ is given in Eq. (F.11). The second one $\left\langle (l,s)j \,||\, (\hat{r}\cdot\boldsymbol{\sigma})\hat{r} \,||\, (l,s)j \right\rangle$ is given as follows:

$$
\begin{aligned}
\left\langle (l,s)j \,||\, \hat{r}\,(\boldsymbol{\sigma}\cdot\hat{r}) \,||\, (l,s)j \right\rangle &= \frac{-1}{\sqrt{j(j+1)(2j+1)}} \left\langle l \,||\, \hat{r} \,||\, l' \right\rangle \left\langle l' \,||\, \hat{r} \,||\, l \right\rangle \\
&= \frac{-1}{\sqrt{j(j+1)(2j+1)}} \times \begin{cases} -(l+1) & (\text{for}\, j = l + \frac{1}{2}\ \text{nuclei}) \\ -l & (\text{for}\, j = l - \frac{1}{2}\ \text{nuclei}) \end{cases} \\
&= \frac{1}{\sqrt{j(j+1)(2j+1)}} \times \left(j + \frac{1}{2} \right) \\
&= \frac{1}{2}\sqrt{\frac{2j+1}{j(j+1)}}.
\end{aligned}
\tag{F.16}
$$

The first equality is derived in a manner similar to the computation of $\left\langle (l,s)j \,||\, \hat{r}\,(\boldsymbol{\sigma}\cdot\boldsymbol{\nabla}) \,||\, (l,s)j \right\rangle$ [see the fourth equality of Eq. (F.3)]. By substituting the reduced matrix elements of Eqs. (F.11) and (F.16) into Eq. (F.15), we obtain

$$
\begin{aligned}
S_A(d_N) &= d_N \times \begin{cases} \dfrac{1}{10}\dfrac{j+2}{j+1} r_{ex}^2 - \dfrac{1}{6}\langle r^2 \rangle_{\text{ch}} & (\text{for } j = l + \dfrac{1}{2} \text{ nuclei}) \\ \dfrac{1}{10}\dfrac{1-j}{j+1} r_{ex}^2 + \dfrac{1}{6}\dfrac{j}{j+1}\langle r^2 \rangle_{\text{ch}} & (\text{for } j = l - \dfrac{1}{2} \text{ nuclei}) \end{cases} \\
&= d_N \left[\frac{1}{10}\left(t_j + \frac{1}{j+1} \right) r_{ex}^2 - \frac{1}{6} t_j \langle r^2 \rangle_{\text{ch}} \right].
\end{aligned}
\tag{F.17}
$$

We then find the expression of Eq. (7.38).

Appendix G
R-Correlation of the Neutron Beta Decay

The decay distribution of the neutron beta decay can be written as

$$\begin{aligned}\omega(E_e, \Omega_e, \Omega_\nu) \propto 1 + a\frac{\mathbf{p}_e \cdot \mathbf{p}_\nu}{E_e E_\nu} + b\frac{m_e}{E_e} \\ + \boldsymbol{\sigma}_n \cdot \left[A\frac{\mathbf{p}_e}{E_e} + B\frac{\mathbf{p}_\nu}{E_\nu} + D\frac{\mathbf{p}_e \times \mathbf{p}_\nu}{E_e E_\nu} \right] \\ + \boldsymbol{\sigma}_e \cdot \left[N\boldsymbol{\sigma}_n + Q\frac{\mathbf{p}_e}{E_e + m}\frac{\boldsymbol{\sigma}_n \cdot \mathbf{p}_e}{E_e} + R\frac{\boldsymbol{\sigma}_n \times \mathbf{p}_e}{E_e} \right] + \cdots ,\end{aligned} \tag{G.1}$$

where the R-correlation is the last term, the triple product of the initial neutron polarization, emitted electron polarization and momentum. This observable is odd under P and CP.

The tree level R-parity violating contribution to the R-correlation of the neutron beta decay is given by the diagram given in Fig. G.1.

The quark level R-parity violating amplitude is

$$i\mathcal{M}_\beta \approx i \sum_{i=2,3} \frac{\lambda_{i11}\lambda'^*_{i11}}{4m^2_{\tilde{e}_{Li}}} \overline{u}(1+\gamma_5)d \cdot \overline{e}(1-\gamma_5)\nu_e, \tag{G.2}$$

where the momentum transfer between currents was neglected. The nucleon level effective interaction is then

$$H_\beta \approx - \sum_{i=2,3} g_S \frac{\lambda_{i11}\lambda'^*_{i11}}{4m^2_{\tilde{e}_{Li}}} \overline{p}n \cdot \overline{e}(1-\gamma_5)\nu_e, \tag{G.3}$$

N. Yamanaka, *Analysis of the Electric Dipole Moment in the R-parity Violating Supersymmetric Standard Model*, Springer Theses,
DOI: 10.1007/978-4-431-54544-6,

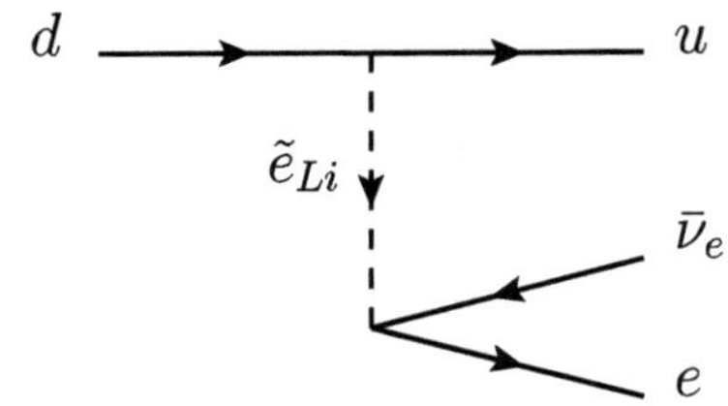

Fig. G.1 R-parity violating contribution to the d-quark beta decay at the tree level

where $g_S \equiv \langle p|\bar{u}d|n\rangle \approx \langle p|\bar{u}u - \bar{d}d|p\rangle$ was obtained by taking the isospin breaking to the first order. We finally obtain the following R-correlation

$$R = \frac{g_A \langle p|\bar{u}u - \bar{d}d|p\rangle}{2V_{ud}(1+3g_A^2)\frac{G_F}{\sqrt{2}}m_{\tilde{e}_{Li}}^2} \sum_{i=2,3} \mathrm{Im}(\lambda_{i11}\lambda'^{*}_{i11}), \tag{G.4}$$

where $g_A = 1.27$.

Curriculum Vitae

Nodoka Yamanaka
Current affiliation, address
Department of Physics, Graduate School of Science,
Kyoto University, Kitashirakawa Oiwakecho, Kyoto.

Former affiliation, address
Department of Physics, Osaka University,
1-1 Machikaneyamacho, Toyonaka, Osaka

City Kyoto
Zipcode 606-8502
Country Japan
e-mail: yamanaka@ruby.scphys.kyoto-u.ac.jp

Appointments

- Post-doctoral Fellow, Kyoto University (April 2013–)
- Post-doctoral Fellow, Research Center for Nuclear Physics at Osaka University (July 2012–March 2013)
- Pre-doctoral Research Assistant (May 2009–February 2012)

Education

- Doctor of Philosophy (Ph.D.) in Physics, Osaka University (2009–2012), advisors Dr. Toru Sato
- Master of Science (MSc) in Physics, Osaka University (2007–2009)

Research Statement
The main interest of my research is the particle physics. I am especially interested in the study of the phenomenology of particle physics beyond the standard model, and in the quantum chromodynamics.

N. Yamanaka, *Analysis of the Electric Dipole Moment in the R-parity Violating Supersymmetric Standard Model*, Springer Theses,
DOI: 10.1007/978-4-431-54544-6, © Springer Japan 2014

MIX
Papier aus verantwortungsvollen Quellen
Paper from responsible sources
FSC® C105338

If you have any concerns about our products,
you can contact us on
ProductSafety@springernature.com

In case Publisher is established outside the EU,
the EU authorized representative is:
Springer Nature Customer Service Center GmbH
Europaplatz 3, 69115 Heidelberg, Germany

Printed by Libri Plureos GmbH
in Hamburg, Germany